Mask Up - The Pandemic

Mark Davenport

Published by Mark Davenport, 2023.

MASK UP - THE PANDEMIC

First edition. May 15, 2023.

ISBN: 979-8223247159

Written by Mark Davenport.

Table of Contents

Forward

It all began as a collection of posts made by its author on social media platforms and a personal blog, which were left largely intact in the book's final version. A chronology of the situation was observed through the evolving awareness and knowledge of the author over the course of the pandemic, touching on a myriad of controversial topics such as climate change, critical race theory, and electric vehicles.

The author, has dedicated much of his time to performing extensive research, analysis, and experimentation using various data analytics tools, such as sentiment analysis and web scrapping, aided by predictive models and Excel/VBA programs. He has been able to form informed opinions based on his findings, which can be found in this book. It is sure to provide readers with an interesting view of the pandemic from the lens of one individual's data and analysis. Whether you find your own opinion mirrored in the book, or you take away different insights, one thing is certain: you're in for an enjoyable read.

At the start, I was only creating simple posts full of my own opinions, occasionally mixing things up with a meme or two. However, as I saw more people looking to my content, I started to recognize the need to up my game in terms of quality. I began to spend extended periods of time researching and fact-checking in order to make sure the posts complied with basic standards of readability and grammar. Moreover, I made sure the posts weren't loaded full of the same tired clichés and recycled content to make sure my readers were getting only the best possible information. With each passing post, I could feel my followers growing as I was now offering content of value which no-one else was publishing. Therefore, with every step of the way, I was learning more about the expectations of my readers and the demands of online publishing.

Recognizing that there was a need for people to understand complex published documents but could be discouraged by difficult language, I set out to rewrite these documents in plain language which could be more easily consumed. The efforts have paid off and hundreds of readers have been able to comprehend what would have been inaccessible information without the rewrite. Now I have a collection of over 300 8.5 x 11 pages and those numbers are always growing with each new edition that I publish. My most recent work is called Post – Surviving the Social Media Battleground and is the fourth book I have written so far in this series. As my articles and social media presence became more popular I felt it necessary to rebrand my work all while staying close to home. I chose to redub my work The Canadian Truth Network in order to gain more traction, and have been pleased with the results. Although I had the idea of completely renovating the book after rebranding, I decided against it in the end as I realized that most readers who would purchase this edition would be new to the content and would benefit more from having a fresh, new read.

It's hard to believe that 2020 has been an entire year since the outbreak of the COVID-19 pandemic. What began as my curiosity over the emerging news quickly shifted to panic and fear as we learned more about the virus and its spread across the globe. As we soon discovered, the virus was much more serious than we initially thought and soon precautions were taken to limit its spread.

From mandating masks in public to restricting movements, myself and many around the world suddenly found themselves adjusting to a new way of life. Trying to stay informed on the latest news and advice, I compiled my thoughts and experiences over the past year into this book, which will serve as a bit of a time capsule of what life has been like living through the pandemic.

Rather than read through the book in sequential order, feel free to jump to sections of interest. As in life, there is not one single beginning and end. While some of the personal reflections may speak more to me, my hope is that it will still be an interesting read to anyone else who founds themself in a similar situation.

While the COVID-19 pandemic has been trying for all of us, it has also been a learning experience. Central to the experience was understanding the importance of proper social distancing, wearing masks, and being informed of the latest news; this will likely be the best prevention for future pandemics or other international crises. In the end, my experience with COVID-19 will serve as an invaluable lesson in the importance of global solidarity and care for one another.

BioNTech and Pfizer falsified key data as part of their submissions to the US Food & Drug Administration (FDA) for securing the emergency use authorization and later marketing authorization approval of their product.

The overwhelming evidence points to fraudulent data being provided by BioNTech/Pfizer about the quality of their novel mRNA vaccines. The quality of the mRNA used during the trials vastly differs from that of the mRNA used for the billions of people who received it. In plain language, the quality, or integrity of the mRNA during the trials, was 78%, and in the batches used for production and sent all over the world was 55%. The issue is, when this was found out, instead of dealing with the issue before full production went into swing, Pfizer and the FDA, as well as the European and Canadian medical safety regulators, lowered the standards that then paved the way for emergency-use only authorization.

The issue with this is quite serious because, again using plain language, if almost half the mRNA vaccines can become fragmented or truncated raises alarming concerns over the rising risk of impurities as there is no way of telling what the damaged mRNA will translate into as they would translate into proteins other than intended spike protein.

A Ph.D. chemist from Hong Kong with mRNA research experience has shown evidence that strongly suggests Pfizer/BioNTech's Western blots were fabricated (copied and pasted). Similar evidence by the same chemist was also presented in a blog by another anonymous expert known as 'A Midwestern Doctor.' Below is a zoomed-in image of a row of the protein bands. Notice the perfect row of pixels compared to the imperfect authentic western blots look like.

Western blots Test

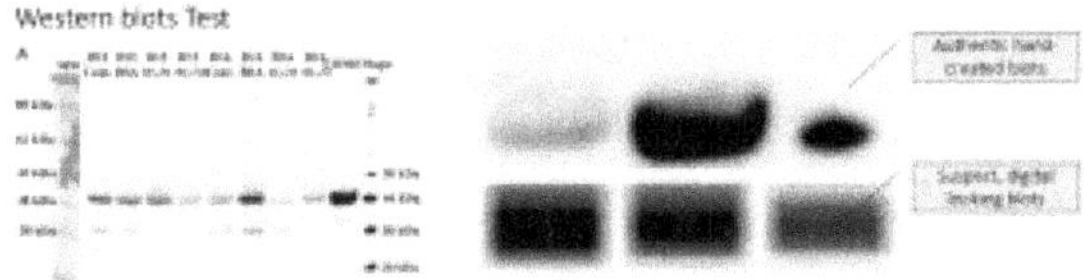

Notice the difference in these two images, the one on the top is an authentic western blot test, and the one on the bottom is highly suspected of being a digital copy (Copy Past), since it seems too perfect and symmetrical used to fool the FDA.

Leaky Vaccines

The discussion surrounding the risk of 'leaky vaccines' has been at the forefront of public discourse in recent times. In short, a leaky vaccine refers to one that doesn't make its recipients immune to the virus, meaning that they could still potentially spread it, albeit with milder symptoms. This has been borne out with reports on mRNA vaccines that have not been effective in stopping transmission. However, this concept is not new, having been extensively tested for decades, notably in relation to Marek's disease. Marek's disease is a virus affecting poultry, which did not cause much harm to chickens in the 1950s, but has become much more severe due to the implementation of leaky vaccines. The use of leaky anti-disease vaccines in poultry has allowed the virus to become stronger and more virulent, to the extent that it can now wipe out an entire flock of unvaccinated birds in a matter of days. There is a concern that the same thing could happen to humans through the widespread use of the experimental mRNA vaccines.

Though the risk of transmission with leaky vaccines is a very real one, it is not necessarily a cause for panic. It all depends on the vaccine, as some are sterilizing (transmission blocking) while others do not provide long-term immunity. People should therefore view the current mRNA vaccines, while they are beneficial to a certain extent, as more of a therapeutic agent. Also, it is worth noting that it is possible for a pathogen to evolve and become resistant to some vaccines, as evidenced by the fact that the World Health Organization recognise four different genotypes of measles globally and twenty-four variations in total. The director of the Centre of Disease Control and Prevention, Dr. Rochelle Walensky, has warned that it is only a few small mutations away from evolving past the effectiveness of the vaccine.

Therefore, there needs to be an honest discussion about what a vaccine is, does and is meant to do. As mentioned before it must provide immunity for it to truly be considered a 'vaccine.' We can look back to childhood vaccines a good example. As a child, I was vaccinated against measles but to this day I have never contracted the virus, nor have I ever had any booster shots. This is proof that the vaccine worked as it should, providing immunity and preventing me from both getting and spreading the virus. I may have escaped the virus, but there are still millions of people who are relying on the vaccine in order to ward themselves or their loved ones from getting Chinese Virus or its variants. In this sense, it seems prudent to at least take the necessary precautions for complete safety and not rely exclusively on a vaccine for full protection.

The evidence seems clear that the so-called 'leaky vaccines' are not entirely successful in what they are ought to do, and should be more accurately regarded as experimental therapeutic agents. As such, it is prudent to be aware of the potential risks to long-term immunization that they pose, while still appreciating their short-term benefits. While it is difficult to predict the outcomes of a vaccine that isn't providing full protection, the case of Marek's disease shows us that there is a risk of a virus becoming more virulent over time due to the pressure exerted by increased transmission. Hopefully, with these facts in mind, we can have a thoughtful discussion on the effectiveness and validity of leaky vaccines and what steps must be taken to make sure that they are safe and reliable solutions to the virus in the long run.

Is it Safe?

As I have been reading through the data and reports about the controversy surrounding the mRNA treatments for COVID-19, I have found myself wondering if perhaps we have made a grave mistake. We were initially led to believe that the 'spike protein' would act as a promising target antigen; sadly, it was not until too late that we discovered that this very same spike protein is, in fact, a highly toxic and pathogenic agent. By moving ahead, and continuing to develop vaccines to target this spike protein, we are, in essence, unwittingly and thoughtlessly administering a toxic compound to individuals who would otherwise be in no way exposed.

This mistake has been a difficult one to grapple with, particularly due to the various complications that typically arise as a result of administering toxins of any kind. By way of example, one study into the effects of vaccinating individuals in a remote farming community in Costa Rica revealed that a surge in neurotoxic reactions were reported by the local population shortly after the vaccinations were administered. Far worse is the fact that similar incidents have been reported across the world and, despite a myriad of research and attempts to address this issue, the exact cause has yet to be determined.

Therefore, it is essential that we seriously consider the implications of administrating this route of vaccination, particularly given our current circumstances with the COVID-19 pandemic. This is no longer a theoretical matter: we must review the potential outcomes and consequences of our actions and make an informed decision based on the evidence we have today.

Some have described this chemical as a toxin, and research conducted by experts such as Stephanie Seneff, Ph.D. and Judy Mikovits, Ph.D. support the claims that it can be linked to cardiovascular damage and infertility. In particular, Stephanie Seneff, a prominent MIT-trained computer scientist and Senior Research Scientist at the Massachusetts Institute of Technology (MIT), has written numerous papers that detail the adverse effects of this chemical on human health. On the other side of the debate, Judy Mikovits, a world-renowned virologist who is widely respected within the scientific community, has also conducted research which has found evidence supporting the possibility of infertility attributed to the use of this toxin.

Overall, the research conducted by these two renowned professionals serves as a strong indicator that this chemical indeed has the potential to be highly problematic for those exposed to it. While the full scope of its effects is yet to be known, it is clear that further study of this toxin is needed to ascertain the full extent of its potential health consequences.

The pharmaceutical giant Pfizer failed to adhere to the industry-standard practices of quality management and safety studies in their preclinical toxicology studies, creating a concerning lack of protocol that puts their results in question. Specifically, it was found that key studies did not comply with the regulations of Good Laboratory Practice (GLP), a rigorous set of safety rules that aim to ensure accuracy, reliability and reproducibility of the results accumulated during preclinical testing.

The biggest ethical issue at play here is the potential implications any errors in preclinical testing can have downstream. Inaccurate preclinical research results could potentially lead to the inadequate safety oversight of not only Pfizer's new products but also potentially other pharmaceutical companies as well that may be using similar methods in their own drug trials. If their research results prove faulty, new products may be brought to market with limited understanding of potential side effects in vulnerable patient populations and the effects that this medication could have on the human body.

This case serves as an example of the kinds of safety issues that can arise when manufacturers fail to meet the exacting industry standards of GLP, which aim to ensure that pre- and post-marketing medical data are produced to the highest possible quality with the greatest efficiency. By disregarding the highly specific protocols protecting standard industry practice, we risk the health and safety of these patients relies in the hands of producers who may not meet the procedures meant to protect us from harm. Adherence to these protocols must be taken seriously in order to protect the vulnerable population who will increasingly be relying on drug therapies for a variety of diseases and conditions.

Despite the importance of performing reproductive toxicity and genotoxicity studies prior to authorizing emergency use of new drugs or vaccines, neither of these critical assessments were conducted prior to the emergency authorization of the current vaccines. It is concerning how these significant problems were allowed to be overlooked as it drastically sinks the benefit-risk analysis of the authorizing process. This is not the first time these issues have been pointed out but still, the underlying risks have not been addressed.

Reproductive toxicity studies are paramount in determining the safety of a vaccine or drug when used over a lengthy period or upon multiple generations, while genotoxicity studies assess the potential for a drug or vaccine to cause damage to a cell's DNA. Without these analyses, the risk of drug or vaccine-induced adverse effects is significantly increased and can potentially have irreversible impacts on many populations. For instance, the lack of adequate genotoxicity studies before the introduction of the paterna cavity vaccines in the late 1960s resulted in thousands of thalidomide-related birth defects in newborns and premature deaths. This highlights why safety assessments of new vaccines or drugs must include reproductive toxicity and genotoxicity studies to accurately weigh the benefits against the potential risks.

Recent reports suggest there is speculation surrounding the impact of mRNA vaccines on women's menstruation patterns. It appears that certain blood clotting issues, which are typically seen when a person contracts COVID-19, are being reported in those who have been vaccinated with the Pfizer COVID mRNA vaccine. This has given rise to further considerations regarding the product's genotoxic (causing genetic mutations) and reproductive toxic (causing damage to fertility or development) risks.

Allergic reactions to the vaccine have been noted in some cases, as have other side effects. However, the existing research pertaining to the aforementioned issues is still in an early stage and there remains much to be explored.

Most of the claims about the risks associated with the Pfizer mRNA vaccine remain unsubstantiated. In one report, a group of gynecologists in France indicated that some of the women who received their shots had irregular bleeding patterns and disturbances in their menstrual cycles. They suggested that further research is necessary to understand the implications for the Pfizer vaccine on female health in the long-term.

While the evidence of risks associated with the mRNA vaccines is still being studied, it's important to note that the benefits of vaccination far outweigh any potential risks. With these vaccines offering the best chance of protecting individuals from contracting severe cases of COVID-19, it is important to weigh the benefits against any potential risks. Ultimately, it is recommended that people discuss their individual concerns with their doctors before receiving the vaccine.

In order to determine the potential risks of the commercial vaccine (BNT162b2), it was necessary to conduct studies that complied with accepted empirical research standards; however, such studies were not conducted. Moreover, while conducting studies to gauge the vaccine's efficacy on whether it stayed confined at the site of injection or if it traveled throughout the body, Pfizer chose not to use the BNT162b2 vaccine but rather used a 'surrogate' mRNA producing the luciferase protein. This decision to opt for a surrogate mRNA instead of the actual network vaccine raised a number of questions relating to the accuracy and reliability of the results of such studies.

In order to ensure that the potential risks of the vaccine are accurately assessed and conclusions are tenable, it is important that Pfizer make use of the actual commercial vaccine or other reliable alternatives that enables the studies to meet the necessary research standards. Similarly, it is imperative that any other research entities or pharmaceutical bodies looking to develop vaccine formulations or medical products adhere to the accepted empirical research standards to ensure the validity of their results. For instance, before the introduction of the novel COVID-19 vaccine, numerous rigorous trials were conducted around the world to ensure its efficacy and safety for public use. Such standards should be maintained by industries in order to conduct successful and reliable scientific research and ensure the safety of end-users.

Recent news reports have put forth some startling revelations which appear to suggest that the American government, along with many other countries, are moving forward with an expansive immunization program which utilizes an experimental vaccine of unknown properties. I have not been able to verify whether a similar program is taking place in Canada, however, I am trying to collect more information on the matter. Moreover, it bears mentioning once more that the research trials that were being conducted to gauge the vaccine's efficacy, which were originally expected to be concluded in 2023, have been compromised by the fact that the control group participants have since been vaccinated, rendering the results of the experiment to be inconclusive.

The release of mRNA-based COVID-19 vaccines to the public under emergency use authorization has been met with understandable urgency and enthusiasm. Unfortunately, it appears that the rush to authorization may have bypassed the routine quality testing procedures that would normally take place, leading to a variety of strange, lengthy, or protracted reactions being identified throughout the world. Reports covering a range of reactions have arisen from countries around the globe, with many of the more extreme side-effects being recorded after patients have completed administering the second dose.

In some cases, this has led to extended periods of fatigue and malaise, with people in Argentina, Turkey, and the USA describing conditions of severe exhaustion persisting for multiple days. In France, healthcare personnel have reported suffering from hair loss and ringing in the ears after their second vaccination, whilst in Spain and South Africa individuals have presented with reactions that result in swollen lymph nodes, nausea, and persistent headaches. The scientific community is currently in the process of investigating these reactions, and attempting to figure out their potential connections with the mRNA-based COVID-19 injections. However, regardless of their link, it is made clear that ample due caution should be exercised by health officials in both authorizing, and providing, vaccines in the future.

It ultimately boils down to this: the assumption that the SARS-CoV-2 spike protein would not enter the bloodstream of humans was a significant and dangerous oversight. Recent studies have discovered the spike protein to remain in the human circulatory system for a period of up to 29 days. This can have numerous deleterious implications on our health, including the ability for the virus to circulate through the body's network of vessels, leading to more widespread infection. Furthermore, the spike protein remaining in the bloodstream has the potential to have adverse effects on the immune system, resulting in a weakened ability to fight off illnesses and disease.

These findings come at a time when we are seeing an increased number of severe cases of COVID-19 and higher mortality rates, which could potentially be connected to the spike protein's decision to remain in the bloodstream over an extended period of time. Healthcare providers are now calling on the public to take necessary steps to protect their health, such as wearing a mask, washing hands regularly, and avoiding large gatherings.

Our misunderstanding of the virus has put us at a great risk, and it is imperative that we act swiftly and responsibly in order to mitigate the impact of the pandemic. Hopefully, by adhering to the aforementioned health and safety precautions, we can minimize the spread of the virus, and ultimately protect our communities from further suffering.

Once in your blood circulation, the spike protein of the coronavirus can attach itself to platelet receptors and the cells that line your blood vessels, initiating multiple pathways of cell entry. Firstly, the spike protein can actually enter the cell, allowing the virus to start replicating its RNA. This allows it to spread to other cells and organs, where it can replicate and cause damage. Secondly, the spike protein can interact with the cell membrane, manipulating it to form an opening which allows a small amount of virus particles to pass through and infect the cell. Lastly, the spike protein also activates certain enzymes, triggering them to produce a certain inflammatory response. This inflammatory response can cause symptoms such as fever, fatigue, and other serious implications, in addition to causing cell damage to the affected area. Ultimately, these spikes proteins are essential in order for the virus to gain access and ultimately cause damage.

The ability of thrombocytes, or platelets, to clump together is an essential factor in how our body deals with blood vessel damage. These specialized cells, which are part of our blood, play a key role in helping to stop bleeding by forming a blood clot when damage occurs. Nowadays, we are seeing an increased prevalence of clotting disorders, both in relation to COVID-19 and to the various vaccines being administered to combat the virus.

This clustering of platelets takes place through a complex process in which the platelets become activated and make their way to the site of injury. Once there, they undergo a series of changes, including a release of granules within the cells that contain ADP, a chemical messenger. This messenger then causes nearby platelets to aggregate and bind together to form the clot, a process which is regulated by the presence of other clotting factors.

Platelet clumping is a hugely important process, as it works to limit bleeding and therefore protect us from serious injury or death. It is also a highly regulated process, with a variety of intricate steps that must be taken in order for it to be successful. However, its complexity poses challenges to researchers when it comes to better understanding the issues that can lead to clotting disorders, the types of treatments that can be used to address them, and the long-term effects of various vaccines.

It can cause abnormal bleeding In your heart, it can cause heart problems In your brain, it can cause neurological damage

The evidence that points to a product that was sorely pushed to market is substantial and quite concerning. Our present results are far below the mark of what we would prefer, and based on the facts at hand, I personally believe that we have yet to experience the full consequences of mRNA technology when it comes to long-term effects. Will there be an increase in abortion cases? Might there be fertility issues, or heart problems among increasingly younger individuals? Only time can reveal the answers to these hypothetical queries.

The consequences of rushing a highly questionable product to market without proper due diligence and a complete understanding of the likely outcomes are severe. Beyond any potential financial ramifications, there lies the frightening prospect of numerous medical issues. Take vaping as a prime example: an industry that insisted on a rapid entry to market, leading to cases of acute lung injury, and horrifyingly, death on an unforeseen scale. We can only hope for a much different result in our own case; yet we should prepare for anything, no matter how unexpected.

In the meantime, we should be aware of the potential future issues that this hastily adopted process might bring. For example, there is the question of whether the unsavory side effects potentially resulting from this hasty product will take time to manifest. Will miscarriages, fertility problems and other heart complications occur? It is impossible to know for certain at the present moment, but it is essential to remain vigilant. In the end, only time will tell the full story of our rushed mRNA technology.

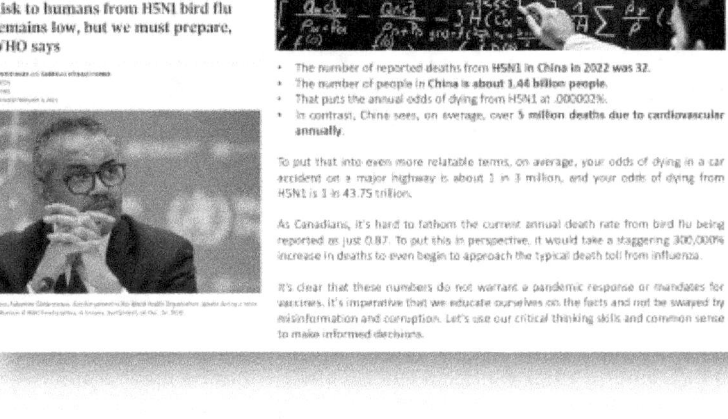

H5N1 – KNOW THE NUMBERS

Back in 2006, President George W. Bush was the first one to declare a "Bird Flu Pandemic," backed by the media, it was declared with apocalyptic warnings of imminent danger that kicked in massive federal funding to develop a new flu shot that incorporated this new "virus."

The outcome was that the pandemic never came, but certain people made incredible amounts of money. Then again, they recycled the dreaded "Bird Flu Pandemic" again in 2015 during the Obama administration. Once again fueled with forecasts of massive deaths, the announcements of a "pandemic" allowed them to take certain measures and create fear among the public. Still, once again, nothing came of it. So now, armed with using the inaccurate PCR test that, for the record, its use halted during COVID due to its inaccuracy, is once again being used to bolster fears of another said to be pandemic. It has also become known that the WHO and CDC have removed annual statistics pertaining to H5N1 and replaced its data with "revised" numbers. But thanks to the Wayback machine, and Wikipedia, the real historical numbers remain there, seen below. Also, contrary to what the media reports, at best using their numbers, the existing flu vaccines do nothing in 98.5% of adults. In 2006, President George W. Bush was the first to declare a "Bird Flu Pandemic," backed by the media. It

FACTS

- To date, there has not been a single death in Canada due to H5N1 and only 457 globally since 2003, which in plain language, puts the risk of dying from H5N1 at 1 in 203 Million.
- Using the highly and proven flawed PCR test to falsely inflate H5N1 stats to raise a level of fear to push a pandemic narrative is unethical and must be called out.

Booster Shots for Everyone!

In the course of my inquiry into the viability of Messenger RNA (mRNA) therapies to combat the novel coronavirus pandemic, I have recently uncovered a multitude of intriguing findings on the anticipated relationship between vaccine immunity and vascular damage. My research reveals that upon receiving a COVID-19 vaccination, an individual is likely to suffer from an alternation of fluid flow in their blood vessels, resulting in 'vascular leakage'.

This worrisome phenomenon has been proven to lead to anemia, thrombosis and other threatening medical conditions. To name a few specific cases, a study published in the British Journal of Clinical Pharmacology found that patients who were administered some forms of immunotherapy have been reported to have experienced a range of associated symptoms, including dizziness, chest pain, hypotension and chest discomfort, along with the vascular leakages.

Similarly, a recent study, published in the Journal of Pathology, has identified antibodies associated with the induction of vascular leakage. Of the 1,827 participant samples, seven were found to have a statistically significant association with an increased risk of developing vascular leakages, reinforcing the notion that vaccination-induced immunological responses may indeed contribute to changes in the integrity of the vascular barrier.

Ultimately, these emerging findings will be of paramount importance as researchers strive to further comprehend how the body's response to the coronavirus may be altered with the introduction of vaccination programs and mRNA therapies. As this very real risk of vascular leakage continues to be more widely acknowledged, further clinical studies will be necessary to enable informed decision-making by those considering immunotherapies, such as a COVID-19 vaccine.

The detrimental effects of continuously administering booster shots are likely to become more severe with each passing injection. Such an action could result in vascular injury at multiple sites all through an individual's body, leading to catastrophic repercussions that could be life-altering. For instance, it has been reported that, in some cases, receiving repeated booster shots can cause a constriction of the blood vessels, leading to serious health complications such as reduced blood flow and oxygen supply to different areas of the body, as well as tissue death due to lack of nourishment. Moreover, this phenomenon can generate a heightened propensity for the development of potentially fatal conditions such as deep vein thrombosis, pulmonary embolisms, and acute renal failure. To illustrate the devastating potential of these effects, the case of a 64-year-old man, whose booster shots left him with a serious vascular injury ultimately resulting in his death, serves as a poignant reminder of their serious drawbacks.

The lack of clinical trials on booster shots and their effect on the immunization-induced pathway has sparked much curiosity among experts and health professionals; leading many to ponder, could we be on the brink of a dawning of an unprecedented landscape of autoimmune conditions? Although no formal proof has been attained, the idea that the introduction of booster shots may heighten one's susceptibility to autoimmune diseases is not invalid. The notion of necessitating booster shots has been recently gaining traction in many western nations with the US Centers for Disease Control and Prevention revising vaccination guidelines to include recommencing first shots or boosters for adults or teens over certain age groups, amid a steady increase in tuberculosis, measles, and other preventable illnesses. In recent years, the medical community has documented a surge in autoimmune diseases, such as Crohn's Disease and Lupus, indicating that the existing immunization-induced pathway may be inadequate to offer reliable protection from many infectious diseases. As such, it may be necessary to modify our current approaches to immunizing to factor in potential combinations with boosters for optimal protection.

It is paramount to consider the far-reaching implications of the booster shot phenomenon, especially in the context of an ever-evolving landscape of autoimmune diseases. It is clear that more research is needed to truly understand the connection between booster shots and the current deficit of protection from many illnesses, necessitated by the existing vaccine-immune system. Until then, we can only hope that the introduction of additional booster shots will not tip the scales towards an environment where autoimmune disease is commonplace.

The potential for clotting abnormalities caused by the recent introduction of mRNA-based vaccines to our society is a very serious concern. Unfortunately, despite the numerous news stories circulating about this phenomenon, it is not making the headlines on main-stream media outlets such as CNN.

The rapid introduction of the vaccine into the bloodstream is what makes this situation particularly dangerous, as clotting may occur even in individuals who have few or no lasting symptoms associated with their condition. The consequences of not detecting and treating these reactions to the vaccine can be dire, with an increased risk of a heart attack, stroke and cerebral sinus venous thrombosis being reported among the many potential adverse events.

Given the gravity of the situation, it is essential that health professionals be acutely aware of the dangers associated with clotting abnormalities and the impact of the vaccine on clotting. In tandem, any adverse event report associated with the vaccination should be investigated fully to better understand the mechanism at play and to ascertain if any further preventative measures need to be taken. This is particularly true for those who may already be at risk for clotting abnormalities prior to the introduction of the vaccine.

Injections of a vaccine are one way to induce an immune response in the body. During this process, the first injection introduces the spike protein into the body, prompting the production of specific antibodies to it. The so-called "Jab," or second injection, results in the body creating additional spike proteins, including those located in the internal surfaces of blood vessels. As the body has already been exposed to the spike proteins, the existing antibodies bind to them, facilitating an attack on the cells. The body's white blood cells are signalled to attack, resulting in potential damage to the vascular walls and increased fluid leakage. This potential for damage and leakage can be seen in autoimmune diseases, like Lupus and Rheumatoid Arthritis, where the immune system mistakenly attacks healthy cells. It is for this reason that understanding the link between these injections and potential damage to the vascular walls is essential for medical professionals to be aware of, as it can help inform their decision-making processes.

As the world sets off on a journey of an unprecedented era with the roll out of multiple vaccines to bring an end to the COVID-19 pandemic, the looming question on many people's minds is what are the long-term consequences of this event? Although the development of the vaccines happened at an incredible speed, the process was done without any in-depth and scientific analysis involving long-term studies. Owing to this, we now stand at a crossroads where we are, in effect, a living lab, who will be the ones to gauge the possible impacts of this pandemic in the years to come.

This is especially concerning considering the amount of risk that this undertaking entails; we have already seen the disastrous effects that this virus has had on vulnerable communities, with the economic toll and medical costs of treating patients with severe cases. Moreover, even now, the implications are far-reaching, with flights being cancelled and mass gatherings curtailed, impacting people's ability to work and continue with their lives. This shows that we are already feeling the impact now, and it is only natural that our concerns over the long-term effects increase further.

Therefore, given the lack of long-term research, and the lack of precedence for such a rollout of vaccines, it is heartening to note that the government and healthcare workers play a significant role in ensuring that we make the most of this situation and strive towards a safe and fruitful future. We can only hope that the seeds that were sown in this endeavour will show their promise and will keep our communities safe and healthy over the years to come.

As scientists, medical experts and the general public eagerly await the distribution of 'vaccines' to protect against COVID-19, we face a new scientific enigma; what will be the repercussions of these potentially lifesaving advances in medicine? Will the introduction of new genes and pathogens inadvertently have an impact on our bodies' ability to ward off disease? One of the most concerning possible outcomes is the emergence of an autoimmune response.

The question remains; what will happen when these foreign agents seep out of their intended vessel – our bloodstream – and into the organs of our body? If the gene uptake is successful, it could result in the production of viral spikes which will activate pre-programmed receptors in our cells and lead to the destruction of potential viruses.

This would, thus, create a "kill or be killed" scenario; the cellular defences become activated, unleashing an army of fast-moving killer white blood cells fleeing to battle the invading force. Could this experience cause our bodies to become confused and perceive its own cells as a threat? If so, could we be at the brink of creating an entirely new world of autoimmune disease?

Only time will tell whether the "vaccines" will successfully ward off virus and prevent serious illness. But we must remain vigilant - by understanding the potential long-term consequences of these treatments, remembering that the consequences of triumph can sometimes be as dangerous as the conquests themselves.

The indisputable truth is that continually receiving immunizations of this sort is fraught with incredible and unprecedented peril - so much so that there is no historical comparison or precedent for us to refer to for advice or peace of mind. Past vaccinations, particularly those employed during pandemics, have always presented levels of risk, however these are minor compared to the novel dangers possible now, due to the technical advances achieved in the past century. Scientific breakthroughs have made it possible to engineer increasingly complex and effective vaccinations which can make bold inroads against devastating infectious diseases, but they also come with the caveat of the largely unknown side-effects that have yet to be identified. It is natural to be scared; the entire world is currently facing an unpredictable and unprecedented situation, and the ramifications of these new immunizations are undoubtedly part of that. Nevertheless, it is important to weigh the source of the fear, and seek to understand the rationale behind making such a decision. Nothing compares to understanding the perspectives of medical personnel who can evaluate the risk value with accuracy, and advise from experiential knowledge.

As Doug Ford, Theresa Tam and the rest of the government medical advisors and experts remain seemingly oblivious to the new medical issues we must face, they continue to rely solely on outdated data and information to direct their decisions. Unfortunately, this has resulted in further harms and casualties. It appears as though we are blindly going forward without fully understanding the potential consequences and consequences of our actions, leaving us with little choice but to cross our fingers and hope for the best.

This lack of awareness and preparedness to the medical implications is particularly concerning as the world continues to cope with the pandemic. Each passing day brings more challenges and potential risks, yet medicine experts continue to ignore the new data and instead double down on decisions made from outdated information.

These issues can already be seen in the case of 'superspreader' events, where the guests were completely unaware of their risk of infection, due to outdated advice given to them by the government and medical professionals. Take for example with the Mississippi wedding which resulted in over 70 cases and even two reported deaths. Most of the guests did not know the risk of attending, with only one guest actively wearing a face covering throughout the entirety of the event. The clear lack of understanding and planning from the medical experts caused this event to become a superspreader, exemplifying how relying on outdated knowledge can come with grave and unforeseen consequences.

In order to create actual, meaningful progress and safety, it is critical that the medical advisors and experts look beyond the old data and adequately plan for the medical implications we may face in the future.

It is undoubtedly concerning that, even when potentially partially realistic, the risk associated with administering the Covid-19 vaccines warrants caution and a thoughtful consideration of the consequences. Therefore, isn't it consequently reasonable to question whether it would make sense to hit the pause button on the vaccine rollout and to conduct a detailed series of tests and review the mRNA used in the inoculations to confirm that they meet all safety standards, as they should have been done in the first place? We find ourselves presented with a quandary so perplexing it is difficult to understand how medical professionals could justify administering multiple doses of the vaccine in good conscience.

It is of the utmost importance, especially with regards to the health of the global populace, that the vaccination process is not rushed, no matter how desperate the desire for an end to the pandemic might be. Doctors must consider not only their own integrity, but also the health and lives of their patients. A 2018 study by the Kaiser Family Foundation revealed that only 63% of American adults reported having high confidence in the safety and efficacy of vaccines; this suggests that such a sentiment is not isolated to the Covid-19 vaccine but rather is held more broadly when it comes to the use of these prescribed treatments. Therefore, in order to increase public trust, stringent steps must be taken to ensure that all vaccines meet the strictest safety requirements and that potential negative effects are determined before the inoculations are administered.

In conclusion, as daunting and difficult as it may be, it is essential to take the necessary steps to guarantee the safety of all sufferers of Covid-19 and those who wish to receive the vaccine. Testing, retesting and verifying mRNA's should be standard procedure and only once a full investigation of the efficacy and long-term implications of the vaccines has been concluded should doctors feel ethically and conscientiously secure in the delivery of multiple doses.

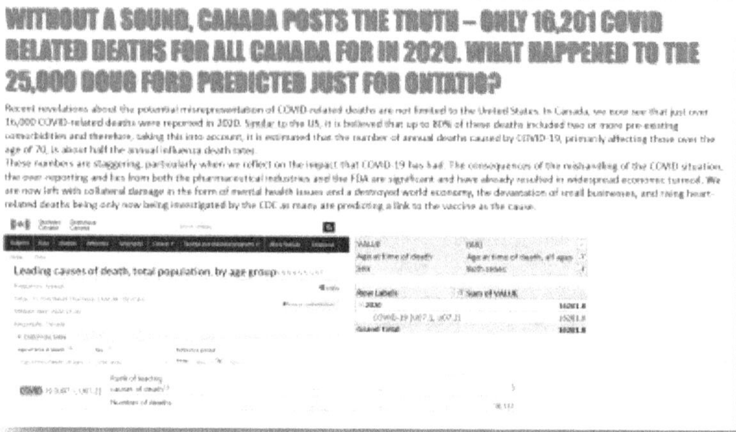

WITHOUT A SOUND, CANADA POSTS THE TRUTH – ONLY 16,201 COVID RELATED DEATHS FOR ALL CANADA FOR IN 2020. WHAT HAPPENED TO THE 25,000 DOUG FORD PREDICTED JUST FOR ONTATIO?

Recent revelations about the potential misrepresentation of COVID-related deaths are not limited to the United States. In Canada, we now see that just over 16,000 COVID-related deaths were reported in 2020. Similar to the US, it is believed that up to 80% of these deaths included two or more pre-existing comorbidities and therefore, taking this into account, it is estimated that the number of annual deaths caused by COVID-19, primarily affecting those over the age of 70, is about half the annual influenza death rates.

These numbers are staggering, particularly when we reflect on the impact that COVID-19 has had. The consequences of the mishandling of the COVID situation, the over-reporting and lies from both the pharmaceutical industries and the FDA are significant and have already resulted in widespread economic turmoil. We are now left with collateral damage in the form of mental health issues and a destroyed world economy, the devastation of small businesses, and rising heart-related deaths being only now being investigated by the CDC as many are predicting a link to the vaccine as the cause.

Leading causes of death, total population, by age group

Diminishing Returns

The question of whether we have gone too far with the protective measures taken to mitigate the spread of Covid has sparked extensive debate across the world. On one side of the argument, some suggest that the beneficial effects of such measures don't outweigh the detriment they cause to various aspects of life, particularly the economy and mental health.

The economic impact of the pandemic is felt in countries around the world. Businesses, small and large, have had to shut down temporarily or close their doors for good, resulting in an estimated global shrinkage in GDP of 3.2% in 2020. Additionally, the economic effects of the pandemic were unequal, with lower-income households facing more severe financial consequences and risks than their wealthier peers.

Mental health has also been significantly affected by the pandemic. According to The World Health Organisation, nearly 1 in 2 people reported feeling worried or anxious, greatly heightened stress, and moderately increased risk of depression, at some point in 2020. The pandemic has also had a substantial psychological toll on those directly infected by COVID, as well as their families. Mental health issues related to the pandemic also disproportionately affect groups such as certain ethnic and age groups, health care workers, and individuals with existing mental health problems.

Given the broad and long-lasting impact of the pandemic, from damage to mental health to major disruption of the economy, it can be argued that the protective measures taken have resulted in 'diminishing returns' and we may have surpassed the point of diminishing returns. It's clear then, that the proper handling of the Covid situation remains a source of contention between those who seek to protect public health and safety and those who see the detrimental effects of overreaching protective measures.

As Ontarians, it is essential to employ the law of diminishing returns when analyzing any situation deemed pertinent to virology. The concept of diminishing returns, envisaged by economists, demonstrates that underlying any system where resources are injected, there will come a point where further input fails to produce significant, or measurable output. This concept is widely applicable to the realm of virology, specifically where the risk of an adverse event can be mitigated in multiple ways. In such cases, the efficacy of every incremental intervention is almost certainly going to decrease as each successive approach is considered.

Moreover, the value of individual interventions is greatly influenced by external considerations such as economic, mental health and educational factors which, if not taken into account, could have an adverse effect on the individual or collective. Furthermore, these interventions require an upsurge in resources including time and energy, which also impacts their overall efficacy. As such, it's important to prioritize interventions which follow a patient-centric approach and are most acceptable given the current resources at hand.

One real-world example of the law of diminishing returns can be seen in the current coronavirus pandemic, in which our increasing knowledge and understanding of the virus, has allowed us to implement a number of strategies to reduce the risk of a widespread outbreak. These actions include self-isolation, social distancing, and the implementation of strict policies and regulations to combat the spread of COVID-19. While each of these strategies has helped reduce the virus' potential damage, their impact was much more significant when first implemented. As we accept these rules and regulations for the sake of safety and prevention, the effect of their implementation slowly begins to wane, unless additional considerations are taken into account.

The integration of the law of diminishing returns into our approach to virology clearly demonstrates its far-reaching implications and widespread applicability. As Ontarians it is essential to consider and understand the limitations associated with any approach taken when facing a viral threat and to provide the necessary resources to implement the most suitable strategies.Such an understanding incentivizes us to prioritize interventions in a patient-centric fashion with greater emphasis placed on the relative impact and acceptability of our implementation.

In medical contexts, when a patient is facing multiple potential treatments or tests, it can be tempting to pursue a 'quest for diagnostic certainty': seeking out the most accurate, comprehensive data and results available. While thorough diagnostic testing and well-executed procedures that provide a large selection of results may seem like a sure method of finding the right solution, this approach can create a false sense of accomplishment because eventually, as one strives for greater accuracy and certainty, the utility of the available data and tests decreases.

To illustrate this phenomenon, an example from the field of radiology can be useful. In certain bone-related illnesses, an x-ray scan is often used to make an initial diagnosis before more invasive and expensive MRI scans or CT scans are ordered, as the latter provide more detailed information and images. While the additional scans may be seen as desirable, they can also provide diminishing returns: as the number of images increase, incremental differences in bone displacement between the images reveal virtually no new information. So, while it's often necessary to get the higher resolution scans for an accurate diagnosis, the utility of these results decrease with each successive image.

In conclusion, it is important to remember that greater diagnostic certainty may not always be attainable, and striving for it can, eventually, lead to diminishing returns when it comes to useful information from diagnostic tests and observations.

In the present situation, the government must carefully consider the potential gains and losses entailed by each and every intervention or step undertaken, such as lockdowns, before deciding how long the controls should remain in place and how far to proceed in honing the patient's health. To do this, a thorough analysis should be made of not only the direct implications on the health of the population, but also the more indirect effects on the broader economy and culture of the society. For instance, a mandatory lockdown can curb the transmission of a virus and reduce the number of illnesses in the short term, but the economic repercussions, such as dwindling job numbers and closures of businesses, can have a devastating impact on the citizens in the longer run. Therefore, a balance must be struck between safeguarding people's health and preserving the health of the nation as a whole.

It is undeniable that the efficacy of the restrictive measures put into place through the use of lockdowns is questionable and should never be considered without careful consideration. The consensus among experts is that lockdowns should only be used as a transient measure and even now, as we witness our society addressing the potential of a Phase 4, the data clearly illustrates that the outlay in terms of testing, activity, and lockdowns have resulted in an only marginal reduction in mortality. In fact, a growing number of experts are beginning to recognize that no amount of measures will completely eliminate deaths in this instance, but instead, the on-going lockdowns and restrictions could potentially only result in a delayed fatalities.

My diagram succinctly highlights the issues associated with the PCR test which has been used as a yardstick to diagnose the presence of the disease. This test has an accuracy of only 65% and more importantly, many of the positive tests have been discovered as false positives as the patient may have only been carrying dead particulates of the disease. Moreover, of the 'confirmed cases', 98% recover with mild minimal symptoms. Unfortunately, the media has largely focused on alarmist news stories related to the total number of positive cases, without adequately interpreting the data to explain the full story; which is why it is essential for both media and governmental bodies to ensure that the interpretation of such data is highly qualified so as to emphasize the convalescence rate and the inaccuracy of some tests.

In light of the potential for a second lockdown, I must ask the government what evidence and criteria they would use to determine whether or not to take such drastic measures. After all, the current state of our pandemic numbers do not appear to warrant such a drastic step. For example, in areas where cases had risen sharply in the past, like Arizona and Florida, the numbers have now dropped substantially. If massive lockdowns were the only answer, why would these numbers decrease? Surely we can find better ways to help keep people safe that don't involve the entire population having to stay home and risk the economic repercussions?

We must also consider why full reopenings are not yet a reality for many parts of the country. Comparing the United States to other countries like Singapore and Korea, it is clear that a fully functioning economy can coexist with mitigation measures such as social distancing, face masks, and frequent testing. We need to make similar decisions here which can help our economy thrive while also ensuring safety. Considering the economic hardship and increased suicides which have resulted from lockdowns, it is vital that there are other ways to keep people protected.

So again, what criterion will this government use if they consider an additional lockdown? Furthermore, why are we not opened fully if the numbers do not support it? Answers to these questions must be provided so that any decision that is made is done so with clear and sound judgement.

Late Night Data Digging

Investigating the current state of the COVID-19 pandemic through thorough late-night research of medical documents and scientific articles has given me more and more insight into how to approach this ongoing crisis. After careful analysis and consideration, I have become increasingly confident in one undisputed fact: Pfizer's vaccine and any other related treatments for the virus are considered experimental until their completion of Phase 3 clinical trials in 2023. Furthermore, the control groups of these trials have recently been administered the vaccine, consequently essentially nullifying their validity. It was this revelation I shared yesterday.

The implications of experimenting with potential COVID-19 vaccines are vast; such measures have the potential to save millions of lives around the world. Additionally, if the trials are not carried out properly, we run the risk of identifying a false sense of efficacy and safety which could ultimately put more people in danger. That is why it is essential that the vaccine development and trials are conducted in stringent and reliable fashion, according to the scientific method. As more and more volunteers are administered the vaccine to verify their safety, therapeutics, and efficacy, we will get closer to the eventual goals of treating the virus, and ultimately ending this global pandemic.

The notion that pharmaceutical companies cannot be held accountable for any adverse effects of their vaccines, is something that is often spoken about, and is worth reiterating. We are currently witnessing, in the last few months, many worrying cases of individuals being exposed to health threats as a result of receiving experimental COVID-19 vaccinations, even though those authorizing, advising, and administering the jabs were likely unaware of the risks these inoculations may pose. Evidence has been mounting, pointing to the fact that casualties and complications have been caused by immunization programs run by the pharmaceutical companies, demonstrating the critical need for better awareness and accountability on the part of those administering and prescribing these vaccines. That is, of course, assuming that these organizations will even accept such responsibility and accountability.

The controversy surrounding the use of mRNA in scientific research has been largely ignored due to the fact that it was originally produced using cells derived from aborted fetuses. Now that the church has been condemned for their abuse of indigenous children, the question remains: will Christianity give the issue of mRNA any consideration? Moreover, what are the moral implications of using a product that was developed with fetal cells?

In the larger context of the topic, it raises concern about the broken morality of using an unethical method to achieve a good objective. On one hand, using mRNA to find a cure for serious illnesses like cancer could potentially save countless lives. On the other hand, any drug or treatment derived from fetal remains should be rejected due to its immoral origin.

The conversation surrounding the use of mRNA has to be taken seriously. It cannot be ignored or dismissed any longer. Despite the advances made in the medical field, the moral implications of this technology cannot be overlooked. Until the whole situation is thoroughly discussed, and a resolution is reached, the use of mRNA should be put on hold until everyone is comfortable with a solution that takes into account ethical and moral considerations.

The concept of utilizing an mRNA synthesis reagent as an inoculation on people is an ultimately new concept, one which has never been tried before. The Food and Drug Administration, in accordance with the very nature of this untested inoculation protocol has not released information regarding the effects that this agent would have on inoculated individuals, leading to the realization that no long-term studies can be conducted and we cannot be certain of any conclusive results until 2023.

The implications of this are immense. Because the safety of the inoculated population cannot be guaranteed, any medical practitioner utilising this untested agent in the name of medical services runs the risk of subjecting their patients to a high degree of uncertainty. What may seem like a miracle breakthrough for a pharmaceutical company may in fact prove to be a disaster for their target demographic.

In addition, the uncertainty of the effects of this agent also impacts national health systems as a whole. While it is true that some countries such as the US are ahead of the curve in terms of introducing novel medicines and technologies to the public, other countries may not have the same luxury. Policy makers in places such as sub-Saharan Africa or South-East Asia would be faced with the difficult task of determining whether to introduce the agent to the public.

Ultimately, the reluctance of the FDA to release information or grant clinical trials until 2023 is understandable. It is of paramount importance that a suitable level of safety be established before administering this untested agent in order to preserve the health of the inoculated population and indeed, the public.

The COVID-19 alpha vaccination is an mRNA that carries a genetic code from (DNA) to the body's cells. This mRNA, containing the instructions for producing proteins from one part of the cell to another, is injected into our body's immune cells and 'reprograms' them to produce unique virus molecules that our bodies then use to produce antibodies. This is intended to be our body's way of identifying the virus and producing the necessary antibodies to fight it off. However, it appears that the mRNA instructions are not entirely sufficient to provide our bodies the protection they require, against the various variants of COVID-19. There is yet no study to prove that this mRNA-coded immunity could enhance our body's natural ability to defend against other diseases and viruses, such as the flu or a common cold. Will this vaccination enhance or impede our body's ability to fight off viruses and bacteria, in the long-term? There is currently no evidence to demonstrate what long-term effects this mRNA-coded immunity will have on our body's abilities to defend against other infectious diseases. In the absence of any long-term studies to back up the vaccine's effectiveness in providing immunity against other diseases, it is important that we remain cautious and aware of its potential impacts on our bodies' natural immunity.

The recent pandemic of the COVID-19 virus has left people in fear, uncertain of how to protect themselves from the virus and its various adverse reactions. It has come to light, however, that the truth of the matter is that there is much information that is being withheld from the public by those in power and the mass media. Governments, the Centers for Disease Control (CDC), and the majority of governing medical bodies are not providing full transparency regarding the potential adverse reactions from the alpha vaccine. This lack of honest communication has become a major source of frustration for many, as the risk of deciding to get vaccinated can be far greater if there is an information void.

These potential side effects are not trivial matters. Reports of swelling, fatigue, and pain at injection sites have been circulating worldwide, as well as more serious reactions such as Bell's Palsy and Guillain-Barre syndrome. It is important for people to be aware of the side effects that are associated with the alpha vaccine and the possible long-term impacts that the vaccine may have not only on the physical health, but also on mental health. Knowing this information can help people decide whether they would be comfortable getting the vaccine or not. Additionally, these reports need to be taken seriously and followed up on with more thorough research so that the vaccine can be as safe and effective as possible.

The media's role in all of this is equally concerning. Rather than address the reports of adverse reactions, they have chosen to be complicit in keeping the truth from the public by giving the issue little to no attention. This results in many unnecessary cases of people getting the vaccine only to then experience the unfortunate consequences. Every person has the right to know about potential side effects and the risks, or lack thereof, that may be associated with getting vaccinated. This way, individuals can then make their own informed decisions about protecting themselves from COVID-19.

The spread of this virus and its adverse effects must be taken seriously by all. That is why full transparency from those in power and from the media is incredibly important when it comes to matters of public health, especially during a pandemic. This way, people can make decisions about their own health without going through unhealthy or unnecessary risks.

In recent years, there has been an upsurge in public concern over the potential risks associated with being vaccinated. Overwhelmingly, the scientific community is in agreement that the rewards outweigh the risks of vaccination for every population group, however, there are also numerous studies based on public health data available - compiled and maintained by the United States Health and Human Services (HHS) - that detail the associated negative health effects and deaths of those who have been vaccinated. These can be accessed via the Vaccine Adverse Event Reporting System (VAERS) website (https://vaers.hhs.gov) after consenting to several requirements and disclaimers.

Having personally accessed this data, I was astounded to find - from over 450 thousand downloaded records - the full extent of these negative effects and corresponding fatalities. These alarming statistics are unfortunately often either not reported on or tucked away, despite the likelihood of them being higher than expected. This has been especially highlighted recently when the CDC was caught deleting and editing data hosted on the VAERS website.

Moderna and Pfizer's recently approved mRNA shots have recently been labelled or referred to as 'vaccines' among news agencies and health policy leaders. However, the actual patents for both of these injections more accurately describe them as 'gene therapy', which is something else entirely. To support this, here is a direct link to each respective patent. It is true that mRNA shots are a revolutionary method of preventing disease, as they use messenger RNA to train the body to create antibodies without utilizing a pre-existing dead or weakened virus, which is what normal vaccines use. It is an exciting new form of science-based medicine and gene therapy, and it is important to truly understand what it is and its relationship to traditional vaccines. In contrast to vaccine shots, mRNA treatments are not made from weakened or dead viruses, rather they contain a small fragment of genetic code that acts as instructions for the body on how to make a specific antibody. Furthermore, upon injection, this genetic code infiltrates the body's cells and instructs them to make copies of the virus which the body's immune system can use to fight illnesses. mRNA gene therapy is much more sophisticated than a traditional vaccine and while they are both effective, they are very different in terms of how they function and what they do.

So Many "Whys"

With the discovery of a vaccine against Covid-19, many people around the world are quickly jumping to get vaccinated, regardless of whether or not they possess any underlying health conditions. Even young adults in their twenties and thirties, with no existing pre-existing conditions, are eagerly waiting in long lines for "the jab". The need to get protected from this deadly virus is so strong that it is forcing even the fittest of individuals to take the necessary steps of getting the vaccine as soon as possible.

The urgency of the situation is made further clear when looking at the consequences of not being vaccinated. With already over two million deaths from Covid-19 in 2020 alone, it is clear that the virus is no joke, and can have serious repercussions on anyone, regardless of age, health, or any other factors. Even previously healthy people, who may have shrugged off the virus in the past, are now paying close attention to the danger that Covid-19 poses. The importance of being vaccinated is not just in the prevention of getting the virus, but in some cases, even keeping asymptomatic carriers from unknowingly transmitting the virus to high risk groups such as the elderly, pregnant women, and individuals with pre-existing conditions.

The rush to vaccine is also driven by the fact that the availability of doses is limited, meaning individuals need to act quickly if they want to increase their odds of getting the injection and protect themselves against this deadly virus. With vaccine manufacturers working overtime to keep up with the demand and some countries facing serious shortages, the vaccine process will take time, making the need to get it sooner rather than later even more dire.

The necessity for a vaccine is also driven by the fact that, as a society, we have yet to find a solution to stop the spread of Covid-19, as we have with other contagious diseases. While proper hygiene, mask-wearing, and social distancing are still important tools to avoid the spread of the virus, they are still not enough to effectively stop it. Therefore, the vaccine is now being seen as a way to take back control over the virus and push it into submission.

Undoubtedly, the Covid-19 vaccine is a huge boon to humanity, and it is no wonder why so many are jumping to get their hands on the injection. The urgency of the situation and the importance of being vaccinated is a driving factor for those who wish to stay healthy and protect themselves from the virus, as well as those who are looking to do their part in ending this pandemic.

It appears that many individuals are not taking into account the possible risks of contracting the Virus, even though it has a 99% survival rate for those not in the high-risk group. Furthermore, even if an individual were to become infected, there are successful treatments available which could prevent any further complications, boasting a 99% success rate. In contrast, comparatively, gene therapy involves injections with unknown long-term consequences, since it has yet to be studied in depth.

The disparity between the two options is stark. The Virus is a naturally occurring phenomenon thus its effects are understood, and with careful management, many lives can be saved. Gene therapy, on the other hand, was not naturally occurring, and therefore its effects are less understood. Consequently, it is important to weigh the risks of both treatments before deciding which route to take. Before opting for gene therapy, it is essential to consult with a knowledgeable medical professional, to gain advice on both the advantages and disadvantages that could result.

In summary, the Virus presents a viable option for people not in the high-risk group, as its effects are well understood and the success rate is high. Gene therapy, however, should be approached with caution, and if chosen, should be done so with the correct direction of a professional.

In recent years, gene therapy has gone from being a largely theoretical concept to becoming a more realistic medical option. However, certain gene therapies are yet to be approved by the US Food and Drug Administration (FDA). This has raised some ethical questions, particularly around why people may be willing to pursue "gene therapies" that are not only not approved by the FDA but are also still in the clinical trials phase.

The notion of taking a treatment that has not been approved by the FDA or even still in clinical trials may seem alarming to some, yet others may feel that the potential benefits outweigh the risks. In a world where there is no one-size-fits-all answer to illnesses, some individuals may decide to pursue a course of action that the FDA has not yet seen fit to approve. This is especially the case for those with serious or life-threatening conditions, for whom the benefits may feel worth the risk of a potentially unknown outcome.

For example, last year a three-year-old suffering from an especially aggressive form of a rare cancer underwent newly-developed gene therapy, bypassing FDA approval. Her tumor was found to have gone into remission and she became a symbol of hope for those suffering from similar serious illnesses. After her recovery was deemed a success, concerns were raised about the ethical implications of taking unapproved treatments.

The weight of the decision to take a "gene therapy" not approved by the FDA and technically still in trials should never be taken lightly. There may be potential risks involved and these should be discussed with a medical specialist before any decisions are made. Yet, for individual with specific needs and no alternative options, taking a potential path that FDA has not yet approved may be an option worth exploring and considering.

The trials that were supposed to bring an end to the dreaded Coronavirus pandemic by 2024 have been completely shattered due to the spread of the virus. It is both surprising and disheartening to see that the media is not covering much news about these trials and the progress of their completion.

These trials were extremely important as they represented hope for the medical world and humanity. If these trials had been successful, it would have meant that medicines, treatments and protocols could have been developed that were able to eradicate Coronavirus for good. Unfortunately, this has not been the case, and due to its rapid progression, we are still no closer to finding a permanent solution to the virus.

According to NPR, long-term studies of coronavirus are being conducted in order to develop vaccines and medications that could help to reduce the number of cases, but this will take some time to take effect. It is disheartening to see that the media has not devoted an adequate amount of coverage to these trials, as the science world would now be far better prepared in the fight against the virus if information about the trials had been made more available.

What is even more concerning is that this lack of media coverage regarding the major trials may have caused the pandemic to spread to far more countries than it was when it started. As it stands, it will take a much longer time to end the pandemic and ensure that the world is a safe place once again. Thus, it is essential that the media make an effort to cover such important trials and studies in order to make people more aware of the possible solutions and treatments that could give us a much better chance of defeating the virus.

Why is there is no talk about the therapeutics that in fact work?

Why have we stopped talking about herd immunity which is by far the best scenario we could hope for?

The truth is, many individuals remain unaware that even those who have been vaccinated against COVID-19 are not 100% immune to the virus. Instead, the mRNA vaccine shots simply reduce the clinical symptoms of the virus, though it is still possible for a vaccinated individual to spread the virus to others. This means that vaccinated individuals can actually become 'super-spreaders' of the virus, unintentionally infecting many others, even if they are asymptomatic themselves.

As an example, consider the United Kingdom's highly successful mass-vaccination campaign during the summer of 2021. Despite the widespread uptake of the vaccines and the fact that it is proving incredibly effective at preventing severe illness, the UK Health Minister and the Independent SAGE Advisory Board have raised warnings that public behavior and adherence to the coronavirus regulations must remain as disciplined as ever, especially amongst those who have been vaccinated.

So, why is it that there is still so little acknowledgement of the fact that those who have been vaccinated are still capable of spreading the virus? Perhaps it is because the general belief is that this is such a remote possibility that it is safe to ignore. To everyone's dismay, however, the potential of the vaccinated to become super-spreaders cannot be taken lightly, and should not be disregarded. Instead, we should be discussing the potential risks and preventive measures available in order to mitigate it.

Overall, we must all come to accept that, even if you're vaccinated, you can still spread this virus and should act according to the applicable health and safety regulations, no matter what your vaccine status is. Only then can we successfully reduce the transmission of this virus, keep each other safe, and prevent any further harm caused by COVID-19.

Good medical practice dictates that the priority should be given to those individuals at the highest risk of contracting the virus, such as people over the age of 70 with pre-existing medical conditions. It is essential that we identify these high risk populations and provide them with early access to new experimental mRNA treatments as soon as possible, to ensure that they can receive the best care available. This is not just a matter of safety, but of fairness, as those at most risk should be the first to benefit from these new treatments.

We already make special provisions for people with pre-existing health conditions and the elderly in many areas of medicine, from the provision of free flu vaccines to free prescription services for those over 65. It is important that we continue this commitment to vulnerable populations when it comes to providing access to experimental treatments as well. By providing access to the experimental treatment before the general population, we can ensure that those people who have endured the most hardship, or who have gone through the most difficult times due to their health issues, are given the chance to benefit from the latest scientific advances.

Furthermore, these treatments can also be beneficial to society as a whole. By targeting those at the highest risk, we can ensure that those most in need of medical support and care can receive treatment before it's too late. By targeting those individuals, we can also ensure that the spread of the disease is minimized among high risk populations, and thus benefit the wider population.

In conclusion, it is essential that we identify those at highest risk of contracting the virus and provide them with early access to the untested mRNA shot. Not only is this an issue of safety, fairness and justice, but also an issue of minimizing the spread of the virus, and thus benefiting society as a whole.

The COVID-19 pandemic has fundamentally shifted the world as we know it. Despite the monumental changes that have taken place over the last year, many governments and health organizations are persevering with the same strategies for managing the virus. One such strategy is the development of vaccine passports. But the efficacy of these passports raises questions, as the vaccine has not been proven to fully stop the spread of the virus.

In theory, these vaccine passports are supposed to encourage people to get vaccinated by lending them some semblance of normality. The mindset is that, by getting the vaccine, individuals can eventually move off the 'lockdown' track and get back to a more traditional lifestyle. The logic is that if enough people are vaccinated, it would ultimately help stop the spread of the virus.

Unfortunately, this is not scientifically proven. Several recent studies indicate that even people who are fully vaccinated are capable of passing the virus on. This means that, although a vaccine passport may be a helpful tool in certain circumstances, it is no substitute for other prevention methods such as staying at home and wearing a mask.

Ultimately, vaccine passports are a flawed approach to managing the virus. Rather than relying solely on them, governments and health organizations should be putting more effort into developing new strategies that are better suited for the current landscape. For example, instituting more stringent travel restrictions and quarantining requirements, as well as devising stronger contact tracing protocols, may be more effective than relying on vaccine passports.

At the end of the day, it is encouraging to see that governments and health agencies are still fighting hard to mitigate the spread of the virus. However, it is important that they recognize the limitations of existing strategies such as vaccine passports, and strive to develop innovative new solutions that are tailored to today's COVID-19 landscape.

Why is it that the mRNA can have more deaths than 27 vaccines distributed in the US in 2008 and not a word from the media?

...and finally, why the hell are we calling the Pfizer and Moderna treatment a vaccine?

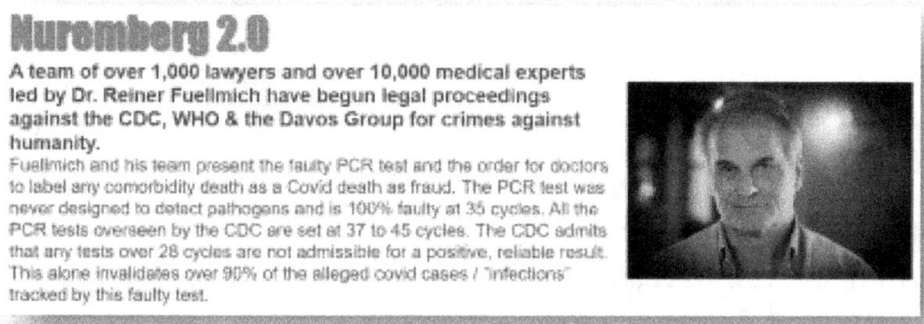

Nuremberg 2.0

A team of over 1,000 lawyers and over 10,000 medical experts led by Dr. Reiner Fuellmich have begun legal proceedings against the CDC, WHO & the Davos Group for crimes against humanity.

Fuellmich and his team present the faulty PCR test and the order for doctors to label any comorbidity death as a Covid death as fraud. The PCR test was never designed to detect pathogens and is 100% faulty at 35 cycles. All the PCR tests overseen by the CDC are set at 37 to 45 cycles. The CDC admits that any tests over 28 cycles are not admissible for a positive, reliable result. This alone invalidates over 90% of the alleged covid cases / "infections" tracked by this faulty test.

MRNA The So-Called Vaccines

The recent events surrounding the US election and the current coronavirus pandemic have raised alarm bells over the censorship issues which have been arising. In particular, the use of mRNA therapies being given the title of vaccines has caused widespread concern. To emphasize this, a huge study conducted in Europe recently reported a series of concerning figures. In relation to the gene-edited mRNA vaccine from Pfizer-BioNTech, the total number of reported deaths and injuries were 5,368 and 170,528 respectively, representing almost half the total for the four vaccines examined. Similarly, the Moderna mRNA vaccine produced a grim 2,865 deaths and 22,985 injuries. Certainly, these statistics make for alarming reading and indicate the potential dangers when it comes to the use of gene-edited mRNA therapies.

This has led to debate on the guidelines surrounding the use of such therapies, in order to ensure that they are used safely. It is essential that governments both inside and outside of the US update their policies and work together to ensure that these gene-edited therapies remain as safe as possible. Only through such co-operation between countries can the potential dangers of mRNA therapies can be mitigated and the risk of death and injury prevented.

The latest statistics from the European Medicines Agency (EMA) indicate that the Pfizer-BioNTech and Moderna vaccines, which are the two of the four currently in use in Europe, account for nearly 78% of all vaccine-related deaths in the region. Specifically, the EMA states that of the 10,570deaths reported from Covid-19 vaccines to date, a staggering 8,233 cases – around 3,400 for Pfizer-BioNTech and 4,833 for Moderna – have been attributed to these two products alone.

Of this alarmingly high number of fatalities, the main causes appear to include blood and lymphatic system disorders, cardiac disorders, musculoskeletal and connective tissue disorders, respiratory and mediastinal disorders, and vascular disorders. In terms of Moderna's mRNA vaccine specifically, these deaths stem from conditions related to the blood and lymphatic system, cardiac disorders, musculoskeletal and connective tissue disorders, and disorders of the central nervous system. All of these problems are significantly more prevalent among those given the vaccine, as opposed to baseline rates seen in the general population.

What is even more alarming is that, despite Europe leading the rest of the world in its investigation of vaccine-related fatalities, the result of these studies seem to be largely ignored. This could potentially be because of the tremendous sums of money to be made surrounding vaccines, but moreover, it is a dire situation that cannot simply be swept under the rug with ignorance. At the very least, it is worth investigating further, particularly when there is even the slightest chance that these distant signals of trouble have a basis in truth.

The fact that the United States Centers for Disease Control (CDC), often thought of as a political and corrupt organization due to their for-profit ties to vaccine makers, has reported an astounding number of "adverse events" related to COVID-19 vaccinations is deeply concerning. According to their Vaccine Adverse Event Reporting System (VAERS), this time period from December 14, 2020 to May 14, 2021 has seen 193,000 adverse events, 4,057 deaths, 2,475 permanent disabilities, 25,603 emergency room visits, 11,572 hospitalizations, and 38% of reported deaths occurred in people who became ill within 48 hours after vaccination. This data includes adverse events related to the two mRNA vaccines, Pfizer and Moderna, and the far less prevalent J&J Janssen vaccine.

The frightening numbers reported by the CDC have prompted many people to become wary of taking the vaccine, and it's a reasonable response considering these risks. However, the majority of adverse events are minor in nature and can be treated with minimal medical intervention according to the CDC. With millions of people vaccinated in the US, this small amount of adverse reactions still serve as a wake-up call; we must remain vigilant in monitoring potential signs of unusual reactions to vaccines, especially when new technologies and methodologies are involved. Vaccines can save lives, but it is equally important to be mindful of the potential risks they pose.

In the United States alone, the death toll of individuals who have contracted a vaccine-related illness is far greater in just five months than the reported vaccine-related deaths from the past two decades combined. A stark reminder of the immense struggle faced during this global pandemic, the sheer number of individuals taken due to the risks posed by the number of different vaccines on the market is staggering.

From the elderly in care homes, to the younger generation willing to take the risk and receive the vaccine, the potential dangers of the different variations of the vaccine have presented a difficult and heartbreaking situation. The cumulative number of individuals in the United States who have succumbed to vaccine-related illnesses is estimated to be above 10,000, and it is expected to keep rising as more and more vulnerable people are receiving the vaccine in an effort to gain protection against the virus.

As the vaccine continues to be supplied to more and more individuals, the vaccine manufacturers must undertake additional steps to ensure that the risk posed by the vaccine is mitigated to the greatest extent possible to ensure the safety of all those who receive it. Until an effective preventative vaccine or cure is made available, increased testing and safety protocols must be implemented to prevent any further avoidable tragedy.

The outlook has become increasingly dreary with regard to the development of antiviral vaccines that can both reduce the risk of severe illness leading to hospitalization, and protect against infection and thus interrupt disease transmission. Unfortunately, the current batch of mRNA vaccines developed in response to the ongoing global pandemic fails to meet these expectations. In fact, so confidently does Moderna's chief medical officer assert this, it has been reported in the media that their trial results will definitively not demonstrate prevention of transmission. This reality has created a great deal of uncertainty about the efficacy of the available vaccines and makes it difficult for public health officials to assess the best course of action.

Though it is easy to feel unreasonably pessimistic about the situation, there are positive developments worth noting. To begin with, the vast majority of those being vaccinated have reported experiencing no harmful side effects. Furthermore, the existing vaccines do have the potential to make almost all of those who inoculated much less likely to suffer severe complications if they do contract the virus. The data clearly shows that this applies across all demographics, although those over the age of seventy-five remain more at risk of developing a serious case of the illness.

All in all, while the current mRNA vaccines may not stop transmission of the disease, they do offer strong protection against it and can thus still play an important role in suppressing the current outbreak. Further, experts continue to strive to identify more effective treatments, with the hope that before too long an antiviral vaccine will become available that can both protect against severe illness and disrupt transmission of the virus.

Despite the numerous discussions regarding the mRNA therapy, it's difficult to predict the long-term effects due to a lack of understanding about this specific type of mRNA delivered through the vaccine. This is because similar forms of mRNA have never been observed in nature, opening the possibility of far-reaching, possibly even permanent, unforeseen consequences. Such a colossal alteration of the code of life can have extremely negative repercussions when unanticipated. Even though there were no long-term trials before the vaccine was administered to millions of people, we are currently witnessing the trials in action. It is therefore vital to continue examining the implications on the long term and ensure that each step of the process is conducted with the utmost approval. This includes doing extensive research on the potential side effects of introducing this foreign mRNA, as well as following up to all of those who have received the vaccine, to ensure that there are no negative implications.

It is beyond dispute that we are rolling out potentially risky, untested experimental gene-edited mRNA vaccines to people all over the world, with little to no knowledge of their long-term effects. This puts us in a precarious situation - the purported benefits of these vaccines are heavily praised by the World Health Organization, Bill Gates, Dr. Anthony Fauci, and other notable figures in the field, however, they may not be as miraculous as they initially seem, and instead could prove to be disastrous. The consequences of this could be far worse than any damages caused to humans by the coronavirus itself, and thus must be considered with utmost caution.

This is not to say that the mRNA vaccines have not had promising results thus far. Indeed, some of them have been successful in protecting the human body against the virus, providing some immunity without the risk of infection. However, the unknown consequences of these vaccines could still be catastrophic, as evidenced by the numerous side effects that have already been reported when such treatments are administered. For instance, the Pfizer-BioNTech vaccine was found to cause several adverse reactions, including headaches, nausea, chills, joint pain, and fever. In some cases, people have developed allergic reactions, and there have been reports of anaphylaxis, a severe and potentially deadly allergic reaction, in response to the jab. In some cases, patients have even developed blood clots or had to be hospitalized after vaccination.

Given these potential dangers, the human race should not be so quick to embrace experimental mRNA vaccines as a potential panacea. Until long term studies can be conducted, the full extent of risks involved with the usage of such treatments must remain unknown. In light of this, the implementation of mRNA vaccines should be carefully examined before rolling out on a larger scale - the safety of ourselves and future generations is worth more than any economic gains.

It is undeniable that the recent numbers have exposed a dire truth - more and more people who have not been vaccinated with a Covid-19 vaccine have been dying as a result or in relation to the virus. Why is this the case? The answer is much more complex than most may realize. It is not just a matter of individuals not obtaining the vaccine; there is often a larger societal and economic context to consider.

Take, for instance, the case of the elderly population. Many of these individuals may lack the resources necessary to procure the vaccine easily. Limited access to transportation, financial costs and the digital divide can all prevent them from getting the vaccine - leaving them incredibly vulnerable to the virus. This problem is becoming increasingly apparent in countries around the world, whether it be due to an inability to access the vaccine from a healthcare provider or due to the disproportionate deaths in minority communities that are already experiencing mistrust in the healthcare system.

Alternatively, consider the issue of vaccine hesitancy. As more information about the virus becomes available every day, some people are left feeling confused and unsure about the potential harms associated with the vaccine. Additionally, false or unflattering stories about the vaccine have made some people wary of vaccination altogether. These individuals may feel that it is too risky to get the vaccine, leading them to put their health at jeopardy.

It is important to understand that there is more than meets the eye when it comes to the relationship between unvaccinated people and Covid-19 death rates. A number of factors must be taken into account, from economic inequalities to religious opinions. It is only by taking a deeper dive into this issue can we really understand the underlying issues and work to develop effective solutions.

The issue of how PCR tests are being conducted has been largely ignored by both Canadian and American governments, and it is an issue of considerable importance. While initially, the prescribed Cycle Threshold for PCR tests was recommended to be between 20 and 28, results of up to 35, 40, and even 48 have been reported. This is deeply concerning, as it has resulted in millions of false positives. The PCR test is not detecting the presence of an active disease, but instead is detecting a specific RNA pattern - a critical factor to consider. When Cycle Thresholds are pushed up to 40 or higher, a majority of positive results become clinically insignificant, as they cannot make someone sick nor be passed onto others. This is why it has been called a 'casedemic' - individuals with 'positives' that involve only dead particles of the virus, which do not cause infection or illness, still count as a positive case. Consequently, it has become ever more critical to ensure Cycle Thresholds remain adequately low, in order to counteract the unreliable amount of false positives and prevent data from being misinterpreted.

The Centers for Disease Control and Prevention (CDC) and other organizations around the globe have been deliberately hiding information from the public. Most citizens remain unaware of the significance of Cycle Threshold settings and harness this ignorance for their purpose. It is believed that the Cycle Threshold setting, formerly set at 40, has been lowered to below 28, so that all vaccinated or unvaccinated persons can be tested. Though this may appear to be helpful for the public, there are two main objectives behind implementing this tactic. Primarily, this will reduce the number of false positives due to testing, thereby conveying a greater effectiveness to the vaccines and enticing more individuals to get vaccinated. Secondly, this will also heighten the amount of "unvaccinated" deaths reported, affording a higher incentive to get vaccinated.

Typically, many scientists suggest that Cycle Threshold settings should be kept relatively high, over the range of 35-40, to ensure that incorrect results from testing are not calculated. Nevertheless, the CDC and other divisions have grossly overlooked the importance of this aspect, thus compromising the integrity of the results. Behavior such as this only furthers the mistrust of the individuals and organizations responsible for administrating and enforcing vaccination protocols. After all, it is the obligation of the CDC to provide citizens with accurate and dependable information concerning the Covid-19 vaccine and its potential risks. Unfortunately, people cannot trust the data they are receiving and in turn, may not be willing to follow the recommended safety measures.

Many people who have obtained the mRNA jab feel an uneasy sense of betrayal, having been directly told that it would protect the people they love. Unfortunately, that does not seem to be entirely true; despite popular misunderstanding, the mRNA jab is not actually a vaccine in the traditional sense of the word. When it comes to vaccines, we usually think of an injection of either a weakened version of the virus itself, or a related one that can generate antibodies to protect against the actual disease if it is contracted. Yet, that is not the case with the mRNA treatment. Instead, it is merely used to educate the body's immune system to enable it to fight the virus, thus alleviating some symptoms, but this does not make the individual completely immune to the COVID virus. Furthermore, if someone is infected, they may be unaware of it and thus, are potentially still spreading it to others. This is why it is imperative to always adhere to social distancing precautions, even when vaccinated.

At the end of the day, what it all comes down to is the utilization of positive (or at least presumed positive) PCR tests to define 'cases' of infection. And, it has been increasingly more common, especially when looking at unvaccinated persons versus those that have been vaccinated, to employ higher cycle thresholds. However, this method of distinction will ultimately exaggerate the true positive presence in the unvaccinated, while simultaneously underestimating it in the vaccinated. This reckless approach of assumption as fact is what leads to the false claims that so often arise in the public discourse about the efficacy of the vaccine.

To put this concept into more tangible perspective, we can look at an example from earlier in 2020 when the United States was in the midst of widespread restrictions and lockdown due to the novel coronavirus (COVID-19). At that time, many people were subjecting themselves to PCR tests to assess their likelihood of carrying the virus, despite having not been exposed to anyone who had tested positive. The accuracy of many of these tests, however, relied heavily upon the utilization of high cycle thresholds, meaning that it was highly unlikely that a test would return a positive result. This subsequently led to many innocent people being declared 'safe' despite coming into contact with someone else who had tested positive in the future.

Ultimately, relying upon positive PCR tests in well people and high cycle thresholds to attempt to gauge the presence of the virus has created false claims regarding the efficacy of the vaccine - claims that could have been avoided had more efficient, accurate and calibrated methods been employed in the measures to assess an individual's likelihood of carrying the virus.

This is key and many don't know it

As each day passes, my suspicions regarding the way the ongoing COVID-19 pandemic is being reported and monitored grows exponentially larger. After delving deep into numerous reports and resources, I have started to form a precise narrative underpinning the pandemic's 'lockdown' measures, triggered by a web of oversights and omissions. Months of extensive research have lead me to discover a plethora of findings, ranging from experiences and opinions, to facts, raw data and other verifiable information. Upon reflection and analysis of this material, I have come to an incredibly saddening and dissatisfying conclusion. I sincerely apologize for the lengthiness of this piece, but what follows is necessary for a full understanding of my thoughts.

I have delved into the topic of the PCR test on numerous occasions, primarily looking into how the implementation of high-cycle thresholds was creating false positives and other incorrect results. Undoubtedly, the utilization of such thresholds was necessary in order to detect mere traces of the virus, as the research done on the genomic extracts of the Covid 19 virus often came up with inconclusive results, meaning the virus was just not successfully isolated and purified. That being said, the studies did uncover that only small fragments of RNA with a length of between 37 to 40 base pairs in length, were present. This is an essential point to consider, as one must bear in mind that the average viral genome typically varies between 30,000 to 40,000 base pairs. When this is put into perspective, it can certainly explain why there has been such outrage surrounding the administration of the PCR test. Looking at the current state of the world and all the reports of inaccuracies that have come to light, it is no doubt that this has been the major contributor to what is frankly deemed as a catastrophic 'Casedemic.'

This raises a crucial and pertinent question: how is it that it seems as though no one in any laboratory research institution in the world has been able to successfully isolate and purify a complete version of the virus? After all, if evidence of the virus can be found, shouldn't it be possible to extract it? This has been a source of puzzlement and consternation among the global scientific community and the general populace alike.

What the scientific research indicates is that, despite the best efforts of researchers, they have only been able to locate small pieces of RNA that can be identified as a Covid case. This goes against the established principles of science, which is based on empirical evidence, and runs counter to basic common sense. It also calls into question ethical principles and integrity.

To illustrate this point, it is important to understand that a viral disease such as Covid-19 can only be confirmed via an oral swab, a throat swab or a nose swab – none of which are likely to capture a complete version of the virus. In addition, the processes used to detect positive cases rely on methods such as polymerase chain reaction (PCR) – a laboratory technique used to amplify segments of DNA – or antigen testing, which uses pieces of the virus to detect an infection - neither of which are capable of isolating and purifying the entire virus.

In conclusion, it is not surprising that no research laboratory in the world has been able to isolate and purify the virus in its entirety, as the evidence suggests that this is only possible through methods that are unable to capture the complete virus.

It has been widely reported that the SARS-CoV-2 virus, which has been identified as the cause of the COVID-19 disease, was first isolated from a laboratory test back in January of 2020. However, while the Polymerase Chain Reaction test, commonly known as the PCR test, was able to successfully identify the virus, it only detected portions of the virus, and was not able to detect the actual pathogen in its entirety. In other words, the PCR test is only able to detect fragments of the SARS-CoV-2 virus, and not the entire entity.

This has been substantiated by public health organizations such as the World Health Organization (WHO), which has warned of the difficulties in detecting the virus as a result of the limitation of the PCR testing. As experts indicate, even defining snippets of the virus is vastly complicated and may take several tries in order to find a fraction of the entity. Furthermore, in cases where the viral load is low, even portions of the virus may be difficult to identify.

The shortcoming associated with the PCR testing for identifying SARS-CoV-2 has led to the rise of alternative methods being used for virus detection. These include antibody tests, which are becoming increasingly available, as well as various antigen tests. It has also been speculated that new breath tests, which would make detecting the virus easier and more accessible, may soon become available. As such, while the PCR test is still reliable and widely used, it is likely that it will be supplemented by other more effective testing methods in the future.

It is important that we are clear on this fact; when a PCR corona test is performed on someone with an immune system, it is not the virus that is detected. Rather, it is small, fragmented pieces of the viral genome that will show up as a positive result. Even if the infectious traces of the virus are no longer present in the body, the corona test is likely to come back with a positive result because of the fragments of the viral genome. This is the reason why the Cycle Threshold has caused a stir amongst those who have taken it upon themselves to learn about it, myself included. This testing regime can be blamed for contributing to the spread of the 'Casedemic'. Instead of detecting the presence of the virus, it instead only tests for the smaller, broken particles of the viral genome. To actively combat the 'Casedemic' it is essential that reliable tests are conducted, ones that demonstrate the presence of the virus and not just tiny pieces of its genome.

The relationship between the media, government and public is an essential, yet complex one and it is no different in times of pandemics. We have seen how across the world the government and the media have been spinning the narrative to spread fear and uncertainty and confuse the public with what are essentially lies in order to further their own interests. By creating a sense of panic in the population and convincing them of the need for draconian lockdowns, strict social distancing rules, stringent travel bans and a myriad of other restrictions, it has become evident that the government have sought to amplify the situation and exaggerate the case numbers.

The underlying driving forces in this endeavour have been significantly flawed data combined with poor science and a lack of understanding of the molecular biology behind the virus; instead, PCR tests have been extensively used for this purpose despite the fact that the inventor of the test, who warned authorities against employing the test in such a manner, has been side-lined and ignored.

Thus, one could conclude that the 'wave' and 'case numbers', which are in reality entirely unsubstantiated, have only served to further the agenda of a plandemic, a notion which unfortunately has been all too successful in its efforts.

The unreliability of the Polymerase Chain Reaction (PCR) testing and its inaccuracy in estimating the number of COVID-19 positive cases has been widely recognized. Despite the high cyclic thresholds ranging from 35 to 48, the test results under the 'waves' of the virus have been almost worthless, making the ensuing lockdowns, economic hardships and corresponding mental health crisis all the more devastating.

The truth is, the hastily implemented measures had no basis in scientific evidence, provoking severe socioeconomic and psychological consequences. Panic in already struggling communities was an all-too-common response and resulted in devastating increases in unemployment and poverty, exacerbating existing inequalities and depriving many of their livelihoods. Moreover, reports contextually link the lockdown regulations to a rising tide of depression and suicide, making the effects of such stringent measures in our society all the worse.

Therefore, humanity is now confronting the stark reality that the seemingly benign policy decisions had absolutely no scientific validation and we ought to reflect on the way such political decisions have been made in the past. Some have even argued that what we ultimately encountered was nothing more than a knee-jerk reaction from successive governments. It is essential that we acknowledge the devastating ripple-effect of such measures, so as to avoid a similar mistake in the future.

Back in March of 2020, I had a sneaking suspicion that this so-called pandemic was more than it seemed. From my research of the data, I learned that when compared to the past 20 years of death rates, as well as the data of past pandemics, there is no evidence to support the claims made by the government and big pharma about the severity of this virus and the need for country-wide lockdowns. I believe that hidden behind the hysteria of the pandemic is an agenda to funnel money to big pharma and give further control to the government.

An example of this agenda's effect can be seen in the U.S., where draconian policies have been put in place in the name of fighting the virus. For instance, we have seen civil liberties being restricted, jobs being lost, businesses being shut down, and massive amounts of money being spent to respond to this virus. If this had been a real pandemic, it's unlikely that the government would have gone to such extreme lengths.

So, while most of the public has accepted the pandemic's story without question, I think it's important that we challenge it by analyzing the numbers and data, and getting to the heart of the issue – the potential money, power and control that it may be creating.

I am not someone who normally puts any credence in conspiracist theories, however, the current situation is certainly prompting a certain level of scepticism in me. If a certain body of information or data is not available to support the claims being made and there has been a profound lack of transparency when it comes to publishing death rates, as has been observed in the past, then it begs the question as to what is really happening here? What precisely is the agenda? Anyone with a modicum of numeracy and access to the figures can quickly ascertain that the numbers do not seem to add up – however, without further clarity and proper explanation it is difficult to reach any definitive conclusions about the situation. Given the far-reaching and potentially devastating impact that such decisions can have on all areas of life, it is of the utmost important that the true situation is made clear and the facts behind them transparent.

It can be difficult to tell what is really going on when all you hear are lies and half-truths. Unfortunately, there are those who will take advantage of people's lack of knowledge or perceived consent for their own gain. As such, the responsibility lies on us to ensure that those who are too easily misled remain informed and aware of what is really going on. This can be done through various methods including fighting for open and transparent communication, combatting irrational fear-mongering, and using reliable sources to raise awareness of the truth.

For example, while false news and rumors can be spread easily and rapidly through social media, organizations dedicated to truth-telling can be increasingly effective in providing trustworthy sources and factual information. Additionally, grassroots efforts by those who care can be a powerful way to reach those who are vulnerable to being taken advantage of, as it can be easier to connect with people on a more personable level which may help them to better trust the facts.

It can be tempting to accept facts without doing your own due diligence, but real knowledge and understanding become available only when we dig deeper into the truth. Left unchecked, the lie will continue to be perpetuated, with people remaining unaware and vulnerable - like sheep being herded in the wrong direction. So, rather than simply accepting what is presented before us, let us commit to staying informed and raising awareness in order to ensure that the truth cannot be ignored.

These days, it has become nearly impossible to ignore the senselessness and futility of the actions some people are taking to prevent COVID-19. We must stand up against these irrational practices and demand that our peers respect the best and most current medical science.

It is simply wrong for parents to be placing masks on their young children who, typically, do not even need to wear one for their own health, far less the welfare of others. We must call out this foolishness and make it clear that this is not a sound or safe practice. Moreover, it should be questioned why someone who has gone through the effort and expense of being vaccinated still feels the need to mask up? Those who have done what is necessary to protect themselves should be able to move forward with their lives unencumbered.

The situation has necessitated us to become advocates. We must take it upon ourselves to speak our minds and be a voice of reason in this pandemic. It is our responsibility to ensure that the public is getting accurate and reliable information and to encourage fellow citizens to do the right thing. It is time to fight back and to refuse to acquiesce to this unbalanced authority. We can start with something as simple as saying "no" to masking our kids and asking those who are vaccinated why they are still wearing a mask.

The horrific impact of the pandemic was not lost on us and we refuse to stand by and ignore the possibility of a fourth wave; allowing such an outcome to take place is simply not an option. We are all too familiar with the crushing effects of the coronavirus, and we must remain vigilant in our collective efforts to contain the spread. The suffering caused by the three previous waves has been devastating and heart-breaking, and we must ensure that we are never subjected to such a situation again.

The reality is, so much of the world has been brought to its knees by this ongoing crisis, with human life, economies, and societies all suffering to varying degrees. Industries have been decimated, international travel has been severely restricted and those who have lost loved ones have been struggling with their grieving process amidst the backdrop of COVID-19, not knowing when their suffering will ultimately end. We all have a responsibility to help prevent the emergence of a fourth wave and to assist those who are still being affected by the three previous waves.

The world must come together to prepare itself and do everything it can to prevent the emergence of a fourth wave. Vigilance, thorough testing, and contact tracing will be essential in helping to protect people against the dangerous virus and hopefully, in time, help end this global pandemic.

Cycle Threshold

As we are all aware, Cycle Threshold (CT) has been a repeated topic in our recent conversations. However, allow me to explain this in my words, so that everyone can clearly understand the significance of this matter. As per the Food and Drug Administration's statement, the exact range of Cycle Threshold used to declare an individual as either positive or negative is not specified by the FDA. Quite evidently, it is up to the commercial manufacturers and laboratories to decide the parameters for CT determination.

To predict the health of a patient based on CT values, a medical professional must take into account the type of test and the threshold set by the manufacturer or the lab. As an example, in the case of reverse-transcription quantitative polymerase chain reaction (RT-PCR), a higher threshold requirement corresponds to a higher specificity and lower sensitivity. On the contrary, when the threshold is lower, the sensitivity is increased, but the specificity is lowered. Going further, the cycle threshold in a Real-Time Polymerase Chain Reaction (RT-PCR) essentially determines the initial amount of a virus present in a specimen. In simpler terms, the lower the CT value, the higher is the virus concentration, hence resulting in positive diagnosis.

In conclusion, it is ill-advised to take CT threshold ranges for granted due to the crucial role it plays in determining the health status of a person, which ultimately depends on the manufacturers and the labs.

The US Food and Drug Administration (FDA) and the Centers for Disease Control and Prevention (CDC) have allotted ample time in emphasizing the need for a highly precise test that involves amplifying minute amounts of viral residues in order to make them detectable. Sadly, there are no regulations set in motion or even any sort of advice available to instruct the test administrators on just how much amplification they should do and the extent they should proceed with, among such critical processes.

The purpose of the amplification process is to increase the volume of the already present fragments to a detectable level, making it easier for the PCR test to ascertain the presence of the virus from the sample. Primarily, the tests utilize a 'Polymerase Chain Reaction' which makes precise copies of the behavioral traits of the virus, intensifying their presence exponentially. Although this is counted as a beneficial process, professionals are concerned about the potential for receiving false negative results if their amplification techniques are not adequately refined enough to discover sufficient traces of the virus. Many tests have already surfaced that have been victimized by such improvisations, such as the batch of antibody tests from China, that claiming to identify the presence of a person's antibodies to COVID-19, when in actuality they did not have any antibodies at all.

Understanding this issue and the disadvantages it presents, it is integral for the FDA and the CDC, along with other similar organizations, to concoct rules and preventative measures to ensure any tests conducted by the professionals should be done with extreme care and precision, ensuring that the amplification technique is safely managed and not over-amplified.

Excessive testing with unrealistically high thresholds can result in an alarming number of false positives - some reports indicate that this false positive rate can climb as high as 60%, and some even suggest it can exceed this percentage. False positives are results that incorrectly test positive for a given parameter, such as mistakenly testing positive for having a certain virus when none is actually present. An example is if a blood test has a very high threshold for a particular virus, even a low presence of it may result in a false positive result. While some false positives can be acceptable, such as having a high number with certain kinds of research studies, too many can lead to incorrect conclusions, such as certain medical treatments being deemed safe or effective when they may not be. With this in mind, it is important to use appropriate thresholds when conducting tests to avoid an abnormally high rate of false positives.

The prevalence of COVID-19 has led to an unprecedented increase in demand for medical testing, as more and more people seek to determine if they may have been exposed to the virus. This has created a major market opportunity for companies providing such testing services, as the more positive test results they are able to generate the more downstream business they will see from individuals who have had contact with someone who tested positive and are consequently concerned that they may have contracted the virus. However, providing reliable testing services is not the only factor at play here; an inescapable truth is that these companies are also driven by financial considerations, in the form of the money and profit to be gained from this situation.

No matter that measures are taken to control the spread of the Coronavirus pandemic, one foreseeable problem that comes with large-scale testing is the emergence of a significant number of false positives. As any test will inherently carry a certain percentage of false positive rate, even with regular tests conducted on a consistent basis, the result will be an alarming number of bogus cases reported every day from the time when the first SARS-CoV-2 test was invented to its utilization for detection of COVID-19 infections.

These false positives can have a lasting impression on public opinion, as people might be deceived into believing that the virus is still running rampant when nothing of the sort is happening. A combination of confusing news coverage and false positives would create an atmosphere of perpetual worry, potentially morphing the current pandemic into an imaginary epidemic that never fully went away or existed in the first place. This could lead to extreme misunderstandings when people project their fears into seeing more infections where there are none, either on the local or global scale.

Unfortunately, the problem of false positives is unavoidable because the accuracy rate of current tests is far from perfect. Because of this, it is essential to create more accurate and reliable tests in order to minimize the number of false positives and to protect the public from being misled about the number of real cases that are occurring. Given that a proper grip must be maintained on the virus, mass testing should be conducted with great scrutiny, sifting through the data with utmost precision to differentiate real from unreal cases. Only then can a true picture of the pandemic be realized, followed by the implementation of the necessary countermeasures that could help contain the virus which would have otherwise have gone undetected.

Most health care professionals and epidemiologists who are familiar with the Polymerase Chain Reaction (PCR) test affirm that the cycle threshold (also known as Ct value) of 28 or lower is likely to produce a disproportionate amount of false positives. This concept has been corroborated by many empirical studies, and the Centers for Disease Control and Prevention (CDC), in an effort to prioritize accuracy, has recommended Cycle Threshold readings of 40 or higher. The significance of this suggested Ct value is immense, as it drastically diminishes the potential for false positive results by increasing the complexity of the PCR assay to a value of approximately one trillion.

The Centers for Disease Control and Prevention (CDC) are being dishonest when they assert that using a sample above 33 cycles will yield a large number of false positives. Interestingly, the CDC's own findings have indicated that detecting live virus in a sample with a cycle threshold of 33 or above is virtually impossible.

This is especially concerning given that certain studies have demonstrated how sample cycling thresholds (CT) can introduce numerous false positive results. To illustrate, a study published in the journal Clinical Infectious Diseases concluded that setting a sample threshold of 40 CT led to between 3.5-5 times more false positive results than those associated with sample thresholds of 37 CT. As a result, adhering to a cycle threshold of 33 may be the only way to reduce the likelihood of encountering these concerning false positives.

In light of the potential danger that false positives pose to public health, it is essential that the CDC be transparent about the real implications of using higher CT thresholds. As such, health professionals and the general public should be able to rely on accurate and up-to-date information from the CDC for understanding and preventing the spread of infections.

Upon further investigation, it became starkly clear that the U.S Centers for Disease Control and Prevention (CDC) were not just struggling mightily to detect live virus particles in Covid-19 samples with cycle measurements higher than 33 - they were utterly failing. In other words, the scientific community had issued testing regulations that their own research indicated would likely lead to misdiagnosis of many individuals not actually infected with the illness.

The examples of false positives are legion. For starters, take the case of a 70-year-old woman in Arizona in April 2020. She received a message from the county health department informing her that her Covid-19 test had come back positive, although her nurse practitioner had reported that her cycle threshold measurement was 34 - a reading exceeding the CDC's own restricted threshold. Similarly, in 2019, a 48-year old teacher in Florida received a positive Covid-19 result, despite being in perfect health, after her test showed a threshold measurement of 45.

The truth is, when measuring viral particles, the higher the cycle threshold, the more likely it is that the test results will be inaccurate. And yet, the CDC had continued to issue guidelines that were not only failing their own research, but also causing confusion and, in some cases, panic.

A recently published study, difficult to locate due to the sensitive nature of its contents, engaged in the retesting of a considerable number of medical patients who had diagnosed with COVID-19 using a widely accepted and more reliable Cycle Threshold approach. To the shock and bewilderment of health care professionals and researchers alike, the modified test revealed that nearly nine out of every ten positive results were, in fact, false positives. This finding is truly monumental, as it has the potential to drastically shift how the public interacts with and understands the severity of this far-reaching pandemic. In fact, one could suggest that this comes as close to a 'smoking gun' as could be expected, as the study's results are downright unthinkable.

Given the various pieces of information we are slowly becoming more and more aware of regarding Cycle Thresholds and the impact that comorbidities can have on Covid-19 mortality, it stands to reason that we must question the accuracy of our current testing methods and the resulting ultimate death toll. After all, despite the virus causing a vast number of fatalities, when comorbidities are taken into consideration it is our understanding that the death rate is actually drastically lower, which casts further doubt on the official figures. Indeed, one recent study, published in the European Journal of Neurology and conducted by the University of Zurich, revealed that the mortality rate of Covid-19 with comorbidities was just 4.7%, compared to the overall mortality rate of between 1.3-3.4%. This therefore raises serious questions about how we are currently recording and understanding the mortality rate associated with Covid-19, and so it is imperative that we consider all of the evidence we are presented with before forming any substantial conclusions.

It seems that the decisions of those in positions of power and influence have been based upon faulty information, false expectations, and inaccurate predictions. This short-sightedness has been catastrophic for so many people, both those in positions of authority and those in the public who have found their lives and livelihoods threatened by the resulting series of events.

The focus of those in charge appears to be based more on personal gain and ambition than what is the best solution for all affected. Politically motivated decisions are often taken over careful and considered steps, leading to situations that have proven extremely damaging.

The financial costs of such missteps can have tragedy consequences, causing hardship to all those involved and arguably an unfair burden placed on the vulnerable. What's more, the power and influence of those in positions of authority can mean that the voices of the masses are ignored as the decision makers cling to a 'business as usual' mentality, regardless of the destructive consequences.

The advice we have received has, sadly, been considered with little regard for the potentially devastating outcomes that can result. Too often those giving this advice have much to gain and nothing to lose, making them inclined to put forward increasingly dangerous and damaging proposals.

It is clear that the situation that many now find themselves in is the result of a combination of factors, such as misguided projections, poor governance and misplaced power. We must reflect and learn from this if we are to better ourselves in the future and protect ourselves from yet more calamity.

I urge you to do your own research, remember to use DuckDuckGo.com as your search engine to avoid the censoring of data.

For too long, we have been obsessing over case numbers in regards to the Coronavirus pandemic. However, there is a reason why raw case numbers should not be our sole focus - we are now testing for the virus much more than we were at the start of the pandemic in April 2020. It isn't logical to compare today's number of positive cases to what the number was at the onset of the pandemic without taking into consideration the dramatic increase in testing that has happened in the meantime. Doing so would make any conclusion invalid because it would not be taking into account the basic concepts of percentage and proportion.

To illustrate this, let's look at a hypothetical situation. Imagine back in 2020 there were 200 tests with 25 of them coming back positive. This would mean that the infection rate was 12.5%. Now, if we double the amount of tests to 400 and come back with 50 positive results, our infection rate is still 12.5% because the total percentage of positive tests has not changed. The 50 positive results only reflect an increase in testing, not an increase in the number of infections. As this example shows, raw case numbers alone cannot be used to track the spread of the virus. We need to look at the percentage of positive test results in order to understand the true infections rate.

It is essential to keep in mind that the data surrounding the Coronavirus pandemic is constantly in flux and therefore, it is vital that we take into consideration all sources of information available to us in order to come to an accurate conclusion. Focusing on raw case numbers alone will provide us with an inaccurate picture and as such, is not a reliable approach.

Monitoring the rate of hospitalizations and deaths from COVID-19 should absolutely be a focus when considering the current state of the pandemic. Fortunately, the most recent data available is portraying an extremely positive trend. In fact, when examining the most recently reported figures in comparison to the historical data, it becomes evident that the fears surrounding the virus have been drastically overstated. The numbers reveal a remarkable fact: in spite of an increase in overall infection rates, there has not been a noticeable surge in deaths in recent months. This has been further confirmed by examining Canada's records and available scientific data.

These figures should offer a great deal of comfort to those affected by the pandemic and illustrate the importance of looking at the latest statistics so not to succumb to fear-mongering or other exaggerated reports. It is often the case that certain media outlets choose to release overly tense information in order to increase viewership, however when considering the overall numbers, this casts doubt on their accuracy.

It is thus paramount that public and government institutions work in tandem to ensure the most up-to-date data is available and accurate. This would grant concerned citizens an increased sense of trust in our government efforts to combat the virus, while peace of mind would have a remarkable effect in reducing the overall stress of the pandemic.

The sheer rise in the number of people walking around with coronavirus in April 2020 was absolutely staggering. It's likely that six times more people had the virus at that time compared to previous months. This staggering statistic may indicate that both the vaccine roll-out and the effectiveness of protective measures such as masks are improving, as well as the physicians' ability to manage the effects of the virus through therapeutics.

To give an idea of the magnitude of the impact here, it's estimated that over 6 million Americans were infected with the virus by April, and the situation only continued to get worse as the months passed. Despite this bleak outlook, medical researchers and pharmaceutical companies worked hard to develop and eventually distribute the first approved vaccines, which people of all ages began to receive in the early months of 2021.

Moreover, minding public health measures such as wearing masks, avoiding large crowds, and staying home when possible began to become common practice, and have been proven to have a significant impact on reducing the spread of the virus. In addition, doctors further refined their treatment practices and began to understand how to most effectively deal with patients who have the virus, especially when it comes to helping them recover and manage their symptoms.

Therefore, it appears that the increased number of people walking around with coronavirus in April 2020 can be largely attributed to progress in the fields of medicine, mass vaccine production, and public health awareness. It is a promising sign that, despite the mounting number of cases, people are learning how to contain and prevent its spread, as well as learn how to manage and treat its effects.

The current diagnosis of Covid-19, the disease caused by the novel coronavirus, has presented us with a conundrum. In this 'casedemic', positive tests do not necessarily indicate an actual 'case' of the virus. All it indicates is that viral RNA was found in the sample. In reality, the vast majority of individuals who test positive will not even show any symptoms, and in extremely rare cases, may only exhibit mild symptoms. Many of them don't even have enough of the virus in their system to spread it to anyone else. Nevertheless, we have all been counseled to follow the guidelines of self-isolating for at least two weeks after receiving a positive result, no matter the severity of symptoms. This is because medical professionals have to remain extra vigilant and err on the side of caution to protect everyone's well-being. While it is unfortunate that many individuals must go under self-quarantine without even having the chance to exhibit symptoms, it only serves as a reminder to us all of why it is so important to take safety protocols such as social distancing, face-mask wearing, and handwashing seriously.

It is concerning that government officials like Doug Ford and the media are reporting such inflated numbers that are leading to unwarranted panic and a further governance of preventative measures, such as the mandatory wearing of masks and lockdowns. It is understandable to anticipate that people will be quick to highlight any statistical increases in cases after the long weekend, often compelling the government to unreasonably invoke another lockdown instead of deepening their understanding and research of the facts. Something which is worth mentioning is the opioid crisis, how it was overlooked and still remains largely undiscussed. It is inevitable that if we continue relying on a sole source of information without conducting our own research, then this same pattern of misinformation is likely to appear and have an arguably deleterious effect.

There has been an alarming trend in government inaction regarding the use of PCR tests to detect the presence of disease-causing RNA - the alarming part being that the recommended Cycle Threshold of 20 isn't even being followed. Instead, much higher levels of 35, 38, and 48 being recorded have been reported. It is concerning, because increasing the Cycle Threshold levels can have a major impact on the number of false positives being identified. To fully understand the effects of adjustments made to the Cycle Threshold, it is important to have an understanding of what the PCR test itself is designed to do.

It is important to remember that PCR tests are not designed to detect diseases, but rather specific RNA patterns. This is why it is so important to keep the Cycle Threshold as low as possible to avoid unnecessary false positives from appearing. When the Cycle Threshold is raised to 25, for example, it is estimated that 70% of positive results identified by the test do not actually pose a threat of illness, due to the over-inflated nature of the positive results.

The PCR test is an essential tool for helping to mitigate the risk of communicable diseases, however, it is equally important to recognize the consequences of using Cycle Thresholds that are too high. If the Cycle Threshold is not kept low, countless false positives can be identified and acted upon, leading to misinformation and unwarranted panic. Similarly, if the Cycle Threshold levels are kept too low, the test results can be unreliable, potentially missing cases of disease that should have been identified. Unfortunately, both scenarios can potentially have disastrous consequences, and it is the responsibility of both governments in Canada and the US to ensure that the appropriate Cycle Threshold levels are used to ensure accuracy.

It is important to look beyond the news headlines when it comes to something as serious and significant as the Covid crisis. When researching the true facts and figures relating to the pandemic, reliable information should always be sought from trustworthy sources. Reading peer-reviewed scientific papers from renowned universities and hospitals across the globe is an excellent way to stay informed and educated on details relating to aspects of the virus. Additionally, consulting Canadian government websites that provide statistics on Covid cases and mortality rates is another source of potentially invaluable information. The Centers for Disease Control and Prevention's website may also house lesser known, yet important facts that are worth examining. By delving deep into the presence available data, there may be a chance to uncover inconsistencies or conflicts between different sources of information.

For quite some time, I have been referring to this situation as a 'Casedemic' and I can assure you that I am not making any kind of jest. As I continue to investigate and learn more about the issue, it is becoming increasingly clear that this is a catastrophic error and has brought forth widespread and long-term destructive consequences that will be felt for many years to come.

It is already evident that the financial impact has been devastating, as businesses have been forced to close or face drastic cost-cutting measures, resulting in job losses on a scale seldom seen before in history. Additionally, we cannot underestimate the psychological pain that isolation and loneliness have inflicted on people who, for months now, have been left unable to see their friends and family in person, despite living in an era of ever-closer contact.

When we look back after the dust has settled, it is likely that we will be able to appreciate this 'Casedemic' as yet another crucial lesson in the need for greater awareness of the potential impacts that our actions can have. We must strive to ensure that this never happens again, and commit ourselves to making wise choices that take people and the planet into account. Until that happens, we will be living with the consequences of this ill-advised misstep for years to come.

The stark reality is that the occurrence of Covid-19 related deaths in Canada has been particularly concentrated among certain at-risk groups — primarily the elderly living in long-term care and assisted living facilities, many of whom are over the age of 80. These individuals, in addition to those who are obese, have been extensively impacted by the virus due to the increased severity of their underlying health conditions.

This should serve as an important reminder for the need for preventative measures to safeguard these vulnerable populations, as no one should overlook the tragedy of these deaths. In Canada, for example, reports have indicated that over 80 percent of COVID-related deaths have been a result of outbreaks in long-term care facilities. In comparison, obesity has become an ever-increasing challenge that has spread rapidly in recent years, affecting those of all ages in both Canada and the United States.

Statistics like these demonstrate just how important it is to take all necessary steps to protect vulnerable populations and ensure we are taking the right precautions to prevent the spread of this virus, as well as other pandemics.

In Canada, 700 people lose their lives each day to a range of causes, and on the 18th of September 2020, the daily average was slightly lower, with only five deaths linked to the CoViD-19 virus. Of the yearly total of 260,000 deaths, only 9205 have been connected to the pandemic. Even if nursing homes and retirement communities are excluded from the count, the total number of CoViD-related fatalities is still under two thousand - a figure that is the equivalent of little more than three days' worth of the country's normal daily mortality rate. This startling contrast between the deadly virus and the draconian measures taken to combat it cannot be ignored; the disconnect between the risk presented by CoViD-19 and the severity of our response is something that the politicians in charge do not seek to draw attention to. Rather, it seems that they perpetuate a sense of danger and apprehension in order to mask any mistakes they may have made in the wake of the pandemic, and to shield themselves from criticism over the far-reaching implications of the lockdown. Unfortunately, precisely how much damage the 'cure' has inflicted on the nation is impossible to measure due to its ever-sprawling reach. However, the implications of the spread of lies and propaganda may be thought to be even more damaging than the virus itself. It is high time that, in the interests of safeguarding the wellbeing of their constituents, the politicians and journalists who have been instrumental in the management of this novel and unexpected situation start being honest and factual without the bias that their masks of ignorance may have otherwise brought.

ONTARIO COVID STATS

It's astonishing to me that there is still so much fear and panic surrounding the latest numbers from Canada's statistics. It's a stark contrast to what I see when I actually look at the numbers, and it leaves me feeling frustrated and somewhat angry. I don't understand why the Government continues to push measures that don't seem to have any impact, but at the same time, I understand why people are scared, as news outlets have been spinning the numbers in their favour for their own benefit. However, my advice is for everyone to do their own research instead of relying on someone else's interpretation of the numbers. When I take the time to look into the Canadian statistics myself, I'm looking for that aha moment - the moment where everything comes together and I understand what I've been missing. Sadly, this has yet to happen and I'm still left wondering why more people aren't doing the same. For example, a recent study showed that only 13% of Canadians agreed that the Government has done a good job of handling the pandemic. If more citizens took the time to reach their own conclusions, I'm sure that number could be much higher.

/

THE PHARMA COMPANIES

Pfizer is one of the world's leading pharmaceutical companies and its rise to the top of the industry is an interesting one. The company is based in the United States and has locations across the world. For years, Pfizer has been one of the most successful drug makers in the world and their drugs have been sold globally.

In recent months, their primary directive has become even more important; the development and rollout of their Covid-19 vaccine. The vaccine has been approved in many countries and has been praised by healthcare leaders and patients alike. The vaccine has been the subject of much excitement, but Pfizer is still a business at heart and its primary aim is to make money for Wall Street.

A closer examination of Pfizer and its business practices over the last few decades further highlight this. They are not a company that puts ethics or morals first. Instead, they have been accused of price gouging and getting away with different misdemeanors and even crimes. All this indicates that the real motto of Pfizer is to put profits ahead of everything else.

The mere mention of Pfizer's name conjures many different feelings within us. From reverence and admiration, as they have consistently been at the forefront of the battle against Covid-19, to doubt and mistrust of the company's ulterior motives. In order to make an informed decision, let us delve into the precarious history of Pfizer and examine if it really does hold up its acclaimed medical ethics.

Pfizer is a multinational corporation that has established itself at the global market, dealing in drugs, and most notably now the Covid-19 vaccine. But astute observers of Wall Street understand Pfizer for what it really is - a cash cow. The prime directive here is to make profits, rather than prioritizing the health and well-being of humanity. This is undoubtedly apparent from the numerous misdemeanours, fines, and lawsuits it has encountered over the years.

In 2009 for example, Pfizer was found guilty of extreme and negligent marketing practices, leading to a fine that was later the highest financial settlement at the time in the United States. One of their biggest scandals was the off-label marketing of their epilepsy drug, Neurontin, which again saw the company warned by the US food and drug administration (FDA). The US government also accused Pfizer of overcharging Medicare and Medicaid in one of the largest cases of civil fraud settlement in US history and has been fined overseas multiple times.

So, in conclusion, is Pfizer really a reliable source of healthcare and medicine? Does it have the integrity and moral code to back its words and actions? The facts speak for themselves. Granted, providing a successful Covid-19 vaccine may be a significant achievement, but one must still consider their history of criminal activities before trusting them with their life.

In a landmark charge, the American pharmaceutical giant, Pfizer, was found guilty of mis-promoting two of its major medicines, Bextra and Celebrex, and paying kickbacks to doctors who were willing to prescribe these drugs for uses for which they were not approved. This resulted in Pfizer receiving the biggest fine ever handed down by U.S. authorities, with the company eventually agreeing to a hefty plea deal amounting to $2.3 billion.

This 'misbranding' of the painkiller medication, Bextra, was conducted in contravention of U.S. Food and Drug Administration regulations which stipulate that medicines need to be prescribed solely for their approved uses only. Despite current regulations, this was clearly not adhered to, leading to the courts taking action.

This should serve as an important lesson to all companies involved in the manufacturing and sale of pharmaceutical drugs. Not only must they adhere to all local regulations, but they should also be aware that, in cases of flagrant abuse of the regulations, significant legal action - and hefty fines - may be taken.

In the 1990s Pharma Troubles

The pharmaceutical giant Pfizer was embroiled in a serious and deadly scandal involving defective heart valves that led to the deaths of more than 100 people. It was later revealed that Pfizer had maliciously and willfully misled government regulators about the potential dangers associated with their product. In response to the accusations, the company agreed to pay a hefty sum of $10.75 Million to the US Department of Justice in order to settle the charges of deliberately misleading regulators.

The tragedy serves as an exemplar of how defective and dangerous products can have fatal consequences when not effectively monitored. Furthermore, it raises serious questions about the integrity of large companies, who may prioritize their profits over public safety. With so many lives at stake, it is paramount that corporations are held to the highest standards of ethical action, take the necessary steps to protect the public, and assume responsibility when their practices result in devastating losses. Clearly, the negligence of Pfizer, in this case, was inexcusable.

Pharmaceutical giant, Pfizer Inc., recently settled a lawsuit for an astonishing sum of $60 million related to their diabetes medication, Rezulin. Numerous patients had tragically passed away as a result of using Rezulin, which had caused them to suffer from catastrophic acute liver failure. This class action lawsuit was settled in a court in Texas, and created a landmark legal decision that has echoed through the medical industry across the globe.

The story of Rezulin, while deeply concerning, is unfortunately not an isolated incident. In recent years, Pfizer has been involved in a number of lawsuits and settlements due to their products. In the case of Rezulin, the company had failed to properly warn local governments, health professionals and consumers of the dangerous side effects of the drug.

As a result, many medical professionals, institutions and patients were unaware of the drug's dangers until it was too late. The legal battle to hold Pfizer accountable for their negligence was an arduous one, lasting for more than a decade. Ultimately, the court found that Pfizer had not acted in good faith and had not taken appropriate steps to protect patients from the potential risks of their medication. The $60 million payout was the result of that ruling.

The situation regarding Rezulin has been a deeply saddening one for many individuals and families, but it has also served as a reminder that we must remain vigilant as consumers and medical professionals when it comes to recognizing the potential dangers associated with certain medications. The outcome of the Pfizer lawsuit is an important reminder to us all that no matter the size of an organization, when negligence or recklessness is caused, the consequences will be significant.

In the United Kingdom, Big Pharma giant Pfizer has been charged and fined €90 million for shamelessly and unscrupulously overcharging the National Health Service (NHS), an organisation funded by the taxpayer. As part of the deal negotiated with the NHS, Pfizer allegedly upped their prices by an incredible €48 million every year, a figure which should not have exceeded €2 million per year. This alleged breach of the contract damages the NHS's already-strained budget and causes the taxpayer to pay more for healthcare than necessary, and at a greater expense than other countries across Europe.

The impact this had on the NHS, and on the British public, cannot be underestimated. The NHS is heavily dependent on taxpayers' contributions and makes sure all citizens have access to a wide range of treatments and services. Overcharging by private companies only serves to increase the costs borne by taxpayers, while delivering little or no value. This hefty fine is intended to serve as deterrent against similar tactics that could be used in the future. Hopefully, this punitive action will be a wakeup call to pharma companies considering implementing such practices, that they will not be tolerated in the UK and that the law will take its course.

In 2004, Pfizer, a leading U.S.-based pharmaceutical company, came under fire from the United States Department of Justice when it was accused of engaging in illegal activities to illegally promote its anti-seizure medication, Neurontin. According to the charges, Pfizer had bribed health care providers including doctors and pharmacists to persuade them to recommend the fast-acting drug for conditions and treatments outside of those officially approved by the United States Food and Drug Administration, such as depression and migraines.

To settle the criminal charges, Pfizer, under the advice of its lawyers and financial advisors, agreed to pay a sum of $430 million in restitution and fines to the United States government. This was not only the largest settlement of its kind at the time, but it was also seen as a warning to other large drug makers that they should ensure full compliance with all FDA guidelines.

The criminal case against Pfizer highlighted the dangers of so-called "off-label" advertising, which occurs when manufacturers market drugs and treatments that have not been thoroughly evaluated by the FDA and do not carry the agency's official approval. Such tactics are deceptive to consumers and can have serious long-term health implications. The Pfizer case demonstrated that the government is committed to cracking down on companies that violate safety regulations and engage in unfair practices.

In 2011, a jury found Pfizer committed racketeering fraud in its marketing of the drug Neurontin. Pfizer agreed to pay $142.1 Million to settle the charges.

Pfizer Inc., one of the world's largest pharmaceutical companies, recently made a startling announcement - they had funneled around $20 Million to nearly 4,500 doctors and healthcare professionals in exchange for them to deliver speeches and talking engagements on Pfizer's behalf.

These payments have become something of a controversial issue over the years, as some feel the payments might encourage a monetary incentive for healthcare professionals to recommend Pfizer brand medicines over competitive alternatives. Big pharmaceutical companies are facing more scrutiny lately, leading some states to take action by requiring more transparency. A new law in the U.S. now requires drug companies to disclose any payments they make to healthcare providers of more than $10 - whether in the form of speaking engagements, research funding, or consulting.

While speaking engagements are certainly not uncommon in the health care industry, and often serve a valuable function in communicating new medical research and treatments, the nature of these payments raises important questions. Is this simply a case of a drug giant leveraging its influence to influence medical practices? If so, it could be distorting the normal push and pull of the industry and giving Pfizer an an unfair market advantage.

Of course, Pfizer has been quick to deny that any kind of "quid pro quo" exists between the company and the healthcare professionals who have been paid to speak on their behalf - but in the current climate, it's clear that more transparency is needed to ensure the trust between healthcare providers and the companies seeking their services is fostered in a transparent and ethical environment.

In 2012, the United States Securities and Exchange Commission (SEC) declared that they had made a settlement with pharmaceutical giant, Pfizer, to the tune of $45 Million. This meant that the organisation had managed to resolve the charges which were made against the company's subsidiaries, who had allegedly been paying off healthcare professionals, such as doctors, in foreign locations to enrich their sales.

This startling series of events highlights how the incentive to increase sales and profits can lead organisations to make ethically dubious decisions. Indeed, in many cultures and countries, bribery is illegal and can have dire consequences for all involved. Revelations about this kind of illegal behaviour have seen a number of other companies being held to account, including Siemens and Alstom who, in 2008, had to pay out in excess of $800 Million to resolve bribery charges with the SEC.

Ultimately, this episode has served as a stark reminder of the importance of abiding by the legal regulations of your home country, no matter where your business operations may be taking place. Organisations in the U.S. must ensure that they operate within the boundaries of the law and refrain from making decisions which could result in being held accountable by the SEC.

The American pharmaceutical giant Pfizer has been sued in a United States federal court for conducting medical tests on children in Nigeria, without consent from their parents. Reports indicate that Pfizer used experimental antibiotic drug Trovan as part of their testing, and became the subject of a landmark $75 million settlement at the Nigerian court, where the company agreed to pay a sum to the affected families of the children. On top of that, it is also known that Pfizer paid an undisclosed amount in the United States to settle similar charges.

The situation created a form of public outrage, as parents of the children who were used as guinea pigs naturally felt that their trust had been violated, and did not want the same situation happening with their children or any other young person. There was even a demand to unpick the entire incident, so that similar cases do not take place in the future and similar unethical practices were exposed. This demand was especially pressing as the company had been accused of conducting a number of clinical trials in the past, on vulnerable communities around the world, without consent or proper due care.

Press coverage and protests ensued, and as a result, members of the U.S. Congress and others around the world have felt the need to step in and act, in order to prevent anything similar from taking place in the future. This is in addition to the established court system, which is the only place in which a party can be held accountable for mishandling the treatment of these vulnerable children.

In the end, the case has been settled, however, it serves as a cautionary tale for big corporations, showing them the consequences of unethical practices, as well as illustrating the need for greater care and attention when using human subjects for clinical trials.

The pharmaceutical giant Pfizer has recently come under scrutiny for its past actions, as evidence has emerged that it violated international law during the Nazi regime. According to reports, Pfizer was complicit in experiments carried out on prisoners of war, against their will, in contravention of the Nuremberg Convention, which was established in 1945 as an affirmation of the highest standards of medical ethics and human rights following World War II.

These experiments included horrific acts of torture, as well as other disturbing forms of illegal medical experiments. One of the most notorious examples was the Sulfonamides Infection Program, in which thousands of prisoners were deliberately infected with diseases, in order to test the effectiveness of different treatments. This program was in direct violation of the Nuremberg Code, which was intended to prevent the exploitation of human beings in experiments.

In light of these revelations, it is crucial to hold Pfizer accountable for its past indiscretions and to ensure that modern medical research and experimentation are conducted in compliance with the highest ethical standards and in respect of human rights.

Amid growing outcry and severe condemnation from around the world for the unfair practices of some drug companies in charging exorbitant prices for medications in developing countries, pharmaceutical giant Pfizer announced that it would donate a staggering $50 million for a lifesaving HIV/AIDS drug to South Africa, a nation struggling to contain the spread of the epidemic. Their decision was commended by international aid organizations and those working towards greater global equity in health care and the reduction of poverty.

Unfortunately, Pfizer subsequently was not true to their word, rather than making their promised donation, they reneged on their commitment and failed to honour the promise they had previously made. This left many people in South Africa feeling betrayed and unfairly let-down in the face of a worsening public health crisis. While it is uncertain why they decided not to provide the money as initially pledged, this example serves as a stark reminder of the ongoing issue of drug company profiteering at the expense of the poor and needy and the lack of transparency and accountability of these corporations.

Also note, Pfizer is on track to make over 30 billion dollars from this vaccine in 2021 alone.

Before making a decision about getting the Pfizer mRNA treatment for coronavirus, it is wise to consider getting a simple antibody test to check whether you already have T-cell immunity, as this could influence the vaccination choices you make. Whether or not you have the antibody can mean the difference between adequate protection against the virus and complete protection, so it's worth knowing if you have it in your system before making a decision. Furthermore, it's important to do your research and use reliable, non-biased sources such as trustworthy medical publications and studies, rather than making assumptions. Also, it's useful to use search engines other than Google, such as DuckDuckgo, to look for information as Google is well-known for censoring certain results. DuckDuckgo is renowned for producing more well-balanced results, and can be used for finding impartial and reliable sources about the virus and potential treatments. Ultimately, it's important to take all the available information into account and make an informed decision that is based on your own research.

500 New Cases but what they don't tell you

Yesterday, while I was enjoying the pleasant day outdoors, I overheard a comment that gave me pause. After I had commented on the large number of families out and about, I heard this in response: "We should be savoring this beautiful day now, because there are 500 new cases and a fourth wave may be 'round the corner."

This conversation came to me at a time when I have been increasingly more vocal about the need to better understand what the true public health picture of the pandemic looks like in Ontario. It is critical that we know the whole story, break beyond the limited news cycles, pay attention to the dataset in its entirety and be informed by informed, not filtered stories with only selected data being reported. This is an important issue for the safety and well-being of Ontarians.

To address this, I put together an illustration to display how the 500 new cases, reported on Ontario's own website, translates into a more realistic interpretation compensated for all the data. This information isn't based on opinion, but on the data from its official source which I've verified.

In an effort to make sure that people are well-informed and aware of the true public health situation in Ontario, it is understandable why I feel the need to ensure that data is presented in an accurate, realistic manner. It is only when we have the whole story that we can trust our decisions and those of our leaders.

Numbers Don't Lie

The data presented in the graph is clear, highlighting the amount of time and resources spent on testing people who were not at risk of becoming infected or sick with the virus. In fact, the media reported increasing case numbers without accounting for the fact that over 98% of those tests provided false positives, were eventually recovered, or had mere remnants of the virus in their system that did not render them sick or transmittable. Although the graph may appear to be complex initially, looking through it reveals a pattern. During all the peaks in case numbers, the death count remained relatively the same, which is why I must emphasize the importance of no longer focusing on case numbers, but instead the rates of sickness and mortality. If anyone is uncertain about the data or the information provided in this article, please do not hesitate to message me and I will do my best to respond. The source of the data used can be found in the link provided in the text.

The evidence presented through this data is clear; instead of merely summarizing the events, a deeper look into the prevalence rates of the virus shows that the magnitude of the situation is not being accurately depicted in the media. Thus, it is essential to start concentrating on the grave truth of the situation; which is the count of those who are sick or deceased due to the virus. Without a proper representation of the statistics, the public may not be able to recognize the full scope of the pandemic, which leads to a lack of preventative measures and further spread of the virus. To accurately garner an understanding of the severity of the situation, it is necessary to take an investigative approach, analyzing the data and being mindful of misleading information.

COVID NUMBERS

After downloading the latest data from the Ontario website, I was astounded to discover that there were an impressive 553,000 records associated with it. To get a better understanding of the information, I decided to correlate the data by age, deaths and cases before overlaying the charts in Excel. By doing so, I was able to plot the deaths in each major Canadian city and create a graphic representation of the results.

The outcome of this visual analysis was eye-opening. From the data, we were able to ascertain that while fewer deaths were attributed to the younger population, a greater number of cases was observed in that same demographic. This brings to light the notion that the amount of positive cases reported by Doug Ford should not be taken as the end-all-be-all, and left in isolation, as that ignores the full scope of the issue. Rather, it is vital to note that the more visible mortality rate ultimately presents a truer picture of the virus, not just the number of active cases.

Alberta's COVID Stats – Why Lockdown?

The past several weeks have seen a flurry of news media attention focused on the lifting of COVID protocols and the subsequent decisions made by provincial governments. In particular, the lifting of masks has drawn criticism, with public health expert Dr. Theresa Tam taking a stance of caution in conjunction with the decision. In an effort to dive into the data deeper, and to see if these decisions may be considered irresponsible, I decided to look into the statistics and do some research.

To begin, I examined Alberta's death rates for any given year and compare them with the death rates of Covid observed in 2021, using trend lines and available data. With this data it became readily apparent that Covid is, in fact, not a major cause for concern in comparison to yearly death rates. The data itself is publicly available and can easily be accessed by anyone wishing to fact check the information. Moreover, with the use of simple trend formulas, it's clear that Covid does not pose a much of a significant risk as some have purported.

Ultimately, the data does show that the decision to lift certain protocols, such as the abolishing of masks, is supported by the evidence which can be seen with a quick glance at the statistical trends. In the end, it's a call for everyone to look into the facts and information at hand and consider the statistics, as opposed to relying on media headlines and 'pundits' to make such a precipitous decision.

It is indisputable that the amount of data released in recent months has led to a widespread realization that the worldwide lockdown was, unfortunately, not justifiable when looking at the available numbers. This is precisely why I have been positing the notion of a 'casedemic', as opposed to a real epidemic; although the tragedy of this farce has spanned multiple areas in the public sphere, it simply should not have happened in the first place.

It is difficult to look beyond the immense disruption and disruption caused by lockdowns amongst many other regulations, and instead focus on the numbers, which paint an accurate picture of how this period has really been little more than a prolonged display of human folly. To take just one example, many states in the US imposed restrictions without any visible change in the graph lines of confirmed cases, or other indices which could be expected to move if the measures were having a real effect. The numbers for something as insignificant as public gatherings during the post-lockdown period have screamed in comparison, actually showing a certifiably measurable decrease in direct correlation with restrictions being put in place.

This highlights why we must take the time to fully understand any data presented to us as a society, and recognise that it is only when we look with a clear eye, rather than simply accepting the case at face value, that we can make more informed decisions which have a tangible good in the long term for public well-being.

Know the Facts: The Odds of Dying

It is essential to be as clear as possible in presenting data and statistics, and in this case, the importance of spelling it out cannot be overstated. I wanted to illustrate the difference in risk levels in the population of Ontario, so I compiled some figures from their own data, plus the death rates statistics taken from the United States, along with the respective populations for each. I used standard formulas, easily found on any reliable sites that deal with mathematical and statistical theory. It is important to remember, however, that these numbers reflect the odds for a single year, so you must take that into account when interpreting the chart. In other words, the chances of perishing over the course of one year for people not thought to be at risk of dying prematurely is slim, so needless to say, the risk is negligible in such cases. An indisputable fact that must always remain true is that one plus one will always equal two, regardless of how difficult it may be to accept.

The figures provided by the renowned Centers for Disease Control and Prevention demonstrate that the number of deaths caused by coronavirus alone is likely much lower than initially suggested. As much as 164,280 deaths were attributed to the virus, however, only 6% (or 9,857) of those were solely attributed to COVID-19. The CDC also released further findings, which indicated that 94% of those marked as having died from coronavirus in fact died with an average of 2.6 additional conditions or causes. Such pre-existing conditions were commonly and quite notoriously dangerous health issues, including but not limited to heart diseases, respiratory illnesses, influenza and pneumonia, organ failure, cancer, obesity, and cognitive diseases.

Just why the CDC resorted to changing their methods and obscuring the details of these figures must be further investigated, especially considering that some speculate that the revision was conducted to bolster vaccine promotion by unjustifiably bringing down the number of reported COVID-related fatalities. The Canadian province of Ontario serves as a telling example. Originally, 25,000 deaths were reported, however, when evaluated further, only around 1300 of those were found to have been caused by coronavirus, and the majority of those deceased were of advanced age. This data was made much more difficult to find by burying it deep within the CDC's webspace, making it hardly visible to search engines and inevitable for the masses to access. However, the resourceful CDC website contains the intercepted report - a testament to how thorough and committed the organization is to promote transparency, even if methods may be regarded as questionable. https://www.cdc.gov/.../covid19-comorbidity-expanded....

All in all, the problem of uncovering the true extent of the carnage caused by the novel pathogen is still very much real, and the overreaching mission of untwisting the knot of this pandemic is still very much in its infancy. It is the CDC's obligation to broaden its commitment to the public good by revisiting its classifications, ensuring that all government-sealed truths are unravelled for the assurance of everyone's piece of mind.

Symptoms of Vaccine Deaths

Through rigorous research of reading medical reports and scientific studies, I have been increasingly gathering more and more evidence to develop an informed assumption on this COVID epidemic. Firstly, the Pfizer treatment and other similar vaccines remain as experimental by nature and have not yet concluded until 2023 when studied in Phase 3 trials. I wrote about an issue observed in the trials yesterday, that all participants of the controlled group had taken the vaccine, consequently blocking the trials from advancing further. Moreover, pharmaceutical companies cannot be held accountable if any damage or harm is caused by the vaccines through administration.

Recently, awareness has been brought to the concept that the mRNA used in relation to the COVID-19 vaccine is based on research that previously borrowed cells from aborted fetuses. Although this causes moral controversy, people are also questioning why there is a lack of discussion surrounding this. Christianity has recently encountered disputes in relation to its stand on the aboriginal struggles, yet this moral conflict seems to become overshadowed by the opposing parties. Until now, mRNA has never before been an unresearched vaccine for humans and the Food and Drug Administration (FDA) does not have information available concerning the prolonged state of health for those committed to vaccination. This mRNA consists of a genetic code from DNA, an instruction guide for the cells located in the body. When the messenger RNA is injected, it attaches to the cells and modifies the user's defense system by introducing virus molecules and antibodies. Unfortunately, all the evidence suggests that the manual, which is carried out in the body, lacks enough protection from from the variants and does not existing information on the wide-scale impact of the code, on the human body's natural defenses.

Not enough is mentioned in terms of the side-effects of the vaccines and the hazard it presents to those receiving the shots. The United States Health and Human Services (HHS) Vaccine Adverse Event Recording System (VAERS) accrues the appropriate data and statistics of those who were otherwise harmed through the injection. On this page, after confirming some disclaimers, unfiltered data can be accessed, regarding injuries and fatalities which were once unrelated to the vaccine, however share a stark relationship with COVID-19 alpha. More alarmingly, the Centers of Disease Control (CDC) has been caught deleting and amending the data from this platform, yet it is available on other reliable sources. What is more, the Pfizer and Moderna mRNA shots have not been medically and systemically tested for a lengthy duration and are only able to exist as therapies, rather than vaccines; although it may have been previously suggested that a single dose is sufficient. The link to the patent consists of more elaborated information if people wish to further their understanding.

Comparing Vaccine Deaths – Shocking!

In recent months, the world has seen a surge in the number of people eager to get vaccinated against COVID-19. Even healthy young people with no pre-existing conditions are among those lining up for their jab without considering all the factors at play. It's important to remember that for most people, the risk of catching the virus is relatively low as the global survival rate hovers around 99%. Moreover, effective therapeutics are available to treat the virus very effectively in the event of an infection. With a 99% success rate, why are so many people choosing to take a gene therapy vaccine not yet approved by the FDA, the so-called trials for which were slated to continue until 2024 before being compromised?

Equally perplexing is the lack of discussion about the therapeutics already out there that have been clinically proven to work. Instead, focus is placed on the prospective outcomes of the vaccine rather than the known outcomes of the therapeutics – notably herd immunity. Herd immunity is perhaps the best solution to the virus we can collectively aim for, and yet it seems to have been all but forgotten in the hand-wringing over the vaccine. Ultimately, it is essential that we weigh the potential risks and benefits of all the available tools available to fight the virus, remedies prepared through rigorous trials and the far-reaching implications of herd immunity to ensure an optimal solution.

The confusion and lack of clarity surrounding the mRNA treatments being termed as vaccines has many questioning why we are seeing such a push for their widespread use when there is no evidence that indicates that those who have had the mRNA shot are immune from contracting or spreading the virus. In addition, why are those individuals at higher risk, like those over the age of 70 and those with pre-existing health conditions, not being prioritized for the experimental treatment? When we look to the data, 27 vaccines distributed in the United States during 2008 had more deaths than the two mRNA treatments yet the media coverage has notably been lacking.

These unanswered questions beg the even bigger inquiry of why are governments and health officials continuing to focus on old data and pushing for vaccine passports while not taking into account the implications that the vaccine does not prevent the spread of the virus and could potentially make those who have had the vaccination "super spreaders" of the virus due to the fact that it only intends to reduce the clinical symptoms. It appears on the surface that the narrative around the efficacy of the mRNA treatments is not being thoroughly explored in public discourse and more research needs to be conducted about their true implications.

In conclusion, it is essential to ask why the mRNA treatments are being termed as vaccines when their efficacy in preventing and stopping the spread of the virus has yet to be fully proven and why this lack of exploration in the media is continuing to persist.

DON'T USE THAT ARGUMENT ON ME

Whenever I get into conversations concerning vaccinations, a remark I often hear is the intuitive assumption that those who remain unvaccinated are being irresponsible and placing an undue burden on the healthcare system. More often than not, this comment is accompanied by emotionally charged criticism of those who have chosen not to get vaccinated. It is worth noting that this view fails to consider the nuances of this hotly debated public health issue.

The reality is that the decision to get a vaccine is an incredibly personal one, based on a variety of factors, including religious or philosophical beliefs, medical history, and of course, legitimate safety concerns about potential side effects. Everyone does not have access to the same information or resources, and some do not have the financial ability to pay for the cost of vaccinations. Furthermore, some orders of magnitude of disparities exist in countries regarding distribution of the vaccines and vaccination rates, particularly in low and middle-income countries.

Ultimately, it is essential to remember that this is a complex issue. Rather than resorting to over-simplified judgement or criticism of those who choose not to get vaccinated, our focus should be on understanding why certain individuals or groups make the decisions that they do and how we can provide them with the accurate information and resources to make informed choices. Furthermore, it is this information that is needed to shoulder the tremendous burden of successfully immunizing the majority of the population against the effects of deadly infectious diseases.

The idea that unvaccinated individuals should be completely excluded from society is not only wrong - it's completely unfounded and unsupported by any fact. I have spoken up against this erroneous notion on multiple occasions, providing evidence to explain why it is so misguided. If readers would like to learn more, I invite them to take a look at the link I have provided in the comments section of this post.

One of the main points I have tried to make is that, based on the available science, it is not accurate to assume that unvaccinated individuals are inherently or solely responsible for the spread of infectious diseases. In reality, numerous factors contribute to the transmission of illnesses, including, among other things, population density and density of contact, hygiene standards and availability of clean water, access to healthcare resources, environmental conditions, and genetic factors.

Furthermore, it is also important to note that some individuals may not be able to receive vaccinations - due to weakened immune systems, difficulty accessing vaccinations, and for a variety of other reasons. Unfairly punishing such individuals and ostracizing them from society does nothing to reduce the risk of contagion, and is both unethical and inappropriate.

The decision whether to be vaccinated is a very personal one, and should not be dictated by social pressures or fear-mongering. It is a choice that needs to be made with the appropriate information and facts in mind. Ultimately, only education and awareness can lead to a more informed and healthier public.

The crucial question we must ask ourselves is, will we see a similar pattern of moralizing and outcasting applied to other people or groups in our society whose cost to the healthcare system is thought to outweigh their benefit? Are we willing to undermine these individuals and make them feel that they are an unwelcome burden? We know that this has already been done with vulnerable communities, such as migrants and refugees who, due to language and cultural barriers, often lack access to healthcare, as well as to previously incarcerated individuals as they struggle to reintegrate themselves into society. In such cases, the message sent is that the medical and financial needs of this population are secondary to their responsibility in, supposedly, overburdening our healthcare system. This message, which is clearly unmerited and unwarranted, only reinforces existing disparities and creates a division between 'the deserving and the undeserving'. It is troubling to think that similar approaches may be used in the future as a way of controlling access to healthcare.

It is important to approach this topic with the utmost sensitivity, however relevant to my point that needs to be made, we must look at the issue of obesity. While I understand I need to get out and walk more, there are people who are truly struggling with being overweight and it can be a difficult thing to accept. These people often face judgement, commenting and even harassment for struggling with their health, which can be quite triggering for them.

There are many theories and social dynamics around why people struggle with obesity. It can be attributed to a combination of environmental, genetic, and metabolic disparities, among other issues. For example, food deserts in cities often mean that people don't have access to healthy food options and are more likely to purchase unhealthy food items due to cost or availability. Health conditions such as hypothyroidism, Cushing's Syndrome and insulin resistance can also influence weight gain. It's also important to note that external pressures play a role - media can be incredibly impactful when it comes to creating and upholding unrealistic beauty standards which can be incredibly detrimental to self-perception.

In conclusion, it is important to recognize the challenges of obesity, consider the variety of factors at play, and approach the issue with sensitivity. We must realize that people struggling with being overweight were not necessarily always this way and need to recognize that there is an issue there rather than shaming them for it.

In 2006, the estimated total cost to the province of Ontario for the prevalence of excessive weight was astonishingly pegged at six billion dollars. Strikingly, that number is thought to have surpassed nine billion dollars by the present day, leaving each individual afflicted with this condition shouldering an estimated average expense of more than $1,400.

This is a significant expense for anyone, and places a heavy financial burden on those with excessive weight. We can illustrate this with a use-case: take Fred, a 30 year old man from Ottawa who has a Body Mass Index (BMI) of 29.9, firmly placing Fred in the morbidly obese category and facing the dilemma of the associated medical costs. In addition to the physical, mental, and emotional tolls this can have, Fred is now faced with a big financial burden that he must bear through understanding and investing in the health and wellness specializations required to strive for a normative BMI. These expenses can vary from physician appointments, tailored physical therapy plans, nutritionist consultations, and other such treatments and advice necessary to ensure Fred's longterm health. Without the necessary financial resources, Fred may never be able to recuperate from his excessive weight, and this is a harrowing thought.

This is by no means an isolated or unique case – obesity continues to plague those across Ontario, and the trend is reflected in the large monetary sum of which we spoke. However, committing to a healthier lifestyle and seeking out appropriate help and advice can go a long way towards not only saving Fred and others like him money, but also saving their lives.

Nationally, research has revealed a startling finding - the economic burden of Obesity in the United States surpasses that of Tobacco Smoking by 27%. This equates to a colossal $23.3 Billion attributed to Obesity compared to the comparatively lower $18.7 Billion figure linked to Smoking.

To better illustrate the magnitude of the economic burden associated with Obesity in the US, it would equate to losses across all fifty states combined in terms of gross domestic product (GDP). For instance, in states like California, where the GDP for 2018 was $2.7 Trillion, it translates to a staggering $7.56 Billion economic burden being attributed to Obesity - equivalent to 0.28 percent of the total GDP figure. Likewise, the state of New Hampshire, with a GDP of $71.6 Billion, faces an economic burden of $151 Million due to excess weight, representing 0.21 percent of the state's annual GDP.

Excessive amounts of both obesity and smoking have been identified as persistent public health concerns in the US and this resurgent data serves to compound the already significant burden of, and cost associated with both. A concerted effort is needed by public health advocacy groups in all fifty states to reduce the incidence and prevalence of Obesity and Smoking.

In recent times, conversations about public health initiatives have brought about a critical issue for us to consider: what about those who do not comply? In the midst of a global pandemic, those who have chosen to not be vaccinated have come under heavy scrutiny. Unfortunately, some have gone so far as to classify these individuals as selfish and stupid, which begs the question—what implications could this have on other social based health issues?

As we all know, obesity and cigarette smoking are two of the biggest public health concerns of our time. Will we begin to regard those who suffer from obesity or smoke as being selfish or stupid, in the same way as those who are unvaccinated? Could those in the health community even consider introducing mandatory diets or banning people from purchasing cigarettes if we go down the path of punishing individuals who have refused to conform to health expectations?

It is impossible to ignore the potential harm incurred by these practices; not only are these actions unjust and ethically questionable, but they could also discourage those seeking help from coming forward, effectively putting them in an even worse position than before. At the same time, we need to remember that, for the most part, we are all capable of making our own decisions and must accept responsibility for our choices. Punishing those who do not fall in line with conventional thinking is arguably against the law of personal liberty and could further contribute to an environment of judgement and mistrust.

Ultimately, we need to be careful with this sort of thinking since it could have far-reaching consequences. While it is important to uphold public health standards, taking measures to punish those who fail to comply without providing comprehensive support and education may be deemed unjust and could have serious ramifications.

Do we block some from purchasing pop and chips by not allowing their bank cards to work for those types of purchases?

As the debate over mandatory vaccinations continues to be at the forefront of public discourse, many people have grown concerned over the potential of restrictive regulations being imposed on those who choose to remain unvaccinated. While proponents of these policies have argued that they are needed in order to protect public health and safety, the implications of imposed restrictions have been met with a great deal of scrutiny as the implications for individual rights and freedoms become more apparent.

A recent opinion study conducted among a random sample of the population showed a clear divide between those who felt current regulations were too restrictive and those who felt exemptions went too far. It is clear that the issue is not a straightforward one and that the various arguments existing on either side must be taken into consideration when assessing the risks and benefits of the measures being proposed.

It is important to remember that while some regulations may be necessary in order to protect public safety, it is also important to consider how they impact the basic right of individuals to make health decisions for themselves. This has been seen in some European countries, where despite large public outcry, the government has chosen to impose fines and other penalties on those who choose to opt out of vaccinations. In the United States, controversy has surrounded attempts to restrict access to public spaces and services to those who are unvaccinated, despite the potential negative implications of this policy on the well-being of those engaging in activities such as attending school, engaging in trade, or exercising other fundamental freedoms.

Given the current tensions surrounding the issue, it is clear that a reevaluation of the proposed restrictions is necessary in order to ensure that individual rights are respected and public safety is not put at risk. It is only by engaging in a meaningful and informed discussion that we can ensure that the policies enacted are fair and just for all affected stakeholders. Therefore, it is essential that we rethink current regulations in order to ensure that we are protecting our societies, our freedoms, and our basic human rights.

Oops, they did it again – CDC makes an error

For several days now, mainstream media outlets have been making a continuous loop of reports claiming that 99% of COVID-19 fatalities have been among those who remain unvaccinated. While there are not any actual data sources to back this claim up – instead, the evidence suggests that the opposite might be true – these reports have, nonetheless, prompted an onslaught of insidiously degrading propaganda that paints those adults and children who have declined to receive the experimental COVID-19 shots as "variant factories" and "incubators" of disease. However, what is lacking from these reports is any mention of the dubious science that underlies the data that has been used to create this narrative.

The type of science referred to as 'Covid Science' is a rapidly developing field of research and evidence gathered through real-world studies and experiments is just beginning to trickle in. What's more, the experimental vaccines authorized for emergency use which have been manufactured and approved in response to this novel coronavirus pandemic lack the necessary decades of safety data and studies about long-term effectiveness that we would expect to see for traditional vaccines. Consequently, to depend on data from studies relying on, for example, antibody levels as a basis for drawing conclusions about the efficacy of a specific vaccine is, at best, precarious.

Rather than relying on biased and often irresponsible media reporting to make decisions about health and safety, it's important that we take it upon ourselves to seek information from unbiased and reputable sources, as well as thoroughly research any data that media outlets or other sources claim to be valid. Doing so is the key to lasting, individual freedom and collective safety – both in times of pandemic and outside of them.

In light of the Provincetown study, which included people who were partially vaccinated or whose vaccination state was unknown, questions have been raised regarding why such data was used to form the narrative that has been presented to us. This data was taken from a time when many were still not fully vaccinated, and thus did not include the most up-to-date information which could have resulted in drastically different findings. This begs the question of whether the data from the Provincetown outbreak should have been considered a reliable source in forming the narrative. It also raises potential concerns about the potential for the results to have been manipulated for the purpose of creating a certain narrative.

Moreover, there are more issues with the Provincetown study than can be addressed in this article: for example, how did such an oversight occur in the first place? Was it simply an innocent mistake or was it a calculated move to distract from the more up-to-date data that could have potentially been more damaging to the narrative being advanced? It is no misnomer to state that we have seen far too many instances where particular narratives are emphasized while any contradictory evidence is swept under the rug in pursuit of a larger agenda.

It is of paramount importance that people take the time to conduct their own research and become discerning consumers of information so that they can resist these tactics and ensure that they are getting the whole story. People should look closely at the techniques used by news outlets and organizations, to ensure that such institutions are not advancing their own interests by sacrificing facts and misrepresenting the truth.

Buyer's Remorse

Even though mRNA vaccines may promise to be a promising solution for combating the global spread of COVID-19 alpha, there are justifiable concerns about the overall timing of their introduction, the current efficacy of the vaccines, the safety, and the risks associated with them. Despite the assurances from the government that those who were vaccinated were provided with the necessary antibodies to protect against the virus, serious questions regarding these matters must be addressed.

The swiftness of the vaccine determinations and how quickly the mRNA-based vaccines were made available internationally has raised questions about the safety testing protocols that were followed. With hundreds of millions of doses already distributed throughout the world, there is no time to waste to answer such questions. Furthermore, many are equally concerned about the efficacy of the vaccines and whether the mRNA variants themselves are effective in eliminating the virus. In addition, there are serious worries about the long-term side effects that could arise from the vaccine, as well as potential safety issues that have yet to be discovered.

While these types of vaccines are seen to be a major step forward in addressing the global pandemic, it is crucial to analyze the potential risks of mRNA-based vaccines and develop mitigating measures to address them. Moving forward, it is important to ensure that the latest research evidence is taken into account and that the focus is on building confidence in the public regarding vaccination programs and mRNA in particular.

Vaccinated people have died from COVID-19 alpha.3.

The widespread availability of the COVID-19 vaccination is certainly a positive step forward in combatting the SARS-CoV-2 virus, however, even with the vaccine, it is not necessarily a guarantee against infection by the Alpha, Delta, and Lambda variants of the virus. To prepare for any further mutations of the virus, booster shots will likely need to be created and administered, much like with the annual influenza shots. What makes the approach to combat SARS-CoV-2 unique is the introduction of genetic instructions into the vaccine. Never before in medical history has such a process been deployed at an international scale, and while it is uncertain what the long-term effects may be, it has unquestionably played an integral role in the decrease in cases of the virus around the globe.

After the discovery of the delta variant, the government increased the expectation for all people to get the vaccination.

The use of an untested mRNA synthesis reagent has never been attempted on human subjects before. This means that the Food and Drug Administration (FDA) has no concrete data or evidence on the potential risks and side-effects the vaccine could have on people in the long-term, which is why the trials for the mRNA vaccine are continuing until 2023. Moreover, even if any of the participants experience severe allergic reactions, deadly side effects, or any other medical abnormalities as a result of the vaccine, they sadly would not have any legal recourse against the government or the pharmaceutical providers. In simpler terms, the volunteers and participants of the trials, who are receiving the COVID-19 alpha vaccination, are essentially the 'test subjects' of the experiment. Additionally, any medical advancements or discoveries made concerning the MRNA vaccine, or any product made or derived from it, would be the property of the government or provider with no remuneration to the inoculated persons. This explains why the majority of trial runs and tests are done on non-permanent volunteers, as taking part in testing such a vaccine exhibits a high level of risk with no potential rewards.

The COVID-19 alpha vaccination is based on mRNA technology, which involves injecting a specific genetic code from DNA into the cells of our body. When these mRNA molecules are inserted, they bind to the cells in our immune system, transmitting a set of instructions which reprogram these cells to create virus-like entities. In theory, this should trigger the body's immune system to identify them and build the necessary antibodies to fight off the virus. Unfortunately, the effectiveness of this mRNA-based vaccine technology on different virus variants remain unknown, and there is currently no scientific study to tell us how the coding of these mRNA molecules impact our body's natural capability of fighting off other illnesses such as flu or colds in the long run. Will this process increase or diminish our bodies' capability to combat other viruses? We don't have sufficient evidence to answer this question.

The truth about the undeniable repercussions of the COVID-19 Alpha vaccinations is not being communicated in a completely honest way from our governments, the CDC or any other medical regulatory group. This negligence of accountability has caused many individuals in our society to become unaware of the numerous negative health impacts and deaths that have been experienced by those that have been administered this vaccine. Although this information isn't easily accessible, there are official government statistics available to the public that document these incidents. If you visit the United States Health and Human Services (HHS) Vaccine Adverse Event Reporting System (VAERS) and agree to the disclaimer, you will come to understand the magnitude of this problem. After meticulously researching this data, I have personally downloaded over 450 thousand records and performed my own analysis, which (unfortunately) confirmed the very serious nature of these adverse reactions. It is absolutely essential that these findings are shared openly, without apprehensions regarding conflicts of interest, in order to protect the health of the public.

Currently, the CDC was caught deleting and editing data on this source. One would ask why?

The efficacy of nanoparticles in the human body has been rigorously studied and documented by a number of researchers, with science and academia noting the range of effects these particles can have. For example, one study from Texas A&M University found that nanomaterial can travel through the human body through different pathways, with the ability to cross cell membranes, the blood-brain barrier, and travel through mucus, saliva, and sweat. These results further suggest that, as nanoparticles move through the human body, they can trigger different cellular activities and might even lead to changes to existing tissues and organs.

Additionally, a report from the National Center for Biotechnical Information outlined how these nanoparticles can interact with living tissue. This includes the ability to attach to certain molecules within the body, influence how cells form and interact with one another, and affect the immunological response of the subject. Furthermore, researchers were able to observe how nanomaterials can enter biochemical pathways and affect physiological systems, indicating their versatility to alter existing cellular functions.

In short, research and documentation supports that nanoparticles are capable of moving through the human body and influencing cellular activities. This opens the door to a range of potential implications and applications, with researchers now beginning to study and explore how nanoparticles might be used in therapies, detection systems, and other treatments.

The potential that mRNA technology holds has been widely lauded, and with good reason; it holds the promise of providing emerging treatments and even cures for life-threatening diseases. But it is not only the potential of the technology that has been brought up; many skeptics, like myself, have expressed concerns about the current implementation, timing and safety of the delivery of mRNA treatments. The efficacy, as well as any possible coverup related to it, are also part of this debate.

Unfortunately, calling out these concerns has often been met by misguided criticism that we are 'anti-vaxxers' or conspiracy theorists. However, this misses the point. An informed and honest discussion about the validity of the questions being raised is necessary for the public good. It is not acceptable to simply raise an eyebrow and ignore the mounting evidence from reputable virologists, doctors, professors, universities and scientists.

Ignoring this growing chorus of voices or pressing ahead blindly will lead to a collective awakening-and it may not be a pleasant one when that ultimately happens. We must become informed and talk about these issues in order to prevent it.

The Lockdowns have become the Danger!

The severity of the Covid-19 pandemic has become abundantly clear with its various mutations and ever-evolving states. Numerous experts, scientists and data suggest that these pandemic-instigated lockdowns, although well-intentioned, have done more harm than good in the long run. From small businesses in towns across the countryside, to local boutique stores, these establishments have suffered drastically from the economic fall-out. Large corporations however have seen a stark contrast, with Amazon and Walmart being prime examples of entities that have reported a hefty amount of profits as a result of these lockdowns. It is essential that we begin to move away from these restrictive measures, and instead rely on scientific, data-driven solutions in order to counteract the unpredictable nature of the pandemic.

The horrendous consequences of our actions are beyond comprehension; we have sadly brought significant disruptions to the heart of our families and caused mental health issues to surge to levels that no previous generation has ever experienced. Every day, we witness heart-breaking statistics presenting an abysmal picture of people taking their own life, which is a result of our neglect towards remedying the mental health crisis commonplace in our society today. It is not hard to imagine the immense cost this has had on our collective emotional welfare; we must strive to do better in the future and take steps to prevent its reoccurrence.

Moreover, our current course of action has depleted our treasury and racked up a whopping national debt of trillions of dollars, which will leave our future generations with an indefensible financial burden to shoulder long after we are gone. Without remedying the sources that have contributed to this lethally high debt, our children and grandchildren will bear the weight of this burden for decades. It is our moral obligation to ensure that we take concerted efforts to make amends for our current fiscal recklessness and alleviate the looming financial doom awaiting our progeny.

If we do not soon wake up to the reality of the Covid-19 pandemic and become more prudent with our daily habits, there will be serious repercussions if we allow the fear-mongering disseminated by our government and media to persist. If we do not take steps to correct and regulate our movements, the consequences will linger for years to come. Our local small businesses, the backbone of many communities, have been fighting for survival since the outbreak began. Many of these once viable organizations have closed, leaving their employees without a source of income. The ripple effect of this has caused many to take drastic measures, including taking their own lives. Left with no other option, many families find themselves uprooted as unemployed parents are forced to move back in with their retired parents who were not prepared to dip into their retirement savings to support them.

It was promised that the implementation of vaccine distribution and lockdowns would allow us to return to what we knew as normal. However, it appears that the reality is much different than what we were promised; in the most recent reports from the Centers for Disease Control and Prevention (CDC), it has become clear that the data does not support the idea of full immunity for those who have received the vaccine. This news is concerning not only for individuals who have been vaccinated, but for policies seeking to facilitate the return to a pre-COVID world.

Indeed, some health experts have concluded that despite the many vaccinations administered and the lockdowns in place that the pandemic continues to spread, leading to the suggestion that additional measures will have to be taken in order to bring the threat of COVID-19 to an end. What this means is that we can not guarantee a return to "normal" any time soon, no matter how many vaccines are administered or how many times a region locks down.

Instead, a more comprehensive approach must be taken to ensure the safety of all individuals involved. This could include utilizing contact-tracing and other methods of limiting the spread of disease in addition to more traditional interventions such as lockdowns and the administration of vaccines. Each step must be studied carefully and rigorously before decisions on how to proceed are made.

Ultimately, the news that the CDC has now shared shows us that the effects of the pandemic are still far-reaching and complex. This emphasizes the need for comprehensive, evidence-based solutions in order to return to a life that resembles our pre-COVID normal. Without a concerted and collective effort, we will be stuck in a potentially indefinite state of pandemic limbo.

For months I have been trying to get the word out that the mRNA injection so-called "vaccine" has not been effective and many people were deceived by the false claims that it would provide protection to our loved ones. The truth of the matter is that it provides more benefit to the individual instead of providing protection to the group. What is worse, and I am not the only one saying this, is that those who have been vaccinated may inadvertently become the next super-spreaders of the virus. We are witnessing this phenomenon now in certain locations that have had extensive levels of vaccination. The Delta variant is more prevalent in these places showing that the spike protein from the vaccine is creating new mutations of the virus.

It appears that a great many individuals are choosing to completely disregard and categorically deny the validity of conflicting viewpoints, for fear of having to confront the reasoning behind their own choices. Just recently, it seems that those who are strongly in favor of vaccinations have begun to take an adamant stance, exhibiting aggressive posturing to defend their respective decisions. Unfortunately, and although understandable, this sort of mentality can be very damaging in the long run, not just to public discourse, but also to the advancement of humanity as a whole.

By punishing those we disagree with and excluding them from the conversation, we're closing off any potential avenues of compromise, and further entrenching ourselves within our respective ideologies. This sort of behavior creates a zero-sum dynamic that can often times leave both sides worse off than if they'd simply sat down, listened to each other's reasons, and made efforts to find a middle ground. Our society should be built upon the foundations of cooperation, tolerance, and mutual respect, and so when these fundamental precepts are disregarded, it immediately threatens the stability of our collective Union.

The Delta Variant

According to the latest data from Ontario's datasets, the highly-anticipated strain of the virus known as the Delta Variant has not posed the dire consequences that the government had previously warned of. Rather, the spread and impact of the Delta Variant has been far less than had initially been feared. Despite the original warnings of exponential growth, the Delta Variant has only caused a gradual uptick in cases, nowhere near the magnitude of what had been expected. This is a positive sign that indicates that the strict measures put in place by the government are indeed working, and that the Ontario population has been doing their part in helping to contain the virus by following the necessary protocols. Moreover, to help combat the spread of the Delta Variant, the government has continued to implement effective strategies such as mask-wearing, social-distancing and widespread vaccination, reinforcing their efforts to protect their citizens and their efforts of building immunity and resilience. As the Delta Variant continues to remain a challenge to communities all over Ontario, the government of the province can take some comfort in knowing that their efforts to date have produced positive results and have thus far prevented the strain from reaching a devastating level.

Recent reports from Intensive Care Units (ICUs) around the world demonstrate what many have been saying all along: the so-called 'Delta' variant of the coronavirus, while certainly more transmissible, is usually associated with far milder symptoms, and therefore is not a cause for undue alarm or drastic lockdown measures. Unfortunately, much of the media has actively fuelled unfounded fears and panic in the general public with regard to this particular strain of the virus, when in fact this alarmist attitude was entirely unwarranted based upon the current data.

The media must take a more measured, scientific approach in its reporting of the virus and related health issues, and not be complicit in disseminating information which risks sparking undue fear among the public and panic buying. Similarly, national governments should focus on relaying to their citizens the results of all the latest research into the virus, and adopt policies based on the best available scientific advice, not just towing a pre-meditated narrative in order to galvanise public opinion.

By way of example, although the 'Delta' strain was at the time of writing responsible for a majority of new coronavirus cases in countries across Europe, scientific evidence does not support an escalation of restrictions based upon this data alone. In Australia, for instance, the 'Delta' strain accounted for nearly 12,000 new verified cases during June, yet Australian authorities determined that the available data showed these cases were largely mild and did not require further lockdowns or restrictions to be put in place.

The most effective approach to fighting the spread of the virus is based upon the evidence-supported facts, not unchecked fear-mongering - and it is incumbent upon both the media and the government to ensure this message is relayed to the people in an honest and open manner.

Kids and COVID

When I arose this morning, I was shocked and disheartened to see the costly push for vaccines for children going on in spite of the data and science at our disposal. This notion runs completely contrary to what is right and proper, and it goes against the very data and science that are meant to serve as our guide and protect us from making potentially harmful and risky decisions.

For instance, recent findings have attested to the fact that childhood vaccinations can have detrimental side effects for the physical and psychological health of a child, especially when vaccinations compound on one another a pediatrician's recommended natural immunity-building protocol. Though the purpose of vaccinations is to protect children from disease, the side effects of them can lead to substantial, far-reaching issues, from systemic fatigue to severe allergic reactions, as well as an increased risk of autoimmune and neurological disorders.

Though the intent of some individuals may be to protect and keep our children safe, I believe it is critical that we look at the full spectrum of the possible outcomes vaccines can cause and closely examine the pros and cons before making any decisions regarding the mandate of vaccinations for our children.

As the virus spread across the globe causing unprecedented disruption to our societies, many countries were left with important decisions to make about how best to contain the virus. In Canada, the provinces implemented strict lockdowns and restrictions, including the closure of schools and the suspension of teachers. But now, many are questioning if these measures are still necessary; whether there is a better approach.

At the heart of this debate is the wellbeing of children. Numerous paediatric health officials have raised the alarm about the long-term impacts long lockdowns are having on our children – from disruption to their education and social development, to increases in mental health issues such as anxiety and depression. Despite these mounting concerns, governments have continued to take a hard line on lockdowns, unwilling to give an inch.

The latest issue of contention has been the potential vaccination of children and infants with the mRNA gene therapy. This has sparked controversy due to the lack of long-term studies around the safety of the vaccine in these age groups. It is a valid point, given that children under the age of 10 rarely catch Covid-19 and if they do, are usually asymptomatic.

What is the rationale for this – for vaccinating children who are not at risk of Covid-19? Despite vocal opposition from medical experts, governments continue to insist that there must be stringent measures in place to protect our children – measures which are doing more harm than good to the very people they are intended to help.

It's imperative therefore, that we put an end to this 'war on our children' and turn our attention to solutions that are based on scientific evidence. Reopening schools, creating safe spaces for our children while they learn, ensuring that teachers return to work and importantly, that we base our decisions on facts and research – this is the way forward. We have been in this for a while now, and it's time we listened to the experts and gave our children the safe, secure and normal life that they deserve.

The push to vaccinate infants, toddlers and children as young as six months of age, alongside the advancement of experimental vaccines that genetically modify the body's cells makes for a concerning mixture. This thought of altering the genetic code of the cells to create a mock virus strikes many as potentially being an unethical practice, perhaps even being a form of child abuse. Unfortunately, the media doesn't always cover stories on the tens of thousands of people who die every year as a result of the flu, so this horrifying reality is lost on many of us. The harsh reality is that over the past three years, influenza-related fatalities consistently exceeded 50,000 people annually, a figure that's been drastically dropped since, with the number of fatalities vanishing completely by the first quarter of 2020. For these reasons, it's truly mind-boggling why there is the push to vaccinate so many at such young ages.

The consequences of the novel coronavirus (COVID-19) amongst children are not as understood as those of regular flu outbreaks. As with the influenza virus, which is highly transmissible amongst children and can lead to serious health complications for this demographic, the closure of schools has been an effective way of reducing the spread and numbers of community outbreaks seen. However, this solution does not appear to be as effective with COVID-19 as research has shown that children generally do not exhibit the receptors required for the virus to take hold.

It is important to note, then, that the strategies adopted to stem the spread of the virus need to be different depending on the target population—those more likely to contract and spread COVID-19 are different than those who are more susceptible to regular flu outbreaks. For example, public health and government officials may advocate for the widespread use of face masks and hand sanitizer in areas where the elderly are particularly prevalent, while taking different steps such as promoting social distancing and remote learning in areas where there is a larger youth demographic.

Indeed, the science surrounding COVID-19 is still developing, and the severity of the virus on children still lacks the scientific data present for regular flu outbreaks. As such, further examination is required to ascertain the best possible ways to protect this vulnerable population and stop the further spread of this virus.

The concept of quarantining healthy individuals is a new development in human history. For centuries, lockdowns were only enforced during periods such as the Middle Ages, with few positive results. This was largely attributed to the fact that overcrowded, confined conditions proved to be fertile ground for diseases to spread amongst healthy individuals, notably seen in the devastating epidemics of plague and tuberculosis, which caused mass casualties. As a result, the practice of quarantine was largely abandoned, in order to protect the health of populations.

However, in more recent times, the spread of infectious contagions has necessitated the re-introduction of this practice. In attempts to contain their spread and protect the wider public, governments around the world have created comprehensive quarantine strategies, with mandatory isolation for any individuals who are suspected of being infected, or who have come into possible contact with someone who is. In some cases, lockdowns have also been imposed on healthy populations, as a preventative measure, and not just a solution. This has led to unprecedented levels of disruption, and has been an unwelcome feature of the landscape to which society has had to rapidly adjust.

Nevertheless, the need for such measures has been deemed essential to ensure the safety of human health, while governments and medical authorities work tirelessly to develop treatments and vaccines. Time will tell whether or not the world has been successful in its containment of this global pandemic, and whether such wide-reaching quarantine strategies have paid off.

- Does "the science" support continuing the lockdowns and social distancing?
- Did it ever support these devastating policies?

Maybe we need a teenager to come up with a better idea for another science project, wait, hold on school has been cancelled.

Comparing the Vaccines

After spending many hours researching, verifying and compiling the data on both COVID vaccine death rates and Influenza vaccine death rates, I have to come to the conclusion that the reported numbers of deaths related to COVID vaccinations could in fact be much higher. This is mainly due to the fact that the COVID vaccines have not yet been approved and that the reported data may not fully reflect the overall situation.

My current question then is why are none of our health officials talking about this? As reports of injuries related to the COVID vaccine grow, with many even discussing the notion of halting the rollout of the injections, it is essential to consider the consequences of doing such a thing.

Just look, for example, at Russia, who decided just two months ago to halt production of their Sputnik V vaccine for two weeks due to a safety issue, whilst the United Kingdom soldier on with their mass immunisation program, having already administered the vaccine to over 11 million individuals.

Clearly there is a resource of knowledge to be drawn on here, and yet, in the face of high-risk, we seem to be silencing the bigger picture discussions, allowing and even promoting the continued use of the vaccine without any discussion of the safety implications of doing this.

It is, therefore, paramount that we have a full and frank conversation on the subject of vaccinations, examining exactly what the risks are, by comparing and contrasting flu vaccinations with the current COVID-19 vaccines.

Lockdowns Based on Kids School Project

Coming across this interesting piece of knowledge was astonishing to me. It's amazing the information that can be unearthed if one looks hard enough. The recent lockdowns experienced around the world are unprecedented; never in the entire history of humanity has something like this been done, until recently. It all started with a school project of a 14 year-old girl who used a computer program to explore how flu could be contained if people interacted less with one another. By simulating a hypothetical viral outbreak and using social distancing as a countermeasure, her simple simulation showed that limiting contact could lead to a significant reduction in the number of people infected. Instead of an estimated 5,000 individuals being infected, the amount was reduced to a mere 500.

Her impressive findings, although initially disregarded, have ultimately made a lasting impact on the world. Her research revealed the power of social distancing in halting the spread of a pandemic, resulting in the implementation of unprecedented lockdowns. The events of this pandemic have thus shown us that information can be hidden away in the most unlikely of places, but with enough dedication and persistence, invaluable discoveries can be made.

When Dr. Drew Pinsky, an experienced professional in the field of infectious diseases, received word from the father of a young girl who had created a computer model, he brought rise to the project at a national level. With the endorsement of Dr. Anthony Fauci, renowned for his work with infectious diseases, the model began to gain traction. However, predictions made by the original models based on Dr. Fauci's advice, largely concerning the extent of the COVID-19 pandemic, were revealed to have been hugely inaccurate and unreliable.

The young girl's computer model, which was uploaded onto a higher-level computer, appeared to achieve more beneficial results. It proposed that instead of completely halting activity and mandating a global lockdown, the focus should be placed on "hot spots" and to encourage regional closures, a policy that numerous countries such as

Japan and Germany undertook to great success. Unfortunately, government leaders disregarded the economic, legal, and political consequences of an extended shutdown were overlooked, and, despite the girl's work being misused and grossly misrepresented by various news outlets, a worldwide lockdown was still enforced. This has caused untold suffering, as hundreds of millions of people now find themselves in an economic abyss without jobs or income.

The effectiveness of the science project surrounding COVID-19 is undermined by a few glaring flaws. Perhaps the most pressing of these is the difference in how Covid-19 impacts children compared to influenza virus. With influenza, which typically spreads amongst children and harms them the most, it was possible to reduce the community-wide outbreak by closing schools; however, this measure is fairly inopportune with respect to Covid-19 since children don't seem to contract the virus or pass it on because they lack the necessary respiratory system receptors.

Whereas influenza is a contagious respiratory condition that is highly transmissible between children, Covid-19 does not seem to be as contagious between them, due to it primarily targeting cells with a particular kind of receptor that are often not present in such a young demographic.

This can be seen in various countries around the world where the impact of Covid-19 on those below the age of 20 has been far less severe than the effects on other age groups. For example, according to figures released by the CDC, only 0.1% of Covid-related deaths in the United States have been in people under the age of 20. This stands in stark contrast to seasonal flu where very young children (ages 0 to 4) are amongst the most commonly hospitalized due to respiratory problems associated with the virus.

This enhances the need to still find and implement suitable Covid-19 prevention strategies in schools, such as creating and practising social distancing protocols, teaching regular hand and respiratory hygiene, and enforcing strict rules on protective wear. While it might not be as effective as school closure would be for influenza transmissions, it will help to contain the spread and ensure the safety of the children.

The current pandemic is an unprecedented event in human history, one that has never before required us to impose restrictions on healthy people in an effort to stop the spread of an infectious disease. In the medieval ages, quarantining of healthy people was commonplace as a way to contain the spread of diseases such as the Black Plague and tuberculosis, yet tragically this had little success in stopping the transmission of these deadly pathogens. The reason for this was that by isolating potentially sick people in confined spaces, transmission of the disease was actually accelerated rather than arrested. In fact, this was so prevalent in medieval times that Romans referred to such efforts as 'bringing in hay for the horse after it had bolted from the stable', in other words, attempting a solution to a problem once the damage had already been done. This gives us an indication of how serious the implications were for those communities which lacked access to effective medical systems and treatments. The lesson to be taken away from this is that even with the advances in medical science and technology, we must remain vigilant in protecting the health of individuals and communities.

- Does "the science" support continuing the lockdowns and social distancing?
- Did it ever support these devastating policies?

Maybe we need a teenager to come up with a better idea for another science project, wait, hold on school has been cancelled.

/uestioning Everything about the Vaccine

A recently published case study and report by some incredibly intelligent and knowledgeable individuals, has incited alarm amongst readers. The highly complex documentation and technical jargon was perhaps most likely intentionally incomprehensible to those without a deep academic understanding of the subject, making it an arduous task to digest. I believe that the information contained within the 42 page report is of paramount importance and must be shared, so to ensure that this is comprehended, I have taken the time to break it down into more attainable language, without sacrificing accuracy. This means that readers won't have to stress their minds too heavy in order to understand the core potential implications of the report.

Recently, renowned senior scientist Dr. Stephanie Seneff, who works at the esteemed Massachusetts Institute of Technology's Computer Science and Artificial Intelligence Laboratory (CSAIL), and Dr. Greg Nigh, a naturopathic oncology specialist, have made some rather alarming discoveries and recommendations. Dr. Seneff has conducted extensive research into the effects of artificial intelligence on our health and the environment, and published numerous papers on the topic. Her research has made it evident that artificial intelligence has the potential to cause detrimental consequences such as air and water pollution, declining public health, and adverse impacts on natural habitat among other troubling effects. Furthermore, her work has revealed that exposing ourselves to artificial intelligence is associated with increased risk of medical disorders and cancer.

Dr. Nigh's research has similarly compelled him to recommend minimizing our exposure to artificial intelligence. He has studied the intersection of cancer and nutrition and has come to the conclusion that artificial intelligence can contribute to cancerous cell growth in our bodies. His findings suggest that reducing our contact with artificial intelligence could significantly reduce our risk of developing cancer and other disorders. Dr. Nigh is known for his ability to synthesize nutrition, emotional well-being, toxic exposure, lifestyle, and genetic risk into a comprehensive cancer-prevention program. He has published numerous books and articles about the dangers of artificial intelligence and has been featured in several notable news outlets.

Given the investigations of these two esteemed scientists, caution is urged when it comes to interacting with artificial intelligence. However, their bodies of research have made it clear that the potential risks of AI far outweigh the rewards, and that reducing our exposure to it may be one of the best ways to maintain our health and well-being.

In the past several months, the gravity and scale of a global health crisis has been exposed like never before. Sadly, more than 6,100 people have died due to certain vaccines, according to data compiled by the U.S. Centers for Disease Control and Prevention's Vaccine Adverse Event Reporting System (VAERS) on August 19, 2021. But this may be just the tip of the iceberg as estimates suggest there could be as many as 10,000 fatalities unreported due to complications arising from inoculation. It is a worrying thought, especially considering reports of systematic manipulation within the VAERS Registry itself. In one instance, it was reported that numerous cases linking vaccinated individuals with potential fatalities or injuries were left unaccounted for. This ought to be taken extremely seriously and the relevant authorities should be held to the utmost scrutiny when such reports surface.

Within this article, I will attempt to decipher the findings of a research paper that assesses the likelihood of Pfizer and Moderna's experimental mRNA vaccines leading to negative repercussions in the vaccinated population. To do this, I will explore the many pathways this phenomenon could take.

The study mentions two primary pathways in which the mRNA vaccines could result in adverse effects. The first is through the immune system triggering a reaction to the vaccine's mRNA components, which may result in inflammation, as well as heightened systemic effects throughout the body. Examples of this could be fever, chills, joint pain and headaches. The second pathway is gene editing whereby mRNA sequences become absorbed by cell receptors and mutated, thus altering the genetic makeup of the cells. This could potentially lead to more severe autoimmune diseases such as Lupus, chronic fatigue, and even cancer.

It is still uncertain if these vaccines are solely responsible for any of the documented cases of adverse reactions or if other underlying conditions are contributing. It is important to further investigate these pathways in order to ensure the safety of the population before the vaccine is widely distributed. Researches should consider conducting more in-depth research into the accuracy and efficacy of mRNA vaccines to provide a better understanding.

Ultimately, it is essential to remember that the experimental mRNA vaccines from Pfizer and Moderna could be causing serious adverse effects in vaccinated individuals. That said, it is important to remain vigilant as more information becomes available and as researchers gain a better comprehension of the pathways this phenomenon may be taking.

One of the possible side effects of many vaccinations is antibody-dependent enhancement (ADE), which is caused by spike proteins being produced in the body as a result of the injection of an mRNA vaccine. Put simply, most successful vaccines work by triggering the immune system to generate neutralizing antibodies which defend against viral infection or the spread of disease. Unfortunately, however, not all antibody responses are strong or effective enough. On rare occasions, when a person is injected with a vaccine, their body might generate antibodies which do not protect cells, and instead actually give the virus a better opportunity to enter a cell and further the spread of the disease. Such a phenomenon is known as antibody-dependent enhancement. Because of this rare yet concerning side effect, there is speculation in the scientific community that even the COVID-19 vaccine, which is administered through an mRNA injection, might be contributing to the emergence of new and more serious variants of the virus.

The mRNA vaccines at the forefront of the current pandemic are designed to deliver the highly unique spike protein to 'infected' cells. This spike protein elicits a powerful immune response from the body's cloned antibodies, which the mRNA vaccine seeks to effectively utilize to fend off an ongoing infection. Thus, the idea is to introduce the spike protein to 'infected' cells for the purpose of training the body's natural immune system to produce defenses that target this protein specifically.

Currently, the mRNA vaccine has been widely deployed as an effective method of boosting the body's natural protective responses to this dangerous and foreign spike protein. Under the guidance of its expertly designed blueprint, the immune system can be quickly educated to recognize the spike protein in order to craft a powerful counterattack against the target infection. It's hoped that these newly generated antibodies can provide a wider range of protection beyond that which the vaccine is able to provide, thus engendering a more resilient and effective defense for any future infections.

In recent times, the implementation of mass vaccination policies has been based on the assumption that the benefits resulting from the vaccine will likely outweigh the potential risks by a long distance. This principle has been crucially relevant in the context of the ongoing COVID pandemic, in which governments have launched unprecedented global vaccination campaigns with unprecedented speed. As a consequence, in a sense, people are the experimental subjects in this unprecedented experiment.

This is a worrisome development given the nascent nature of the science behind this pandemic, and the sharp disparities in global infrastructure. While countries with well-resourced healthcare systems can attempt to more thoroughly measure and track outcomes, this is not always the case in nations with more limited resources. Citing the example of India, one can argue that the breakdown of the nation's health infrastructure and the limited access to basic medical services prevent the monitoring of any long-term effects of the vaccine, or even identification of any unreported adverse events.

Another challenge associated with the global mass vaccination campaigns is the lack of transparency in how decisions are made surrounding the vaccine's production and distribution. With the situation so rapidly evolving and the complicated web of vested interests, it is impossible for the average individual to know how safe a given vaccine is and which elements of the risk/benefit ratio have been overlooked.

Overall, mass vaccination campaigns can be a powerful tool in reducing the spread of dangerous diseases and should certainly be viewed positively. At the same time, it is crucial to understand the implications and consequences such campaigns can have and that they should only be undertaken with careful consideration.

Now that the wealth of new data is available, it is time to gain further understanding of this hitherto unfamiliar technology. The current vaccinations that are commonly recommended are inoffensive and not the cause of any adverse events. If any of these vaccines did happen to lead to such problems, they would immediately be withdrawn from the market. Clinical trials of the third phase are specifically carried out to locate serious or common adverse effects, before a vaccine is given the green light and made available to the masses. Unfortunately, these tests are ongoing, with many, many people included as voluntary participants.

The development of novel messenger ribonucleic acid (mRNA) technology presents a unique set of logistical challenges in terms of delivering and administering the treatments effectively and safely. While the term 'vaccine' has been widely used to refer to this technology, it should be noted that mRNA-based treatments are not actually vaccines in the traditional sense.

First, the manufacturing of mRNA treatments requires sophisticated production facilities, equipment, personnel and processes. These must be optimized to ensure newly created mRNA treatments are of the highest quality to meet safety standards mandated by the FDA and other governmental regulatory bodies. It is also necessary to manage the storing of mRNA treatments in refrigerated temperatures that vary depending on the specific product.

Second, the lifesaving treatments must be quickly and accurately tracked and shipped to designated locations. This can be especially challenging in warmer climates and environment when the mRNA treatments must be kept cool and dry. Specialty shipping companies often employ state-of-the-art thermal packaging and monitoring systems to ensure the integrity of these treatments during transport.

Finally, it is also important to consider the well-trained personnel, equipment and protocols required for injection, as the entire administration and delivery of mRNA treatments must be handled with the utmost care and precision. Ensuring proper storage and temperature control is essential, as well as understanding the potential risks and side effects of the medications prior to injection.

Overall, there are numerous considerations that must be taken into account when managing and delivering mRNA treatments, requiring advanced production facilities and processes, specialized shipping methods, and proper administration and delivery protocols.

The development and eventual deployment of Covid-19 vaccines has caused a seismic shift in the healthcare landscape, both nationally and internationally. This marks a unique moment in history, unlike any we have seen before, with a vaccine being made available to the general population.

The magnitude of this achievement is extraordinary; seldom has a vaccine obtained the kind of widespread, multi-governmental support we have seen for this particular health emergency. Numerous governments around the world are putting considerable resources into the development, testing, and eventual distribution of the vaccine, a commitment which will require immense medical and logistical coordination.

This timeline is unprecedented in terms of scale and speed, especially when compared to the more traditional vaccine development process. Whereas vaccine development can typically take up to several years to advance from laboratory testing to market introduction, the Covid-19 vaccine has managed to reach the general population in under a year of development.

Organisations involved in the development process, such as the World Health Organisation and countless pharmaceutical companies, are collaborating to ensure the rapid yet safe deployment of the vaccine. Many of the initiatives being undertaken to advance the development process, for example, the practice of expediting some clinical trial phases, are groundbreaking and beneficial for the future of public health.

The development of Covid-19 vaccines marks an extraordinary moment in history, and the coordination of multiple stakeholders involved in the process of vaccine deployment is an unprecedented effort for a health emergency of this magnitude. With any luck, the efforts being carried out today may well serve as a blueprint for the successful production of future vaccines.

- This is the first time to use mRNA vaccine technology.
- First time Moderna has brought any product to the market.
- First to have public health officials telling those receiving the vaccination to expect an adverse reaction.
- First to be implemented publicly with nothing more than preliminary efficacy data.
- The first vaccine to make no clear claims about reducing infections, transmissibility, or deaths.
- First coronavirus vaccine ever attempted in humans
- The first injection of genetically modified polynucleotides (DNA stuff) in the general population

Vaccine development is a highly specialized field that often requires a deep understanding of the latest scientific and medical practices. One fact that has surprisingly gone under-discussed, however, is the categorization of the types of available vaccines into three distinct areas - Simplicity, Complexity, and Unprecedented.

Simple vaccines, such as the traditional polio vaccine, represent straightforward applications of existing medical technologies. Complex vaccines modify existing vaccine technologies, and often require in-depth medical knowledge. An example of such a complex vaccine would be the HPV vaccine, which utilizes a modified version of the common rabies vaccine in order to be effective against the Human papillomavirus.

The final and arguably most important category is the Unprecedented vaccines. These are vaccines developed to combat diseases for which no suitable vaccine has ever been created. In this regard, the newly developed coronavirus vaccines represent the perfect illustration of an Unprecedented vaccine. This remarkable feat of biological engineering, accomplished through the hard work of scientists around the world, promises to be the saving grace from a global pandemic that has been devastating society for the past year.

In summary, vaccine development has recently become a hotbed of innovation, with categories ranging from Simple to Unprecedented. This wide range of scientific advancement provides researchers with a unique opportunity to solve a myriad of medical challenges.

The development of vaccines can be a lengthy and perilous process, one that involves overcoming numerous hurdles and is confounding to even the most educated individuals. To illustrate, it is thought that a vaccine takes an average of 12.5 years to go through the entire development process and the estimate for a vaccine making it through Phase II trials - where the efficacy of the vaccine is evaluated - is only 5%. Sadly, from this statistic, the estimated chance of emerging from the following Phase III clinical trial, which focuses on assessing population benefit, is only 40%. This means that ultimately, an unprecedented vaccine, one without a counterpart in existing medicines on the market, has a shockingly low chance, estimated at a mere 2% probability, to make it through the final trial. This serves to reinforce the idea that tackling vaccine development is an arduous task, but one that is vitally important to addressing the global community's health-related matters.

On May 10, 2021, Pfizer's clinical trials for their much-anticipated coronavirus vaccine were given the green light to continue after the company reported that despite some seriously worrisome issues reported at the initial screening, there were still good grounds to move forward with the tests. Unfortunately, these issues have cast doubt on the interim efficacy data, which suggest that the vaccine is around 95% effective at protecting against the virus. This high efficacy data has yet to be adequately explained or supported by Pfizer, leaving many people concerned about the safety and reliability of the vaccine.

With the long-term dangers of the virus still very much a reality, and considering the heightened importance of public health, it is crucial that Pfizer publicly address these worries in order to offer reassurances to the general public. Otherwise, it may be difficult to convince potential vaccinees to go through with the process, as any doubts regarding safety and efficacy can easily lead people to look for an alternative method of protection. As such, it is important for the company to outline the reasons behind their decision to continue the clinical trials, in order to clearly demonstrate that the risks involved are not too great and the vaccine will ultimately protect people from the virus.

A comprehensive analysis of the vaccine's release revealed a number of flaws, missteps, and unanswered questions regarding its safety and efficacy. Notably, much of the research was conducted after the inoculation was already in public circulation, sparking public health concerns. It must be considered whether the administration of the vaccine followed established protocols and if the drug was subjected to appropriately rigorous trials prior to distribution. Furthermore, data must be collected on the vaccine's effects to ensure that its use in the population remains safe and effective. Scientists must garner an understanding of how it interacts with other available medications, vaccines, and treatments, as well as any potential long-term adverse reactions it might have. In an era of rapid technological advancement, ethical considerations must be kept paramount in any discussion of scientific breakthroughs. The critical questions raised about the vaccine's safety and efficacy must be addressed in order to ensure public safety, confidence, and trust in the medical community.

- Will Vaccines Stimulate the Immune Response?
- Will Vaccines Provide Sustainable Immune Endurance?
- How Will SARS-CoV-2 Mutate?
- Are We Prepared for Vaccine Backfires?

Recent issues in the news have raised some troubling questions surrounding the efficacy of mRNA design and its ability to successfully unfold as anticipated. This tendency to falter is primarily attributed to the spike protein failing to operate in its expected manner, which is causing a great deal of concern, as this issue should have been caught before the vaccine reached Phase III clinical trials.

The importance of correctly determining the drug's effectiveness prior to Phase III is paramount when it comes to public health and safety, as vaccines are intended to protect those who are most vulnerable and susceptible to the virus. Thus, it is necessary for the drug to meet all scientific standards and expectations before being rolled out for mass vaccination.

In the instance of mRNA designs, scientists have deemed the spike protein to be essential in fostering a protective immune response. However, if this spike protein is not performing properly, the vaccine will be inadequate in countering further mutations of the virus. This means that even if the vaccine is administered, it may be unable to effectively protect individuals from contraction or transmission of the virus.

Ultimately, the current situation has thrust the topic of vaccine efficacy into the spotlight, with many calling for a more diligent approach to conducting pre-clinical trials in the future. This is so that any potential issues, such as the mRNA design's inability to unfold and produce spike proteins as needed, are identified much earlier, before the vaccine is approved for mass vaccination.

The original text discussed a topic that was particularly interesting, spike protein toxicity, and required the interpreter to dive into a great many details and explanations of the science hiding behind those big words. What made this topic even more challenging was the sheer volume of information, with the discussion of the spike protein toxicity taking up no less than five pages, while the total amount of information was fifteen pages. Such a long text was filled with unfamiliar terms, which presented an arduous task of interpretation and distillation that needed to be completed in order to make sense of the information.

Diving into the details, the spike protein toxicity primarily refers to the toxicity of the spike protein found on the membrane of SARS-CoV-2, the virus that causes the coronavirus COVID-19. This spike protein is primarily responsible for the virus's ability to bind to and enter cells, allowing the virus to replicate and spread. The degree of toxicity of the spike protein is of great interest as it could be key to understanding how the virus affects human cells. Thus, various studies have been conducted to better understand how the spike protein affects the human immune system, tissue damage, and other physiological effects.

The findings of these studies are quite intricate, enough so that the topics described within the fifteen pages of information are far more detailed and in-depth than one can summarize in one paragraph. Indeed, further examination of the spike protein toxicity and other facets of the virus's pathology require a great deal of understanding of the science and terminology behind it, in order to make sense of the complex information. Even more so, navigating through the five pages devoted to the spike protein toxin was a tough feat which tested one's comprehension and interpretation abilities to their fullest degree.

The recent emergence of experimental mRNA vaccines has been met with great optimism, as they hold the promise of revolutionary benefits and potentially provide cures for life-threatening diseases. However, with the emergence of such new technologies and the unprecedented speed of their development and application, also comes with it a concerning prospect of life-altering and even catastrophic unforeseen consequences that are now manifesting themselves through scattered reports and posts scattered around the globe. From cases of facial paralysis and anaphylactic shock being reported in clinical trials to cases of Guillain-Barré Syndrome being arising in vaccinated individuals, the potential implications of these vaccines are significant and warrant attention. Although these vaccines undoubtedly carry a remarkable potential for the future of medical treatments, the long-term safety of these technologies must be carefully considered. Thus it is important to take into account all the possible risks, before these treatments are regularly administered, in order to avoid any potentially tragic scenarios.

The development and implementation of mRNA vaccines against the novel coronavirus, SARS-CoV-2, has been met with strong support from both the media and government, turning the vaccines into something of a political weapon. Nevertheless, several significant issues remain cause for concern, many of which don't receive the attention they deserve. To name but a few, the risk of long-term side effects caused by the technology has not been fully researched and explored yet; on top of that, the high cost of the drugs mean that not everyone will be eligible to receive it, with the poor being particularly affected; and the rapid process of development for the vaccines raised eyebrows due to its lack of rigour. All of this, plus much more cause for worry, merits serious deliberation and discussion. There is a wealth of knowledge in the scientific community that should be utilised, and it is the responsibility of those with greater intellectual understanding of the subject matter to bring such issues to the fore.

In order to effectively eliminate any potential side effects, numerous medical experts, researchers and virologists have suggested that basic surveillance measures be put into place. At a national level, in-depth data collection of any unwanted reactions that arise from the use of mRNA vaccines is a must, and frequent testing of those vaccinated to gauge their autoantibody levels is critical. One of the most important measures that stands out is the conducting of in vitro studies to determine whether or not the mRNA can have any damaging repercussions on male fertility.

To further illustrate this point, a study conducted in 2019 by prominent doctors and scientists from the United Arab Emirates University sought to research any influence of mRNA vaccines on male fertility. After conducting several in vitro trials, the results found that mRNA had a minor impact on testicular cells and revealed that it had the potential to reduce the sperm motility of men. However, the study ultimately concluded that the administered dosage of the vaccine was too small to produce any conclusive evidence. Despite this, the study highlighted the importance of continued research into the matter and further emphasized the need for careful monitoring of any potential adverse effects.

The increasing number of adverse side effects reported from recent clinical trials have created a atmosphere of uncertainty. Many have argued that instead of recklessly pressing ahead, the rollout of medication should be halted and all evidence should be examined with a critical eye. Our current situation has been described as a 'wait and see' as we struggle to determine the safest course of action.

As we have seen in recent times, the consequences of ill-advised or ill-thought out decisions can have long-reaching effects - not only in the medical field, but in a myriad of other sectors too. In the political world, we have seen the results of inadequate research and testing leading to poor policy outcomes. In the business world, 'short cuts' can cost companies millions in missed opportunities.

Hence, it is essential that we take the time to properly assess the implications of decisions, particularly those with profound implications, such as the roll-out of new medications. Taking such an approach would act as a safeguard, not only against costly mistakes, but also potentially life-threatening scenarios.

Therefore, it has become incredibly important to determine whether or not we should continue down the current path, or if we should pause and reassess the situation, as some have suggested. Undoubtedly, this will require a great amount of due diligence, however, the benefits stand to greatly outweigh the risks.

This year has seen a dramatic increase in infection rates across the globe, leading many nations to introduce vaccination programs in an effort to combat the spread of disease. With so many people being vaccinated, the risk of those who were not vaccinated coming into contact with those who had, and potentially becoming infected, has become a concern. We could be at the dawn of an epoch where a vaccine passport could keep those who actively chose to avoid getting the 'jab' from being exposed to individuals who have been vaccinated, thus avoiding possible contamination of the unvaccinated person's DNA.

It is a worrying thought that, in 2021, we may be unintentionally spreading a disease that could be extremely damaging to both those who have been vaccinated and those who have chosen not to receive the vaccine. This can have a profound psychological and physical impact on the lives of those affected, not to mention the strain it can place on an already overburdened healthcare system. It is therefore essential that we use a vaccine passport as a way of protecting those who did not get the jab, while also providing a much-needed security net for those already vaccinated.

The vaccine passport has the potential to be a powerful tool in curbing the spread of this disease by allowing easy tracking of who has been vaccinated and who has not. It could also serve as a reminder of the risks to those considering opting against getting the vaccine, allowing them to make a more informed decision. With an effective vaccine passport in place, a vaccination rate of 70 – 80% may be achievable, representing a huge stride forward in the fight against this disease.

It is clear that, if we are to protect both those who have been vaccinated and those who did not, a vaccine passport can be an integral part of this plan. If used smartly and effectively, it could be the way to ensure we limit the spread of this disease, as well as empowering those searching for accurate and dependable advice on whether to get the 'jab'.

US Death Rate – Vaccines

After extensive careful research, I've finally been able to compile the latest available data from VAERs that was obtained a few days ago. After carefully analyzing the information, I've noticed a remarkable trend in the distribution of the vaccine. When the vaccine mortality rates were overlaid on these figures, the findings were truly remarkable.

For instance, in the state of California, it was found that the number of vaccine mortalities was directly proportional to the number of people who received the vaccine. It was also seen in data collected from other states that, despite having lower population densities, the vaccine mortality rate remained relatively constant. This data can be used to solidify previous theories and help medical experts to determine the effectiveness of the vaccine in different areas.

Furthermore, it was observed that the vaccine mortality rate decreased as the number of people vaccinated increased. This could be a result of more individuals being informed of the potential hazards of receiving the vaccine, leading to a more informed decision of whether to receive it. In addition, more individuals receiving the vaccine helps to create a herd immunity within the area, reducing the risk of certain diseases being spread, and ultimately decreasing the mortality rate.

The data that I have collected over the past few days has been incredibly insightful and serves to be a valuable resource for anyone interested in learning more about the vaccine's effects. With the help of this data, it should be possible to continue to monitor the situation and make appropriate changes as needed in order to ensure that the vaccine is being distributed as safely as possible.

The debate over the correlation between injuries and death rates continues to be a significant issue that is often ignored by some outlets. For many years, people have been told time and time again that if there was any action that could potentially save a life – whether it be the use of masks, lockdown measures, or a vaccine – it should be implemented without hesitation due to the 'if it saves one life' mantra.

However, despite this 'if it saves one life' message being heavily fortified over the years, opinions and attitudes now seem to have shifted in light of the deluge of questions regarding the efficacy of the vaccine, with many openly questioning the validity of the 'if it saves one life' claim. For example, although the vaccine is being credited with saving many lives, there remain numerous variables in the equation – such as individual health issues and the decisions of different governments – that make it difficult to fully comprehend the potential long-term outcomes associated with the vaccine.

Moreover, despite the fact that the number of injuries and deaths associated with the virus has only seen a decrease in recent months, the amount of confusion and fear surrounding the vaccine has only seen an increase. As a result, it is important that we remain open-minded when it comes to considering all possibilities when it comes to injuries and death rates, as well as the data and evidence surrounding the vaccine, in order to make an informed decision as to whether the 'if it saves one life' mantra is still relevant today.

Fight the Vaccine Mandate

Vaccines are essential tools that humanity has devised in the fight against preventable diseases. When people hear the words "vaccine" or "vaccinated", the importance of immunization is immediately highlighted. Taken as a whole, a vaccine is an immunological preparation used to induce immunity and protect an individual against an infectious disease. For example, people have been taking the human papillomavirus (HPV) vaccine to prevent the spread of sexually transmitted HPV infections, which can lead to various types of cancers such as cervical cancer. Essentially, the vaccine works by introducing a weakened form of the virus into the body, prompting an immune response and the subsequent production of antibodies that protect against the targeted strain. In layman's terms, this means that when you get a vaccine, you are "taking this" and becoming immunized so that you "never catch that". Vaccines are examples of modern medical advancement at its finest, whereby you can be healthily and safely inoculated against a disease and its transmission is stopped, as it is with most other contagious afflictions.

Despite the common misconception, the Pfizer and Moderna treatments popularly known as the 'Jab' are, in fact, not vaccines. Despite what some might mistakenly believe, these two treatments were never developed to be a vaccine and do not result in immunization. Recent reports from British Columbia, Ontario, England, and Israel all confirm that those who received the 'Jab' are still at risk of spreading the disease as they have not been immunized. This fact has been illustrated in many of my previous posts, and is no longer a matter of speculation or theory. The importance of understanding that the 'Jab' is not a vaccine cannot be overstated. Until the majority realize this, many people will continue to erroneously believe that the 'Jab' has immunized them and that they are not at risk of spreading the virus. Those who have received the 'Jab' should continue to practice the same safety guidelines as those who have not, such as wearing masks and social distancing, to protect themselves and others.

The increasingly concerning situation that employers mandating the Pfizer and Moderna medical treatments to protect staff is equivalent to forcing an individual employee to take Tylenol so the other staff members in their office don't get a headache. It cannot be stressed enough how these treatments merely reduce clinical symptoms and do not provide immunity or prevent the spread of the virus. Unfortunately, many businesses are unfortunately enforcing this policy as a result of ignorance and a lack of leadership from the authorities. It is necessary that steps are taken to prevent this from continuing and we must rally together to try and educate others on the facts surrounding the vaccines.

If you're an employee facing pressure to take the vaccine, ask yourself this one simple question: "How does the Pfizer or Moderna treatment protect others?" If your employer's plan is to mandate regular testing of unvaccinated staff, then make sure they include vaccinated staff too. It is undeniably true that a vaccine has not entirely prevented cases in people who have received this medical treatment, meaning the vaccine has failed. Thus, it is essential that we do not allow ourselves to be bullied into taking something that is both experimental and unsafe. It is vital that we stand our ground and make sure everyone is aware of the risks.

THIS PUTS EVERYTHING INTO QUESTION

Israel has been one of the undisputed success stories in the global fight against COVID-19, thanks to its highly effective and efficient vaccination program. The country has had to contend with a sharp increase in the number of cases caused by the more contagious Delta variant of the virus in recent weeks; however, the good news is that its symptoms are generally milder than those associated with prior variants of the virus. This increase in infection rates has, unfortunately, taken a toll on the efficacy of the vaccine, which health experts originally said would provide up to 75% protection against the virus with just a single dose. Subsequent shots, they said, would provide full protection. However, those predictions have been tested in the face of the Delta variant's remarkable resilience.

In light of the sky high number of reported cases of COVID-19, there are some doubts that our year-long battle against the virus is far from over. As such, our efforts of early containment appear to have been rendered futile, making it essential for citizens to receive an additional dose of vaccination, otherwise known as a 'booster shot'. Studies conducted in the US have shown that for diseases such as measles and whooping cough, booster shots have proven to be effective in providing immunity, as have regular doses of the flu vaccine. The World Health Organisation has also recommended people who come into close contact with infected individuals to get vaccinated, even if they have already received their primary dose of the vaccine. The medical and scientific consensus is, now more than ever, that we should all embrace the idea of a booster shot.

The question of how many more boosters will be necessary before the drug is fully approved remains to be seen. Employers are implementing this mandate, and unfortunately, it has largely been unsuccessful. The safety of this drug still remains an imperative consideration; there have been numerous cases in the United States where its usage has caused serious harm. The unfortunate truth is that the drug has not lived up to expectations.

Take, for example, a prominent case from 2019. It involved a patient in Pennsylvania who overdosed on the drug and suffered from severe short- and long-term health effects. While thankfully this incident was not deadly, it highlights serious risks. It also questions the efficacy of the drug for its intended purpose - to help reduce medical complications caused by an uncontrolled medical condition.

Similarly, a case in Maryland in 2020 resulted in the death of a patient. While this was a rare occurrence, it raises questions about the suitability of the drug and its usage among vulnerable patient populations.

It is clear that we must consider the risks and benefits of using this drug before moving forward with more boosters. We must also determine whether these risks are justified, especially considering the known effects and possible outcomes detailed by authorities in the US. Until then, caution must remain top of mind when evaluating the efficacy of this drug.

This troubling report from Israel's COVID website leaves one wondering what conclusion to draw from these concerning numbers. The data indicates that, as the vaccination rolls out, the death rate increases in tandem rather than exhibiting the expected downward trend. Such findings are especially alarming in light of the rising cases, further fuelling the scepticism surrounding the effectiveness of the vaccine. With serious questions raised about the vaccine's efficacy, some are even asking if it is instead making the variants more contagious and damaging natural immunity. To analyse the potential risks associated with the vaccine more closely, a more comprehensive study is needed to substantiate these worryingly high death rates.

In Ontario, there has been an effort to conceal the full and comprehensive record of COVID cases by age and other criteria. It is understandable why this record has been concealed from the public, for it verifies the vulnerability of our society to this horrible pandemic. The stark reality of these numbers is something that should enrage each and every one of us, for it has put our friends, family, and neighbours at risk.

Take, for example, the data which shows that people aged 60 to 79 are twice as likely to suffer from a severe case of COVID-19 than those who are 20 to 39. There is also research that suggests that among people aged 80 years and older, two-thirds were hospitalized for at least two weeks. This is a harrowing reality and yet, it provides insight into the dangers of this pandemic. We must use these numbers to protect the most vulnerable in our society and to learn from them.

It is essential that all of us understand the importance of protecting those who are most at risk of falling ill to COVID-19, yet this is seemingly not enough. We must use these numbers to become increasingly vigilant in our response to the pandemic, as they are a reminder of the pains we are all facing. The numbers are evidence that this pandemic has caused too much suffering and too much pain; pain that is not just borne by the elderly, but families and friends that have lost loved ones.

Truly, the numbers should make each and every one of us angry as hell, for if we open our eyes and understand the gravity of the situation, perhaps then we can all work together to protect the vulnerable and to work towards the betterment of our society. Ultimately, this is why the Ontario government has concealed these numbers from the public, for the matter is not only about numbers, but about the lives lost to this horrible pandemic.

I scoured the webservers of the organisation in search of the filename I had downloaded before and, lo and behold, I struck gold; the dataset that I had been searching for was sitting quietly on a server contained within a single, seemingly forgotten web page. The discovery of this hidden trove was both startling and fortuitous, given that it was crucial for the successful completion of my project. To say I was elated would be an understatement. As I wrapped my head around this unexpected find, I thought of the countless other digital assets that lay waiting undiscovered and unseen by our ever increasingly automated and hyper-connected world. They were relics of another time, custodians of secrets of untold importance and power, now just as lost as the sands of time.

After meticulously parsing through the data, I drew an inescapable conclusion: the vast majority of deaths are among individuals over seventy suffering from pre-existing medical issues. This, however, was not surprising in the least; a trend noticed as far back as March 2020, that made it all the more sobering. The staid statistics reveal that the probability of dying from COVID-19, according to the documented data, is shockingly, dismally low at a mere 0.004%. It boggles the mind that the lockdown measures enforced beyond necessity and the vaccine passports demanded by state officials are so disproportionate to the actual risk of death. Ponderably, the mundane risks we are exposed to on a daily basis are far higher than that of COVID-19, yet no preventative restrictions are placed on those activities. We should all be protesting this wrongful confiscation of our freedom to our mayors, MPPs, premier and health minister. This awful situation cannot continue, for it was never a pandemic from the start. Our freedom is being unjustly taken away, with no reference to sound judgement, legitimate reasoning or fact-based evidence.

99.6% RECOVER FROM COVID

According to the latest available data compiled by the Government of Ontario between May 1st and August 31st, 2020, it has been determined that a remarkable 99.6% of all cases of individuals under 70 years of age who were admitted to a hospital have fully recovered. This quite frankly astonishing figure is a testament to the immense amount of knowledge and skill possessed by the frontline healthcare workers and medical professionals throughout the province, as it confirms the efficacy of their world-class practices in assuring the safety and wellbeing of those affected by the virus. Furthermore, it serves as a beacon of light that shines through such a dark and uncertain period, and encourages citizens of the province to remain hopeful and vigilant in their efforts to slow the spread of the virus and safeguard their community.

Statistics show staggering numbers in regards to the impact of COVID-19 on individuals with comorbidities. According to a recently released report, more than 90% of all COVID-related deaths were individuals who had three or more pre-existing chronic conditions. Specifically in the United States, it was observed four comorbidities predominated within this demographic.

Given the high number of fatalities attributed to the virus, many countries have implemented preventative measures to limit the spread, including vaccines and vaccine passports. Vaccines have been recommended to those most vulnerable to the virus, i.e. people with underlying medical conditions. In many places like the U.K. and the U.S., having a vaccine passport offers certain benefits, such as ease of travel between countries without long quarantine periods. In addition, countries such as Iceland have implemented a universal national health passport app as a preventive measure against disease transmission. Furthermore, as a way of incentivizing the population to receive a vaccine, the United States is currently discussing offering a vaccine passport system that would allow vaccinated individuals more freedoms, such as entry into a restaurant and theatre.

It's undeniable that COVID-related comorbidity deaths represent an alarming number. However, vaccinating vulnerable populations and incorporating vaccine passport systems is proving to be a successful measure in limiting the spread of the virus and offering extra protection to those with underlying conditions.

/

WHY ALL THE PANIC AND THREAT OF ANOTHER LOCKDOWN

It is simply mind-boggling why our society has arrived at the point where we are turning to vaccine mandates, punishing individuals for wanting to protect their physical health, and establishing a vaccine passport system when the statistics tell a perplexing story. How can we explain, for example, the astronomical number of people over 70 who have three or more comorbidities?

This dilemma is of paramount importance to us all. We know, for instance, that the elderly population is particularly vulnerable to SARS-CoV-2, the virus that causes COVID-19. Thus, any vaccine-related rules or regulations must be sensitive to the various needs of our seniors. This includes exempting at-risk people – the ones with three or more pre-existing medical conditions – from facing pressure to vaccinate.

Furthermore, it is completely unjustifiable to terminate employees who opt out of the Moderna or Pfizer vaccines. It's unacceptable that individuals who believe their own physical bodies are best equipped to fend off the virus should be subjected to such harsh forms of discipline.

Finally, there are repercussions associated with putting into place a "vaccine passport". If a healthcare provider records proof of vaccine administration in a patient's medical file, then what's to prevent other companies or entities from requesting access to that personal information? Ultimately, it could lead to dire consequences for an individual's privacy rights.

Given these realities, we must ask ourselves: why are we in the situation we are in? And more importantly, how can humanity remedy these issues moving forward? These are tough questions, and it's unclear what the answers will be. But the graph is clear: the vaccine mandate status quo simply doesn't make sense.

It is noteworthy that, when examining the reported deaths as a result of COVID, a pattern emerges where many cases showed flu as a comorbidity. This is extremely puzzling, because the death rates of flu have registered at zero during this period. How can this be possible if flu is playing a role in these fatalities? To put this into further perspective, an average of 2,300 deaths per year due to flu have been recorded in the province of Ontario alone. Therefore, one might suggest that these fatalities should be eliminated from the tally of "COVID Related" deaths and labeled as due to flu instead. This apparent discrepancy raises several questions that may evaluate the accuracy of these reports.

This is Horrible – They Do Not Work

It is alarming that the government appears to be withholding this important information from the public. The data coming out of Israel and the UK is quite stark; even those people who have received two doses of the COVID vaccine are becoming infected, transmitting the virus, being hospitalized and unfortunately becoming seriously ill from COVID at extremely concerning levels. This disproves the theory which many had previously held that the virus is exclusive to those who are not vaccinated and demonstrates that it is not simply a 'pandemic of the unvaccinated'.

What makes this story particularly worrying is that the clinical trials performed on the Pfizer vaccine did not explore a large representative sample of populations and therefore did not fully prepare healthcare workers for what has happened. Had such a sample been tested and had the results been as expected for a vaccine, it is likely that the vaccine would not have received the Emergency Use Authorization given by Operation Light Speed. This is why it is so imperative that the government share the facts on this matter with the public, rather than trusting that what occurred under Operation Light Speed will be sufficient.

In the Canadian province of British Columbia, a review of the data collected by the BCCDC (BCCDC Data Summary August 26 2021 – page 19) suggests there is little evidence to indicate that the Covid-19 vaccine has had any substantial effect on the number of deaths resulting from the virus. In fact, the numbers of fatalities between July 27 and August 23 among those in the 80+ age range show a shocking near-doubling among those who are fully vaccinated, when compared with those who have not been inoculated. This finding is similar to those reported in both Israel and England, where the vaccine has failed to demonstrate a positive effect on the mortality rate.

The results of this data analysis constitute a 'smoking gun', representing damning evidence that the Covid-19 vaccine is ineffective and that policy-makers must rely on sound science and valid data when creating and implementing protection policies. It is extraordinary how, even in the face of such compelling research, the value of vaccination continues to be touted as the only legitimate and viable means of providing Covid-19 protection. In contravention of this supposed 'wisdom', this data shows that there is, in fact, no reliable evidence to suggest that the vaccine is any less of a health risk than no protection at all.

It is wrong for our government to attempt to divide our country and foster a sense of mistrust and animosity between different people, instead of providing proper leadership and accountability. We cannot simply blame those who have chosen not to be vaccinated as the main cause of the current state of affairs and must refrain from engaging in such divisive behaviour. Instead, we need to recognise that there are multiple complex factors at play in this situation and give them the consideration they deserve.

It is also important that we are mindful of taking any form of untested or ineffective treatments, as these have yet to be proven and can put our health at risk. We must be particularly cautious when it comes to medications and treatments with unknown effects, and if we are to try them, we must ensure that we are doing so safely and responsibly. We owe it to ourselves and our communities to refrain from taking any form of action that could be detrimental, particularly when it comes to our individual and collective health.

The consequences of our mishandling of the pandemic are becoming painfully obvious. We have gone too far and need to take a step back, assess the damage and ensure everyone is held to account. We must not be tempted by irrational arguments that cannot be substantiated with facts and figures. This isn't 'COVID Logic'; it's a genuine misstep that has caused untold disruption and suffering and it must be addressed without delay.

We have all read the stories of people who have lost their jobs due to the impact of the virus, of vulnerable families becoming destitute, of lives changed beyond recognition. It's a sorrowful, disgraceful situation but one we have all fallen into through carelessness and neglect. We must not let our misplaced optimism cost us any more than it already has. We must start to actively manage this situation according to sound scientific judgement and evidence, or else the costs in lives and livelihoods will only deepen.

This is no time for denial or deflecting the blame. We must look this challenge straight in the face and accept the blame where it falls. Only then will we be able to properly rebuild and revitalise our societies, setting us on a path to more secure and prosperous future. We can no longer turn a blind eye to the truth; it's time to act and bring an end to this senseless chaos.

Real Vaccine Looks Like

It has become all too common to hear the argument that a higher rate of cases and hospitalizations among those who have been vaccinated for a particular disease is not a concern due to there simply being more people who have been vaccinated. This line of thinking and attempts at rationalizing away any potential issues with the vaccine however is not only fundamentally flawed but is also not supported by any reliable historical data. We should be wary of buying into this type of reasoning and instead focus more on determining potential causes for this higher rate of cases and hospitalizations.

What we do know about the correlation between vaccinations and the potential for people to still contract the disease is that, although vaccinations can and do provide a certain degree of protection, there are still some cases where people were vaccinated yet still managed to contract the illness. This suggests that the vaccination may not be effective or may not be functioning as intended, and so further study would need to be conducted to determine the root cause. Additionally, if the population being studied consists of those who were vaccinated, the demographics of that group should also be considered, as any correlations between certain population characteristics and the higher rate of cases and hospitalizations would need to be considered.

Accepting flawed rationalization is a slippery slope and not doing enough to ensure that any claims made have reliable historical data to back them up is irresponsible. It is for this reason that we should not buy into this way of thinking and instead take a more holistic approach when it comes to assessing why a higher rate of cases and hospitalizations are happening among those who have been vaccinated.

The need for the government and pharmaceutical companies to justify their efforts in creating a potential COVID vaccine has grown to be increasingly desperate as the public has been presented with mounting evidence that the vaccine has not achieved the success that was promised. This reliance on the news and media outlets to relay only their view of the vaccine's efficacy, prevents the public from forming their own educated opinion. Consequently, those that research and read the reports and studies around this topic are met with disdain and labeled as 'conspiracists', while at the same time taking the blame for the increasing numbers of infections.

Yet, if this vaccine had been as successful as the measles vaccination, our current reality would be far different. We would be out and about, engaging in our favorite hobbies, travelling freely, and enjoying our hard-earned liberties. Instead, the fear of contracting COVID has caused worry and unrest to reach levels not previously seen in recent history, seemingly out of nowhere.

Vaccines provide protection from a range of diseases and can save countless lives when administered successfully. A key measure of their effectiveness is efficacy, which refers to their ability to protect against a particular virus or illness. To understand how effective a vaccine can be, let's look at the success of the Measles vaccine, first introduced as part of widespread immunization programmes in 1963.

The chart on the left shows that shortly after the introduction of the vaccine, cases of measles and associated deaths dropped sharply and have since remained close to zero. Apart from a brief period in the 1990s, we have been able to virtually eliminate cases of measles and their associated deaths in many parts of the world, thanks to the effectiveness of the vaccine and the widespread use of it over the last 50 years.

This is a perfect example of the protection that can be offered by vaccines and the difference they can make. The success of the Measles vaccine has created a safer world for countless individuals who, without it, could have faced serious illness or even death. Geographically, immunization programmes have vastly reduced the spread of the virus and helped to create a healthier generation of people for us all.

The current situation with the Pfizer vaccine, including the numerous booster shots which have been announced, together with the rising number of cases, causes enormous concern. It is clear that this news is not at the level which we had hoped to observe at this stage in history. Recent reports from British Columbia, England, and Israel illustrate this further - it appears that it's only a matter of time before matters become increasingly complex surrounding the global management of the COVID-19 pandemic.

What this shows is the importance of remaining vigilant and keeping the public informed, no matter how difficult the circumstances may be. It emphasizes the necessity to adopt robust strategies to ensure the safety of people living around the world, and a comprehensive predictive framework to determine the best course of action as new developments continue to arise. Furthermore, it is clear that communication and collaboration between stakeholders is paramount, as information sharing and joint strategies are more important than ever due to the fast-paced and uncertain nature of the pandemic.

Overall, it is important to look beyond the immediacy of the situation, in order to establish what the long-term implications of the facts and proposed solutions may be. Ultimately, responding appropriately and effectively to the pandemic is a responsibility shared between countries, governments, corporations, and citizens alike. In this way, we can all strive towards the same goal: controlling the spread of the virus and giving hope to the world that this situation - however hard to face - will be mitigated as soon as possible.

In an age where there is a heightened focus on the media and its influence on the everyday lives of people, it is important to take it upon ourselves to invest in our own research and analysis. We should be looking for the facts and data ourselves and not treating what the media say as gospel. Each news service, and each individual news anchor, has their own opinions and biases that can often colour the interpretation of facts and data and mislead us.

Not having all the facts means that we can often be unable to make accurate judgements or evaluate situations properly. For example, if we are only told half of the story of a political scandal, then it can be hard to get a full understanding, or to even know what to believe. It is therefore we to our advantage to conduct our own investigations and to research information further. It also shows that we are actively engaged and seeking truth and more accurate answers.

The media can also often oversimplify and misrepresent an issue, particularly when it involves complex and nuanced topics such as international relations, climate change and economic policies. Researching these topics for ourselves is a vital step in being better informed, so that we can engage in debate and discourse on these issues, and ensure that our beliefs are evidence-based.

Ultimately, if we want to truly understand current issues and the implications of recent developments, we should take it upon ourselves to research, look at data, and form our own conclusions. This will help ensure that we are presented with the full facts, rather than relying on the media to interpret or translate them for us. Otherwise, we might not be getting the whole truth.

Personal Choice Discrimination

Recently, the government has made considerable strides to place a greater emphasis on vaccinating more citizens in order to protect those who have already received the vaccine. However, we are now left to grapple with a burning question; if those who have been vaccinated are fully protected from contracting COVID, why does it matter what other people do? This seemingly illogical stance has revealed the truth about the efficacy of the Pfizer and Moderna treatments; that is, that both vaccines offer no immunity or protection from contracting COVID, and do not stop the spread. In fact, the only thing that these vaccines offer is a therapeutic response - one that can help reduce the severity of clinical symptoms should someone contract the virus. Evidence of this is notable in a UK report which states this clearly.

Put simply, then, are the government's attempts to 'protect' those who have been vaccinated nothing more than a bid to simply prevent the symptoms of COVID in those who have yet to receive the vaccine, regardless of the risks associated with it? Moreover, is it fair for the government to demand that people take a specific drug, when there are several other therapeutics available that are both less risky and simpler to take than the vaccine? These are important questions that we need to consider as we move forward.

Pfizer is set to launch an innovative oral therapeutic at year-end 2021. This release has the potential to reduce the burden on our healthcare system and eliminate the associated economic costs, should the unvaccinated population opt to utilize the therapy as a precautionary measure against contracting the virus. It could well be argued that such a move is in the best interests of all parties.

To maximize the efficacy of this oral therapeutic, several objectives must be met. Firstly, the manufacturing process should be geared towards producing a high quality prototype that is readily available to all. Not only must it be safe to use, but it should also be delivered at a competitive price point; permitting universal access. Secondly, the therapeutic should be easy to administer, with a straightforward process for dosing. This will ensure compliance, and remove the risk of contamination or incorrect dosage.

Finally, it is paramount that effective communication strategies are employed in order to ensure confidence in the product. Through highlighting the health benefits, as well as explaining the science behind the therapeutic, the public is more likely to adopt its use. In addition, demonstrating its effectiveness as an alternative to other forms of treatment will provide additional incentive.

In conclusion, should Pfizer successfully adhere to these goals and launch a viable oral therapeutic, then it could help to protect the unvaccinated population and consequently reduce the financial burden faced by our healthcare system and the whole country.

Today, the mounting costs of obesity to the Canadian healthcare system are staggering. Recent reports estimate the nationwide cost of obesity to the nation nearing $8 billion, amounting to a per-capita expense of approximately $1,400 for every citizen of Canada. The burden of treating obesity-related illnesses has created a significant strain on our healthcare system due to the wide-reaching physical implications obesity can have on our body. Reports suggest that over 50,000 deaths are expected to be attributed to obesity this year alone. Unfortunately, the numbers tell us that over half the population of Canada is technically overweight, with 51.1 percent falling into the unhealthy weight category outlined by the World Health Organization.

It's irrefutable that smoking is also a costly menace to society. As officials estimate, the direct healthcare costs of smoking to Canada total $6.5 billion every year, with a staggering overall cost to the Canadian economy of an estimated $16.2 billion. While it is terrifying to comprehend how high the cost of addiction can be, we must strive to make changes to better the health and wellbeing of our population.

Tobacco use is a major contributor to the global burden of preventable morbidity and mortality. This includes all forms of tobacco, such as cigars, pipes, and smokeless tobacco, as well as cigarettes. The cost of smoking to society is tremendous, both in terms of its impact on healthcare and the economy. In Canada, the total costs are estimated to be over $14 billion annually, with key contributors including healthcare costs, lost productivity, and criminal justice costs. Whilst there is also a personal choice element when it comes to alcohol, this should not be discounted when discussing how best to protect ourselves and our healthcare system. A recent report on smoking, found here, details the extent of the problem in greater detail.

By a wide margin, obesity, smoking, and alcohol consumption are the three most prominent "personal choices" that exert a tremendous impact on both healthcare expenditures and the economy in general. Practically speaking, all cases of obesity, alcohol use and smoking are completely avoidable and can be minimized if not eradicated altogether. So it follows that any sort of policies and restrictions that target individuals who choose not to receive vaccinations without equally addressing these other "personal choices" cannot be labeled as anything but discriminatory. It is unfair to limit the freedoms of certain individuals while turning a blind eye to other self-destructive choices.

The repercussions of obesity, smoking and drinking are threefold. The obvious detriment to one's health is the primary one. Obesity increases the risk of heart attacks, strokes, and arthritis, while smoking is known to cause chronic respiratory diseases, and frequent alcohol use can lead to severe liver damage. All of these conditions are expensive to treat and place an unbearable strain on the healthcare system. On top of that, drinking and smoking decrease productivity in the workplace, leading to a decline in GDP and decreased revenue for governments. Lastly, a growing body of research suggests that unhealthy lifestyle choices often lead to depression, social exclusion, and subsequent increase of criminal activities, thereby amplifying the domino effect.

Consequently, it is unfair to impose restrictions on one chosen "personal choice" group while failing to hold accountable those engaging in behaviors that contribute to a negative economic and health outcome as well. We must recognize that any attempt at creating a more balanced and equitable society must take into account all personal choices in order to avoid discrimination.

Discrimination is a pervasive issue in our society, and the restrictive measures currently taken by our government and employers are a clear demonstration of its presence. This unfortunate reality has worsened with the government's recent steps to prevent those who choose not to vaccinate from participating in certain activities within the public sphere. This action has opened a can of worms, as the government now carries the responsibility of protecting the health system and of promoting healthier choices for all Canadians.

It is not just the public sphere that is affected; employers too are making decisions that are steeped in prejudice. For example, certain companies are enforcing mandatory vaccination amongst their employees as a condition of their employment. This is fundamentally unfair, and can make vulnerable people feel like they are being subjected to a form of discrimination. In addition, there are disparities in access to vaccines, with certain demographics being far less likely to be given the resources to get the vaccine than others.

To truly combat discrimination, we must strive to create a society where all have equal access to resources and opportunities. We must take steps to ensure that individuals can make personal decisions without fear of repercussions, and demand our governments protect our health system from the dangers of discrimination.

The government has an obligation to impose the same restrictions on those who elect to drink alcohol, smoke, and are obese as those of other groups. Doing anything else in this instance would be considered a form of discrimination and is entirely unacceptable. I understand that this an uncomfortable conversation to have, yet it is necessary: the current situation is completely ludicrous, and it appears as though there is a specifically targeted assault against those freely exercising their right to personal choice. It goes without saying that double standards should never be allowed, and that any group should never be put in a position of unequal treatment.

To illustrate, consider the example of a county that recently raised the drinking age from 18 to 21. While this is an apt policy to keep young people safe, it is worth noting that the new age limit applies to everyone — regardless of background. By contrast, imagine a scenario in which a certain type of ethnicity or religious affiliation was restricted from drinking alcohol for reasons of personal choice. Apart from being ethically unconscionable, this type of rule-making would send a clear and disturbing message: those belonging to a certain group (or groups) do not have the same fundamental rights.

The above dilemma emphasizes how important it is for governments to ensure that all citizens are treated with the same respect and held accountable to the same rules.Unequal restrictions based on arbitrary characteristics not only promote a culture of inequality, but they also create an atmosphere of injustice.Therefore, the government must strive to ensure that no one is subjected to discrimination or targeted for exercising their right to personal choice.

The facts are abundantly clear and should never be overlooked: The current medical treatment mandated by the Canadian government will not offer anyone protection from contracting or transmitting the COVID virus. This is not a vaccine - a vaccine is, by definition, an immunization against a particular illness - it is a therapeutic. While terms such as 'vaccinated' are often incorrectly used to describe this treatment, it is important to note it is not a vaccine, but rather a specifically tailored medical treatment intended to deliver therapeutic benefits. It is thus entirely misguided to assume that everyone needs to have been vaccinated to protect others from the virus. The efficacy of this treatment has not been tested to determine whether it can act as an immunization, and thus should not be relied upon as a means of promoting public health.

The recent restrictions imposed on individuals who, by their own volition, remain unvaccinated are an affront to the equality rights set forth in the Canadian Human Rights Act. This is a egregious disregard for the basic freedoms enshrined in Canadian law and it is a pervasive injustice that demands to be addressed.

Canada is a multicultural country built upon the foundations of human rights. As such, any restrictions based on whether one chooses to adhere to a particular course of treatment are fundamentally unfair and antithetical to the principles of our society. We cannot allow this kind of discrimination to take place in our own backyard.

The injustice of this decision is further amplified by its implications of religious discrimination. Many Canadian citizens hold personal beliefs that dictate their stances on health-related matters and it is unjust to limit the rights of those who choose not to vaccinate. This is of particular importance to those with conscientious and religious beliefs and values which stand in opposition to the prodigious vaccination campaigns which are occurring globally.

It is now moment for Canadians to actively stand up and challenge these infringements on human rights. Everyone needs to come together to defend freedoms which should never be taken away. It is our responsibility to ensure that Canadians, regardless of their health-care decisions, remain equal and lack any form of discrimination. Together, let us remain vigilant and fight against such injustices.

In my opinion, the government's approach to this matter was misguided and can potentially have deleterious consequences of far-reaching magnitude. A haste to implement their agenda without sufficient foresight has precipitated a situation that, without remedial action, could foster a deep-seated rift in our society. For example, if initiatives are not revised in time, political unrest could fuel a nationwide calamity that could linger in the form of litigation for generations to come.

Furthermore, the opportunity costs associated with continued negligence can be immensely expensive in terms of both economic and social consequences. This is especially true in scenarios where urgent legal aid is necessary to protect citizens' rights and/or ensure an equitable outcome. A bad decision now could lead to a series of interminable court battles that could disrupt both public and private life for years to come.

Thus, it is with this in mind that I feel the government must act swiftly and judiciously to take a hard look at their current policies and reassess the potential implications of their actions. It is only by developing sound and decisive courses of action that the looming crisis can be averted and the public can be guaranteed of a secure future free from contention and strife.

While the majority of individuals can and should receive the vaccine, it is important to recognize that for some individuals there may be medical reasons to avoid the vaccine, and in some circumstances, the same could be said for individuals who are obese. It is important to note that with obesity, as with any medical condition, each situation is unique. In some cases, an individual may be struggling with obesity due to a medical condition or other underlying health issue, while in other cases, an individual may have difficulty controlling their weight despite their best efforts. Thus, it is possible that obesity may be an appropriate reason in certain cases where an individual may not be able to receive the vaccine. For example, if a person is obese due to an underlying health condition, like hypothyroidism, they may wish to consult their physician to determine the safest course of action with regards to the vaccine. Additionally, if an individual is obese due to a lifestyle choice that is not in their direct control, like limited availability of nutritious foods or lack of accessibility to safe spaces for physical activity, this could also be taken into consideration for any medical decisions about the vaccine. It is important to realize that the reasons for someone's obesity are very personal and that medical professionals can determine if it is an appropriate exception for them to not receive the vaccine.

A 99.9% SAFETY RATING IS NOT ENOUGH

When considering the safety of vaccines offered by the Government and Pharmaceutical community, the reported percentages may appear relatively benign on the surface. It is important to remember, however, the sheer volume of people being administered the drug. Take, for example, a drug with a 99.973% safety rating being administered to 10,000 people. The average person may see this as an acceptable percentage. Even so, the reality is that this means that three people would experience adverse reactions or require hospitalization. This is a significant number of people and emphasizes the importance of understanding the numerical implications of what percentages can actually mean in real world scenarios.

The touted 99.973% effectiveness rate of the mRNA vaccine has been causing a stir - and rightfully so. To put that figure into perspective, at the global level, where over 5.4 billion people have already been vaccinated, the number of potentially hospitalizations resulting from adverse reactions to the vaccine stands at an alarming 1.2 million! Such an outcome is simply unacceptable and has prompted experts to consider the establishment of new guidelines when it comes to administering drugs of this magnitude.

When confronted with a medical situation of this magnitude, close is unfortunately not good enough. It is true that the number of individuals in need of vaccination has never been seen before; however, that does not excuse us from providing the highest quality of care and protection for the billions of people around the world who have taken part in the rollout of this vaccine. It is therefore paramount that we continue to consider the potential implications of administering drugs to such a significant proportion of humanity, and set out clear and meticulously calculated parameters for administering such treatments.

It is important to understand the significance of the numbers involved when discussing the effectiveness and safety of a vaccine at a global level. To demonstrate this, let's use the example of Tylenol, a widely-used pain relief medication formally known as Acetomenephen. This medication is administered and taken hundreds of times more than any vaccine, and yet there is an established fact that approximately 31,250 people per year experience serious adverse reactions to the medication that require hospitalization, which is a number much lower that the 1,200,000 one would expect, based on its general 99.973% safety rating. This number is critical due to the sheer scale of inoculation that has been conducted worldwide.

It is clear that there is still a disparity between Tylenol's safety rating and the one a vaccine would require in order to ensure that hospitalization cases are kept at more acceptable levels. As such, in order to ensure public health safety, a safety rating of no less than 99.99999% is required. This would place hospitalization cases at around 45,000 per year, a number still somewhat short of Tylenol's, but a number that is hopefully achievable with current vaccination practices and procedures. Establishing new standards and regulations has become imperative to the success of vaccinating the world's population and reducing hospitalization cases.

Note: Based on known data, Tylenol safety rating is extrapolated to be estimated at 99.999993%

QUALIFYING DATA

There are increasingly concerning examples of reports making use of data which fails to meet the required standards for accuracy in order for people to make an informed opinion. All too often data is not fully qualified, thorough enough or even current enough for people to use it as the basis of an opinion. Without this data being available, people are unable to interpret the truth from the falsehoods.

Take for example a report released by a major news source about global warming. On the surface it would look like this report is providing a comprehensive breakdown of the facts, but on closer examination it can be revealed that certain data points crucial to the debate are omitted or misrepresented. If people were empowered to access the full data themselves and make their own conclusions, instead of relying on an incomplete and potentially biased study, then the public discourse around global warming would certainly be much better informed.

Another example can be seen with the recent outbreak of the coronavirus. Numerous reports have been released on its spread and impact; however, many of these sources of data have been incomplete or not fully qualified. This leaves the general public with unreliable information and a cloud of uncertainty hanging over their heads. To ensure people are in possession of the full predictive data they require, governments and medical experts must ensure that the majority of their reports contain transparent, accurate and up-to-date information. Only then can people form reliable opinions and react appropriately.

All in all, it is essential that data presented to the public is held to a higher standard of accuracy than is often currently seen. People must be able to access the full set of facts for themselves to be able to draw the appropriate conclusions if decisions are to be made wisely. By providing the public with comprehensive, trustworthy and pertinent data, we can start to make a better informed society.

The Ontario Government released the latest data from its datasets, and the Delta Variant has not met the expectations that the Government's ominous warnings had set. This is in contrast with the original fears and concerns surrounding the new strain of the coronavirus, particularly in large metropolitan centres like Toronto and Hamilton. In these areas, the Delta Variant has been more prominent and defiant, causing a much more rapid rise of cases than the original strains of the virus.

Despite this, the data trend shows that the Delta Variant has not had a catastrophic impact on Ontario's overall cases. For the entire province, the case count still remains within a manageable level, with the exception of a few hotspots that have seen a sudden surge in cases. This means that the province is seeing a gradual decrease in the spread of the coronavirus, and that the situation is not as dire as was initially feared.

What this data shows us is that the Government's warnings were in many ways overstated, and that steps taken by members of the public to be vigilant in following protocols for public health and safety have enabled Ontario to keep the delta variant from being a disaster. In the coming weeks and months, the effectiveness of the public in containing the virus will need to be kept under close watch in order to ensure the continued safety of all citizens.

Recent ICU visits have lent credence to the assertions made by many experts and medical authorities that the spread of the Delta variant of COVID-19 is more easily transmittable and highly contagious, compared to earlier variants such as Alpha and Bravo. However, research suggests that the severity of the virus, as well as the symptoms experienced by those infected, is far less severe compared to earlier variants. The vast majority of patients who contract the Delta variant will experience mild symptoms or none at all, with only a small fraction of patients needing medical attention or hospitalization. This stands in stark contrast to the Alpha and Bravo variants, which were far more likely to cause respiratory issues, and were responsible for the implementation of severe lockdown measures in many countries. It is important to note, however, that caution and preventive steps should still be taken, as the Delta variant is more likely to spread, leading to more cases and fatalities. Although this may be seen by some as cause for panic and further lockdown measures, the data collected thus far justifies the stance of medical professionals who believe that the Delta variant is milder than prior strains, despite being more contagious.

The coronavirus pandemic has played out in a particularly worrying fashion in the public eye, with sensationalised media coverage and governments making decisions that often seemed to be rooted in fear and confusion rather than rational analysis of the facts and figures. All the fear and media attention generated about the pandemic has ultimately been unfounded, misguided and needless, and it hasn't helped in our attempt to contain the virus. Perhaps what is most concerning about this is the notion that the media and government may be complicit in inciting this fear in the public.

Government and media figures often seem to be focusing too much on outlining the potential horrors of the virus and not enough on informing the public of the actual numbers - how many cases have been confirmed, what proportion of affected people are displaying serious symptoms and so on. Instead, they rely on a sensationalised version of events in order to grab headlines and manipulate sentiment. Obviously, this isn't helpful: it adds to the fear, panic and confusion, when what we actually need is calm, sensible decision making based on actual data. That is why it is so important for the media, and governments, to start reporting honestly and accurately, basing their assessments and predictions on the most up-to-date science and nothing else. Fear-mongering serves nobody, least of all the public. We need trusted sources of information that are based on facts and data, not unsubstantiated rumours and conspiracy-laden narratives.

/

The Vaccine's Effectiveness

By examining the COVID-19 data from the Ontario region, I was able to realistically predict what the 'anticipated death rate' would have been if the efficacy of the COVID-19 vaccine had been equivalent to that of other historically available vaccines. Traditionally, it has been observed that death rates tend to diminish as the population is increasingly vaccinated, a phenomenon observed in both the Polio and Measles vaccine. For instance, a 2015 study by the Institute for Health Metrics and Evaluation noted that through the widespread implementation of the Polio vaccine, global cases of the virus were virtually eliminated. Similarly, the introduction of the Measles vaccine in the late 1960s is widely credited with reducing the global instances of the disease from around 4 million cases per year to an estimated 500 cases annually in 2016. Clearly, these two examples serve to illustrate the point that vaccinating a larger portion of the population results in an overall decrease in death rates from a specified communicable illness.

It is of utmost importance to establish clarity; the data presented here has been sourced directly from Ontario's data sets. This allows members of the public to access and authenticate the information for themselves. All that is shown here is data, represented through simple line charts and linear regression models - there is no room for opinion or interpretation. Therefore, what is being displayed is an accurate reflection of the raw data which has been collected, eliminating any potential bias or human subjectivity that may arise. This makes it easier for readers to get a truer picture of the information being provided, rather than relying on someone's personal interpretations or conjectures.

Alberta's Death Rate

Recent statistics from the province of Ontario illustrate that in the 19-month period between early 2020 and mid-2021, only 25 deaths related to COVID-19 occurred among individuals age 20 and under. Seeing this, the government has proposed vaccinating this group of young people, yet there is no evidence to support the need for it.

This begs the question: why does the government want young people to get the vaccine? According to the latest medical data, it appears that those ages 20 and under face a miniscule risk of death from COVID-19. The mortality rate among those in this age demographic is extremely low, with only 0.0005% of deaths related to coronavirus in this category. As such, the urgency of vaccinating the young seems unjustified.

Health experts have also advocated against it, arguing that the vaccine's safety and effectiveness in this age group remain unconfirmed. Studies have suggested that the vaccine may prove beneficial in this demographic, but these cases are limited to a few countries and it's still not completely clear whether it will provide an adequate level of protection. In addition, ongoing trials for the vaccine are still needed to provide long-term data regarding its efficacy and side effects.

Given the lack of evidence that this group needs to be vaccinated, it is wise to think twice before rushing the process and make sure that only the most appropriate measures are taken. Thus, further research should be conducted to confirm the safety and efficacy of the vaccine before it is administered to younger individuals.

Alberta – More cases but less deaths?

This chart paints an interesting picture—despite an overall rise in the number of reported cases, we are nevertheless seeing an encouraging decline in the death rate, suggesting that Albertan's health situation is increasingly secure. We must, however, be sure to consider the full picture when seeking to accurately analyse the situation, rather than just pay attention to the superficial figures that can easily cause alarm. So long as the number of fatalities remains relatively low, the stringent lockdowns and other restrictions which have been employed can eventually be lifted.

Real-world examples of the impact that this shift has had can be seen in the providence of Alberta. Recent estimates suggest that, due to a fall in the death rate, various stay-at-home orders have been successfully eased allowing Albertans to go about their lives with a greater degree of freedoms. Furthermore, analysts are suggesting that the observed trend is likely to continue, with more liberalised policies potentially on the horizon in the near future, depending on the evolution of the virus within the region.

/

ou Can't Blame the Unvaccinated

The most recent data is far from promising when it comes to the efficacy of Pfizer and Moderna's treatments. In England, Israel, and British Columbia alike, numerous reports are emerging that cast doubt on the success rate of these treatments. This has left people scratching their heads, wondering what exactly is going on.

Take England as an example: some studies have indicated that the Moderna treatment might be less effective than initially thought. The University of Oxford's interim analysis, for instance, unveiled a problem. While the overall vaccine efficacy rate was 70.7%, the effectiveness rate dropped significantly to 57.8% among older individuals, or those in their seventies and older. Consequently, it has been suggested that other measures, such as social distancing, should be taken in addition to relying on vaccination alone.

In Canada, the scene is much the same. In British Columbia, for example, the Ministry of Health reported that the Pfizer-BioNTech treatment had an efficacy rate of 55.1% over the two-week period following the first dose. These results were considered concerning, and it's feared that other countries, which were initially heralding the treatments as game-changers, might be in for a similar disappointment.

These findings clearly demonstrate how the Pfizer and Moderna treatments may not be as effective as first thought. With vaccine rollouts continuing around the world, further research is necessary to ascertain just how successful these treatments will be in the long-term.

It is true, as has been widely reported, that many of us - in Israel and elsewhere - were led to believe that life could return to 'normal' once certain levels of vaccination were achieved. Unfortunately, data from Israel, which has exceeded these levels, tells a different story. As we enter our fourth lockdown for the country, there is compelling evidence that lockdowns are the only measures that appear to be effectively decreasing case levels; sadly, the vaccine does not seem to have had the hoped for results.

This has led to a great deal of frustration and confusion for many. As the world widely embraces vaccination in an effort to bring the virus under control, people in Israel are now, once again, caught in the grip of strict lockdown measures - bearing in mind that the country was one of the first to launch its mass vaccination campaign. What was meant to be a national act of optimism, a game-changer to prompt a return to pre-pandemic life, has ultimately failed at the goal.

If there is any lesson to be learned here, it is the significant importance of stringent lockdown measures, however unpopular they may be. Although we have all wanted to leave this period of pandemic life behind, if we look to the Israeli case - and similar in other countries that are heavily-dependant on vaccine rollout - it has become increasingly clear that lockdowns remain the most efficient and cost-effective way to stem the spread of Covid-19.

Where are the Pandemic Numbers?

Over several months ago, the Canadian government took a brave and unprecedented step in removing their records of death rates for 2020 and 2021; a move that was unexpected considering the records had been kept and made available to the public for decades prior. After conducting a thorough search in hopes of finding replacement data that showed the influence of COVID-19, I was unsuccessful in my endeavours. However, recently I discovered that Ontario has incredibly accurate death records dating back to 1949. What's more, each record is backed up by an official document or death certificate - a significant advantage over other datasets.

In addition, when I compared the Ontario statistics to those from a paid subscription data service's records, I found an almost direct correlation between the two. Analysis of Figure 1 also supports this assertion and further proves my findings. It is important to keep in mind that not all deaths recorded in Ontario will have a death certificate on hand; however, this does not in any way diminish the importance of my findings.

Despite the fact that there may be certain deaths not accounted for in the official death certificates, these documents represent an exceptionally reliable source of information when it comes to determining death rates for an entire state or province. Utilizing the official death certificates, public health officials and other authorities are able to paint an accurate picture of morbidity and mortality levels in an area. Moreover, examining death certificates can also help reveal patterns that can lead to effective interventions and mortality reduction initiatives.

For example, a few years ago, a study was conducted in Canada that examined death certificate information from the span of time covering 1994-1996. The study found that deaths resulting from injury, violence, and substance abuse had increased significantly during that time period, while deaths resulting from chronic conditions had decreased. These revelations provided authorities with valuable insight into the health of the population and the interventions necessary to reduce the mortality rate in that province.

To sum up, official death certificates are incredibly valuable for determining the death rate of an area and for providing vital information that can be used to implement effective initiatives to reduce mortality.

Figure 2.0 paints a picture of an unexpected drop in overall deaths, which has been attributed to the inactivity of people spurred on by COVID-19. Despite this, the predicted vast numbers of deaths that were previously discussed have yet to materialize. There isn't any concrete evidence to suggest that the deaths due to the flu were mislabeled as COVID-related fatalities in an effort to alter the numbers. However, with the understanding that news outlets would have certainly seized on any overly high fatalities rate to further the narrative and push for further measures against the virus, the lack of any meaningful increase in the death toll does raise some questions.

Many experts predicted that 2020 was going to be a particularly severe flu season, with zero deaths being registered as flu-related being practically impossible. With the standard protocol being to record any COVID-19 related deaths, where are the numbers that were promised? It is perplexing to consider the absence of any meaningful spike in fatalities statistics and prompts us to ask, where are the pandemic numbers? It is unclear what, if anything, the government or media may be withholding, but the effect of this lack of data is plain to see. Any COVID-related fatalities should be noted accurately and given the attention it deserves. Until then, we cannot be sure if the projected numbers of deaths have been omitted or are yet to be seen.

Are Our Bodies Being Edited?

With the current COVID-19 pandemic ravaging the global population, much research has been performed on the RNA structure of the virus in order to better understand and combat its potentially catastrophic effects. This has led to a number of in-depth scholarly papers written by a variety of scientists, attempting to answer the various questions surrounding this emergent virus.

In order to gain insight into the inner workings of the virus, scientists have begun to pore over these papers, reading the dense and complex scientific terminology used to explain the dynamics of the virus and its genome. Those that commit to delving into this body of research have discovered a wealth of information, not only about the virus itself, but also about the potential implications of novel attributes and characteristics that are found within its structure.

These findings are especially pertinent given the man-made origin of this pandemic, and therefore, some of the more peculiar elements of the virus such as its ability to readily mutate, could provide researchers with valuable information on how to prevent similar events in the future. For example, research published in the journal Nature highlights the highly responsive nature of the virus, allowing it to rapidly evolve even with limited genetic diversity which, when compounded with human-to-human transmission, can pose a serious threat to public health.

In addition to the scientific findings, these research papers have also provided general observers with interesting details about the virus and its structure, giving observers the unique opportunity to engage in meaningful conversations with experts in the field. With knowledge of the virus's astonishing evolution as well as its intricate genetic composition, observers can discuss potential treatments, preventive measures, and other potential solutions to the virus's global spread.

While there is no denying that reading these complex, hard-to-read scientific papers can be a daunting task, the possible advancements in our knowledge of the virus are well worth the effort. By examining these papers, scientists can develop a better understanding of the virus and its genome, while general observers can gain a better appreciation of the importance of preventive measures in preventing additional outbreaks.

Moderna's revolutionary work is pioneering a new era of molecular biology, one where mRNA is seen as the 'software of life'. Tal Zaks, Chief Medical Officer at Moderna, described this concept in 2017 as 'hacking the software of life'. This phrase may seem perplexing at first, but at its core it reflects the idea that mRNA injections can be used to alter our genetic code or DNA. This technology is often referred to as 'information therapy': taking our understanding of how information is transmitted in cells, combining this knowledge with medicine, and creating a way to adjust biological processes. You may find yourself wondering how this is possible and what the implications would be if we could one day edit our 'genetic code'. Bill Gates, a strong supporter of mRNA injection technology for coronavirus, stated that it could be used to alter our DNA, a notion that Tal Zaks firmly agrees with.

This highlights the potential of mRNA injectables to interfere with the genetic makeup of a human being, which can have a profound impact on many areas of medicine. For instance, It could possibly revolutionise the treatment of cancers, genetic diseases, and the spread of certain virus or other infectious diseases. Furthermore, it will be used to create new and improved protein-based treatments that could provide relief for many chronic conditions, such as diabetes and arthritis.

By exploring the possibilities of mRNA injection technology, we can unlock a multitude of therapeutic opportunities with the potential to increase human well-being and life expectancy. Let's not underestimate the power of Moderna's groundbreaking work and Tal Zaks' leadership in this revolutionary field of information therapy.

Transhumanism is a concept that seeks to break the boundaries of the human condition by leveraging technology to augment human intellect and physiology. In a modern world where the Covid-19 vaccines are purported to be able to alter human DNA, some may suggest that the progenitor of this transformation, the Fourth Industrial Revolution, has already arrived.

The idea of transhumanism, derived from the Latin trans meaning 'beyond' and humanism meaning 'of the human condition', essentially denotes the use of technology to transcend the bounds of what it is to be human. This ranges from simple, socially accepted concepts such as genetic engineering to more extreme aspects, such as the development of artificial intelligence to entirely replace human faculties.

The impact of transhumanism is already quite visible in modern society as everyday technology blurs the lines of what we could once have only conceptualized. The internet enables us to connect with others around the world instantaneously, biotech and artificial intelligence are starting to make processes more efficient, and advanced robotic prosthetics are allowing those with disabilities to live a more normal life.

At the same time, some do question the potential risks behind transhumanism, such as its potential to create a two-tiered society, where the elite with access to transformative technology would be able to outpace those without, leading to a greater socioeconomic inequality. Additionally, some fear that unchecked transhumanism could lead to apocalyptic outcomes, such as the total superseding of human consciousness by artificial intelligence.

All things considered, transhumanism is an idea whose potential remains both revolutionary and risky in equal measures. Now, we can only watch with anticipation to see whether it adequately serves the promise of transcending the human experience.

In recent weeks, the BBC has declared that rumours claiming the COVID injections alter DNA as 'debunked'. Nevertheless, in decades to come, this article is likely to be cast aside as one of many inaccurate articles that unjustly declared 'fact' despite being wholly inaccurate. The same can be said of their piece informatively outlining the purported need for under 30s to receive the COVID injections. Whether or not this is a coincidence, one cannot ignore the fact that BMGF - one of the BBC's largest donors in 2021 - is wholly invested in its success.

The pandemic has created an ideal platform from which to launch the mRNA future. In order to make the best of this opportunity, we have seen a rapid proliferation of what can only be considered experimental Covid injections - with no regard for the consent of the human subjects on which they are tested. Should this practise continue, we risk seeing a dystopian future in which the rules of engagement are being determined exclusively by a chosen few rather than by the people whose lives it impacts. Not only is it immoral - it is downright irresponsible.

We, as human beings, deserve to have some level of control over our own destiny. Even if we were to accept the purported benefits of the Covid injections and the next generations of mRNA injections, asking for our input and consent is the very least we can do. The question is - will the powers that be acknowledge this and allow us some involvement in the decisions that influence our future, or will they continue to ignore our voices and irreversibly alter the course of history? Only time will tell.

BILL GATES - OBSERVATIONAL

At just 14 years of age, Bill Gates was already demonstrating his technological prowess, uploading a virus of his own invention to a global computer network and successfully bringing it down. His ambition was clear from a young age, and five years later, Gates was propelled to global fame as the founder of Microsoft.

Gates was an innovator, the first to recognize the market potential in the release of flawed, buggy software and to offer up the product for free on a trial basis prior to charging for further upgrades. He understood that, by allowing users to sample his product, he could attract a loyal following as soon as he had his foot firmly in the door, and that there would be an abundance of profits to be made from the successive iterations of the technology.

Though the same cannot be said within the realm of biology, the notion of virulence, the production of a malicious code and the subsequent sale of the antidote, is a trend that has mirrored the origins of both Microsoft, and indeed the recent development of the virus that has caused a stir in Wuhan. By identifying that the demand will always be present for a protective layer to safeguard users, whether they be operating a computer or fighting a pandemic, Gates saw that potential and capitalized on it, capturing a slice of the market and becoming one of the world's leading tech moguls.

The Bill and Melinda Gates Foundation (BMGF) is undoubtedly the most prolific financial backer of vaccine-related initiatives worldwide. By the July of 2020, the foundation had, in their own words, 'donated' over $300 million towards the global response to the 2019-2020 COVID-19 pandemic, making it a significant contributor towards the fight against the disease. In addition, BMGF holds a substantial share of equity in multinational pharmaceuticals companies such as Pfizer, Inc. and BioNTech SE – two organisations at the forefront of the international efforts to develop a successful vaccine.

Furthermore, BMFG have committed US$20 million to the US biotechnology organisation Moderna and their mRNA-based projects, paving the way for additional funds – potentially up to US$100 million – to be directed towards similar projects in the future. All in all, the Gates Foundation continues to demonstrate its commitment to advancing public health worldwide through its capacity to create meaningful partnerships, at a global scale, as is evidenced by its cutting-edge partnership with the Serum Institute of India. Not only is this partnership is key to global vaccine access, but it is also a testament to the foundation's dedication and commitment to fighting life-threatening diseases around the world.

Alberta Being Victimized

It appears that once again, our political and media entities are taking data out of context in order to paint a picture of widespread despair in Alberta due to rising COVID-19 deaths and illness. When taken at face value, there is reason to be worried – however, let's take a moment to examine the situation more closely and make sense of what is really happening.

In Alberta, the total number of COVID-19 related deaths since the start of the pandemic has been 676, which is undoubtedly a tragedy, but the numbers need to be put in context. Over the same period, 4.15 million residents of the province have contracted the virus. That means the death rate in Alberta is approximately 0.0162%, which is significantly lower than the national death rate of 0.0242%. That number, though not entirely satisfactory due to the death of even one person, makes it clear that the situation in Alberta is not as dire as some in the media would like us to believe.

Political and media entities have been known to use this type of "data fudgery" before in order to bolster their agendas and distort the truth. Rather than blindly accept their portrayal of events, it is important to take the time to look at the data ourselves and make our own value judgments. With that knowledge in hand, we are better able to understand the challenges that Alberta has faced since the start of the pandemic and to determine what can be done moving forward to ensure that everyone in the province is as safe as possible.

It is an undebatable fact that the deaths attributed to COVID-19 have been greatly exaggerated by up to 75%. This can be attributed to the fact that many people with multiple comorbidities are tested positive for the virus and if they die,they are accounted for in the stats as COVID deaths. A close example of this being, a patient in a critical condition due to a major heart attack, if tested with a COVID-19 panel and tests positive, it is primarily considered a COVID-19 induced death. Many of these patients had minimalist to no symptoms of the virus yet is still accounted for as a COVID-19 casualty.

Cases such as this distorts the statistics and caused a worrying spike in the reported COVID-19 death toll. This can be understood by talking a closer look at those numbers; it becomes evident that a significant number of the people did not so much die of COVID-19, but due to further complications of other illnesses they had. For example, a patient that was tested positive for COVID-19, but subsequently died due to a lung infection or stroke did not therefore die by the justification of the virus- he died due to other conditions that attacked his already weakened body.

This does not however, imply that the virus has deterred in any way from the horrific death toll it has cost humanity. In addition, this phenomena of false representation demands attention and the statistics should accurately represent the true picture of death toll owing to the deadly pandemic.

Alberta's September report painted an unsettling picture of the impact that COVID-19 has had in the province, revealing that 185 deaths have been attributed to the virus so far. A startling 166 of these deaths occurred in those aged 60 and older with evidence pointing to 75% of such cases involving comorbidities. This increases the number of true deaths attributed to the virus to a much more accurate 41 throughout the province in this age group. A much smaller cohort of those under 60 were reported in Alberta's data with 19 "COVID-19, virus identified". Again taking into account the 75% comorbidity rate highlights the low number of under 60 age group deaths attributed to COVID-19, to be less than 5.

This data gives more insight into the reach of the COVID-19 virus, displaying the severity of the situation for those over 60 and the importance of focusing preventative measures on this demographic to avoid a rapid increase in death rates. It also helps adjust government resources to target those who are most at risk and in need of assistance, ensure the elderly are being cared for and protected to the best of their abilities. Those under the age of 60 should also remain vigilant to prevent further escalation within their age group. Knowing the real impact of COVID-19 on those it is impacting most is of great importance in planning successful strategies towards managing the continued spread of the virus.

There is absolutely no empirical evidence that suggests the hospitals in Alberta are suffering due to the unvaccinated population. The numbers simply do not aid in making this argument persuasive or valid. Sadly, the healthcare system in Alberta has been unstable for many years. It was not long ago, in 2015, that Alberta encountered a widespread and severe flu season, sparking outcry and distinging disbelief from many of the healthcare officials and workers. Despite this instance, the government continues to make desperate attempts to keep people in states of anxiety, forcing them to line up for booster shots and thus, deepening the already present social divisions within the country. This is an unfortunate truth that many people choose to seek refuge from.

The notion that there is no pandemic is patently false, as the sheer number of confirmed COVID-19 cases globally attests to the existence and severity of the disease. According to data from the World Health Organization, there have been more than 24 million confirmed cases of COVID-19 as of October 2020, leading to the death of over 830,000 people globally. In many countries, the number of active cases of the disease continues to rise, necessitating the adoption of draconian social distancing, quarantine and lockdown rules by governments to prevent its spread. It is evident, therefore, that there can be no doubt as to the existence of a pandemic.

Contrary to popular misconception, the evidence supporting the fact that there is a pandemic is far more substantive than simple headlines or tweets. In the United States alone, the Centers for Disease Control and Prevention provides an exhaustive list of COVID-19 data, which includes number of cases, deaths, hospitalizations, and demographic profiles. Similarly, the World Health Organization has released detailed reports and summaries on the spread and consequences of the pandemic globally. Clearly, more reliance must be placed on empirical, fact-based data rather than on untrustworthy media sources or tweets.

In conclusion, although some may claim that the pandemic is a mere media fabrication, the sheer amount of reliable, scientific data shows that there is, in fact, a very real global health crisis which calls for responsible, science-based solutions.

Absolute Risk vs Relative Risk

In today's world, it's seemingly commonplace to hear companies and manufacturers advertising their products or medications as having incredible results. It is only natural, then, that upon reading such claims, people can be easily convinced and tempted to purchase the product or medication advertised in hopes of experiencing these benefits. Such a notion can be particularly enticing when the drug claims to reduce a risk or side effect, as was the example given. Many may assume that a 2% chance of a heart attack being reduced to 1% is a large leap; however, this cannot be further from the truth. When evaluating real-world examples, it can be observed that the absolute difference that would be experienced when something such as a 5-year risk for a heart attack is reduced from 2 in 100 (2%) to 1 in 100 (1%) is only 1%. This seemingly small change may still benefit people, but it is an important step to avoid broad generalizations and painting a false picture of the effectiveness of a drug.

Calculating risks can be complicated and so understanding the differences between absolute and relative risk is of the utmost importance. Absolute risk is simply the difference between two risks, for example, if you have a 2 percent chance of an outcome and a 1 percent chance of an outcome, subtracting the two results in a 1 percent difference in risk. Yet, when understanding claims, relative risk is often used which can be misleading. To illustrate this, let's say you have a 1 percent chance and a 2 percent chance of an outcome. The relative risk would be 50 percent; thus, a 50 percent reduction in risk. Though this sounds very impressive, it is in actuality skewing the statistics in an attempt to deceive.

It is essential to be aware of this if you want to know the truth and have all the honest facts since a false impression can be created if relative risk is not understood. This is especially important when making important decisions and assessing risk as a false impression can lead to an incorrect understanding of the facts which could put you in an undesirable situation. Therefore, it is critical that you understand the difference between absolute and relative risk and be cautious when assessing risk to ensure that you are making the best decisions for yourself.

Staying with me as I guide you through this crucial understanding of reading and interpreting data. Let's use the comparison compensation of relative risk and absolute risk shown above as a practical example. Relative risk has the ability to make the benefits of a product or service appear much more significant, while making the risks associated appear far smaller. This clever use of framing is a common practice amongst a variety of industries, particularly the pharmaceutical industry, and it is crucial that you be aware of it so that you can identify when it's being used.

Although both absolute risk and relative risk are valid forms of data, they can be used in combination to represent a greatly misleading statement. For instance, if you are being sold something the promoter may highlight only the relative risk, exaggerating the benefits while downplaying any risks. Unfortunately, it is rare that this type of behaviour gets picked up when decisions are being made based on the data reported. For this reason, it is paramount that you understand the difference between relative risk and absolute risk and and be alert for any hidden implications.

The COVID19 – Mismatched Framing Reality

When the coronavirus first reared its head just over a year ago, scientists were quick to release information with regards to the virus and unfortunately, it wasn't long before media outlets latched onto the story, stirring up all sorts of mixed media framing. In March, 2020, I recall attempting to explain the true nature of the virus to those around me but sadly, it seemed like too few people were receptive to logical information—the panic was too great. The reports issued at the time indicated a four-percent chance of catching the virus, two percent of those affected eventually ending up in an intensive care unit, and then two percent of those in ICU would die. Unfortunately, these numbers were interpreted as meaning that there was a two-percent chance of dying from the virus, leading to even greater fear and panic. We now know that the risk of death is far lower, more in the order of .00024%, which is a massive discrepancy as opposed to the two-percent figure that was initially thought to be a more accurate reflection of the situation. This is because it is an absolute 4% chance of catching covid, however the two-percent number related to those who end up in hospital and the two-percent those who die is a "relative" number reflecting the percentage of those in ICU who sadly don't make it. A percentage of a percentage, of a percentage.

Why Are We Not Talking About This?

It is puzzling why we are not having an open dialogue about the elephant in the room: comparing the efficacy of the current COVID-19 vaccine to those that have been used for other diseases in our collective past. The evidence is irrefutable; this vaccine is simply not working as intended and expected. We are now faced with the unfortunate predicament of our government restricting access to a variety of potentially life-saving treatments and medications, with the unhindered ability to remove them from store shelves, along with prohibiting medical professionals from prescribing the same. When our government entered into a contractual agreement with the pharmaceutical giant Pfizer, the details of the agreement were made quite clear: our government would not promote nor endorse the use of any other cure for this global pandemic. This is leaving citizens of our great nation without any real options available to treat COVID-19, an ailment that has already claimed so many innocent lives.

Ontario's COVID data presents important data points on the effectiveness of the Pfizer and Moderna vaccine. By modeling an "anticipated death rate" based on the efficacy of traditional vaccinations, a concerning reality is made apparent. Vaccinated citizens are still becoming infected with the virus; why has the government not taken this into account and adjusted their course of action? Many Canadians were promised a return to 'normalcy' for those taking the vaccine, yet we are yet to see any of that follow through.

The real-world implications of the lack of effectiveness of the Pfizer and Moderna vaccinations are far-reaching and cannot be ignored. For example, in Ontario, there are still many who are opting not to get vaccinated because of fears over its efficacy. Those who are vaccinated worry that they may still contract the virus, despite all the necessary rules being followed. And as the vaccine is not as effective as we had anticipated, this has also caused a delay in the reopening of the economy, impacting an untold number of Canadians who are relying on that economic relief.

It is more important than ever to assess the current course of action, and adjust it based on our current reality. There is a need for transparency and a closer examination of the efficacy of the Pfizer and Moderna vaccine, so we can find the best possible solution to resume a sense of normalcy. It's time to review, reassess and adjust the course of action: only then will Canadians have a real chance of returning to normal.

As we approach the dawn of a new era in vaccinations, it's easy to see why so many of us should not be seeing any published outbreaks at this point. Vaccines in the past have been pushed through with haste and with little accounts for the long term implications of their decisions. It's now more important than ever that we think carefully about the effects of our actions today on the health of future generations.

Rather than fix the problems of the past, however, many governments around the world have continued to impose further restrictions on their citizens. For some, this has been out of necessity, whilst for others it has been for political reasons. But regardless of your opinion on this particular vaccine, it's become abundantly clear that the issues we face today are far more profound than many of us could have imagined.

We should all be mindful of this fact when wondering what the biggest concern is right now. Despite our need to protect public health, is it really a good idea to think more about if we have enough propane for the BBQ we're hosting with our friends? It is important to remember the gravity of our situation; of the importance of the work being done to create a safe, effective vaccine. After all, if we don't get it right, then the costs to society could be immense.

While they keep us busy fighting over the vaccine and the lockdowns, much bigger problems are about to go nuclear

Influenza / Pneumonia Deaths vs. COVID19 Deaths

Gathering this information was no easy task, as it appears that Ontario and Canada were not particularly keen on having the flu and pneumonia data compared with COVID data. In fact, extraneous data which was more recently acquired had been removed in full, for one whole year in one instance. Fortunately, I was able to reach out to other reliable sources that could provide the data required for my comparison. It must be noted that the findings line up with what many had predicted for a long time. Examining the mortality figures shown in the accompanying chart, there appears to be an uncanny parallel between the monthly totals of COVID and the flu/pneumonia. It is worth noting that there have been multiple other research studies conducted in the past which have concluded that the mortality rate for influenza and pneumonia is significantly higher than the mortality rate for COVID-19. Thus, my findings may be in line with the rest of the research community. There is no denying the gravity of the situation and the importance of continuing to enhance safety measures and protocols to mitigate the spread of not only COVID-19 but also Influenza and Pneumonia, to save as many lives as possible.

As numerous analysts and researchers have come to realize after examining the associated data, the vast majority of reported COVID-related deaths, approximately 70%, also involve at least two to three other complicating mortality factors. This means that a patient may succumb to a fatal heart attack, for instance, where the presence of COVID-19 is merely incidental. Nevertheless, if the infection is recognized, the death is labeled "Covid Related." To counter this, a far more conservative adjusting methodology has been suggested to accurately discern fatalities "due to Covid." This calibrating approach deducts between 30 and 35 percent from the published COVID death totals, thereby excluding those deaths described as "Covid Related." A sizable portion of the deceased may have had existing conditions that were not considered, leading to a potentially skewed reported figures without the additional adjustments. Unfortunately, heart disease, cancer, respiratory conditions, diabetes, and other ailments have been linked to numerous COVID-associated fatalities, emphasizing the need for greater scrutiny and more exacting analysis of the relevant data.

Recent data indicates that when adjusted and assessed using the same methodology, COVID-19 is not necessarily more dangerous than the annual flu or pneumonia. This brings into question why we have taken the drastic measures of lockdown and unemployment of those who were not vaccinated for the flu in previous years. In fact, the Centers for Disease Control and Prevention reports that influenza any a virus has caused an estimated 9 million - 45 million illnesses, 140,000 - 810,000 hospitalizations and 12,000 - 61,000 deaths annually since 2010. In contrast, the CDC estimates that for the 2020-21 season, about 36 million people were infected with the flu, 490,000 people were hospitalized, and 34,200 people died from the flu. Comparatively, COVID-19 has been responsible for slightly fewer illnesses and hospitalizations, but higher mortality rates.

In spite of this data, we can see the clear inequalities in the way governments have reacted to the two diseases. While the United States and many other countries employed strict lockdowns, travel restrictions, and unemployment sanctions in response to the novel coronavirus which began in late 2019, the same measures have not been taken for the flu. This could be due to the fact that many people become immunized against the flu each year, while a vaccine for COVID-19 did not yet exist.

Therefore, many experts argue that it is unfair and unjust to be implementing different measures for two similar diseases, taking into consideration the amount of people afflicted and the number of deaths resulting from each. It is clear from this assessment that a more unified, consistent way of responding to threats to public health must be developed for the future.

Comparing SARS-CoV-2 natural immunity to vaccine-induced immunity: reinfections versus breakthrough infections

There is a recent study out of Israel that is making waves around the world and providing unique insights into the human body's ability to fight against the many variants of the virus. It has been found that the body's natural immunity is even more effective against the most recent variant known as the Delta variant. While the study itself can be a bit overwhelming with many scientific terms and jargon, it is important to translate these findings into simpler language so that more people can grasp the key concepts and information present. This can be done by providing real-world examples that connect these findings to our everyday lives, and by providing alternate perspectives to help provide a better understanding of the results. From this, we can gain a greater appreciation of the powerful abilities of our natural immune system and better prepare ourselves in the face of potential future variants.

This recent study analyzed the population size of 673,676 individuals who were 16 years or older, who were identified as eligible for the study group. Upon further inspection, it was determined that 62,883 of these individuals had been fully vaccinated. The pool of prospective participants was further narrowed down to 42,099 individuals that had not been vaccinated, but had previously been infected.

This finding is highly significant for many reasons. Firstly, it showcases the efficacy of the vaccination, hinting particularly at how the vaccine has the ability to protect the public from a potential second wave. The data provides evidence that those who have been vaccinated are far less likely to become infected again, thus heavily influencing the country's overall infection rate. In addition, this study also illustrates the importance of expanding access to vaccines, given the relatively large number of individuals that are still unvaccinated and susceptible to the virus. This data presents an urgent call to action for healthcare providers and governments around the world to increase efforts to vaccinate vulnerable populations across the globe.

Another thing to keep in mind this study focused primarily on the Delta variant that was dominant in Israel at the time.

It is clear that the primary motivator for this study was the rising reports surrounding the efficacy of the vaccine. A detailed description of the study's methods requires a thorough explanation; they have reported that the study was a retrospective observational study with three separate groups, where three multivariate logistic regression models were applied. Although this phraseology is difficult to decipher, it essentially means that the research team tracked and observed the individuals in the three groups and used specific mathematical models to segment and better understand the data points. This enabled them to analyse the relationship between the vaccine and its clinical outcomes in more detail, investigating the various factors that impacted its success. Examples of such factors could include the age of recipients, pre-existing medical conditions, the dose of the vaccine administered, or even the environment and diet habits of the individuals. All of these variables play a primary role in the efficacy of the vaccine, which is why this method of research is so comprehensive.

The study was comprised of three groups, all 16 years of age and older:

- Those who had previously contractive COVID and recovered.
- Those not exposed to COVID but had two doses of the Pfizer vaccine
- Those who were exposed to COVID and had only one dose of the Pfizer Vaccine.

The study then looked for four things:

- Infection – How easily did the subjects become infected
- The severity of the symptoms
- COVID related hospitalizations
- Deaths

This study followed a cohort of individuals in Israel who acted as both a control and experimental group spanning the period between June 1st and August 14th of 2021, an epoch in which the Delta variant of the novel coronavirus was the dominant variant in the region. In order to account for any potential discrepancies in results due to outliers, demographic data of the participants was tracked, including age, sex as well as the time at which the first event occurred. To judge the efficacy of these metrics, a follow-up period was included in the study to assess the rate of virus spread between the two groups.

For example, it was observed that the participants grouped within the control subgroup were generally older, compared to the experimental subgroup, potentially diminishing their overall susceptibility to the novel virus. Moreover, despite the fact that the rate of infection was found to be generally consistent between the two groups, careful examination of the results revealed that the older age groups within the control subgroup experienced less infection than their younger counterparts in the experimental subgroup. Furthermore, it is important to note that the difference in infection rate between the two groups varied depending on the time of the initial testing event.

THE RESULTS

The results of Group #2, which received two doses of the vaccine, have highlighted an alarming thirteen-fold increased risk of infection for the Delta variant when compared to Group #1. This is a significant observation that has been backed up with a 95% confidence level, a measure used to assess the reliability of data. Moreover, the probability of these results occurring by chance is highly unlikely, as evidenced by a P factor of (P<0.001) - a scientific term that is thought to be factually accurate and accepted when the value is less than 0.05. Interestingly, it is thought that the effect of vaccines could have had an influence on these results, making it difficult to draw clear conclusions without considering this possibility. Other reports have also found substantial increases in infection rates with the Delta variant, suggesting that it poses a cause for serious concern when taken into account with the findings of Group #2. What is clear, however, is that more research needs to be done in order to gain a full understanding of the situation and to identify effective strategies for containing the virus.

Members of Group #3 experienced a disproportionately greater risk of infection and worse symptoms than those previously infected, leading to higher rates of hospitalization. A tremendous 5.96-fold increase in the risk of infection compared to Group #1 was determined, with a remarkably high 95% Confidence Level, as well as seven-fold higher severity of symptoms. These results are significant as they strongly suggest waning immunity within the affected population, particularly in those who had been previously infected by the virus.

The effects of waning immunity can have particularly concerning affects, especially in areas where the re-infection rate is high. Without strong natural immunity, more people are likely to become infected and a higher number of related hospitalizations will become necessary. In addition to the potential for strained healthcare systems, this could lead to higher rates of death due to the virus and other complications related to infection. Simply put, waning natural immunity could be devastating to public health, leading to a never-ending cycle of infection or, even worse, an uncontrollable pandemic.

This study analyzed the effectiveness of natural immunity versus vaccine-induced immunity and the findings indicate that natural immunity provides a stronger form of protection against infection, symptomatic disease, and hospitalization caused by the Delta variant. This is an important point that has far-reaching implications, as it signifies that natural immunity can offer a more effective defense than taking just a single or double dose of vaccines. This is an advantageous finding, as immunization programs are often not accessible to all populations, whereas natural immunity is more feasible to attain in many communities. For example, in low-income countries, resources and healthcare infrastructure might be limited and populations may not be able to gain easy access to vaccines. In this situation, having natural immunity as a viable option can provide valuable protection against the Delta variant. The findings from this study, therefore, demonstrate that natural immunity can be an efficient tool for safeguarding vulnerable communities against the virus.

The latest analysis has revealed that natural immunity is significantly more effective than vaccinations in providing sustained, robust protection against infection, illness and hospitalization with regard to the Delta variant of COVID-19. Individuals who have already contracted the virus and have received just one dose of the Pfizer vaccine have also gained increased immunity against the Delta variant. However, the long-term protection that may be gained from the recently prophylactic third dose administered in Israel is yet uncertain.

Evidence of natural immunity's superior potency in offering defense against the Delta variant is notable, particularly as the majority of the world has yet to receive a full course of vaccinations. Through natural exposure to the coronavirus, individuals are capable of developing a protection that work best in cases of moderate to severe infection, providing better efficacy than the first two doses of any major vaccine.

In India, for example, many of those infected exhibit relatively fewer symptoms than vaccinated individuals. Moreover, research has demonstrated that natural immunity does not decrease over time, in contrast to responsiveness to vaccines which tend to fade after several months. In this way, natural immunity offers the advantage of staying power in terms of protection against the virus.

Given these findings, the proper application of current knowledge can enable communities to develop an immunity that is tailored to their specific needs. These strategies should include proactive measures such as contact tracing and proper quarantine procedures, along with increased access to therapeutic antiviral treatments.

All in all, this new analysis shines an important light on the understudied issue of natural immunity as a potential way for communities to mitigate the risks associated with COVID-19 in the long-term. In combination with proper utilization of the available vaccines, adherence to public health guidelines and the pursuit of therapeutic antiviral treatments, natural immunity may prove to be the key to helping the world reclaim the safe and healthy society it so desperately needs.

Despite there being some general findings in the study, the long-term effectiveness of the many variants are still inconclusive. As more information is being gathered, reports are starting to appear which suggest that immunity levels may be decreasing, especially in terms of antibody levels in the real world. For example, studies have uncovered that after six months, those who were infected with the SARS-CoV-2 virus developed considerable drops in the number of antibodies present in their system. Furthermore, recent research conducted in Madrid has revealed that for those who received both doses of the Covid-19 vaccine, the quantity of antibodies recorded in their system had also declined after six months. These findings indicate that more studies must be undertaken to better understand the long-term effects of all different variants, as natural immunity may not be as sustainable as initially thought.

The remarkable strength and endurance of natural immunity was demonstrated by the results of the tests conducted for seniors aged 65 or older which spanned seven months. These tests revealed no signs of waning immunity despite the passage of time - demonstrating its effectiveness and reliability. Natural immunity serves as a vital safeguard against illness and disease, providing protection from a variety of potentially contagious infections and illnesses. For example, elderly subjects with healthy natural immune systems may be less likely to develop debilitating conditions such as influenza, pneumonia or bronchitis. Furthermore, it has been demonstrated that overall physical health and mental well-being are positively affected by the presence of robust natural immunity. The results of the aforementioned tests are a testament to the effectiveness of natural immunity, and a valuable resource for healthcare professionals as they strive to promote healthy longevity amongst the elderly demographic.

COVID IS NOT THE BAD GUY

In Alberta and all around Canada, a survey of death records has shown that influenza, pneumonia, respiratory failure, hypertension, diabetes, and cardiac arrest have been the most common comorbidities that have been associated with cases where a patient has contracted and eventually died from COVID-19. Unfortunately, these conditions often go undiagnosed, essentially serving as ticking time bombs that are lurking in the shadows of the population, and thus, making predicting and preventing deaths due to the virus all the more difficult.

With something like COVID-19 happening on a global scale, we face serious questions that may actually force us to look at the bigger picture: Does the virus serve as a test of the general condition of people's health and bring to light the issues that are impacting our society, issues that we've all known about but done very little to change? Is there something else that we are labeling as the 'bad guy', when in fact, it might be something entirely different?

Recent records seem to answer this question quite clearly: roughly 30% of the population is obese and nearly 40% are considered overweight – both of which are considered to be health factors that would strongly contribute to having a weakened immune system. To have a better understanding of the death toll from COVID-19, it would be hugely informative to have statistics that were able to capture the various health factors of each patient, such as body weight, fat levels (BMI), whether they are smokers or drinkers, and their general well-being. With this information, we would be able to determine whether or not a patient's health status is actually having an influence on the severity of their symptoms.

Had proper preventative action been taken to address the trends that were foreshadowed nearly forty-two years ago, the world could very likely have been spared much of the hardship that the current coronavirus pandemic has brought about. In 1978, the world had yet to experience the SARS outbreak and the swine flu pandemic of 2009, however, the inklings of future pandemics were on the horizon. Global scientists had highlighted the looming risk of infectious disease, yet few countries had the foresight to prepare for such a crisis.

Had measures been initiated much sooner to shape a pandemic-ready infrastructure, the effects of a potential pandemic could have lessened drastically. However, even in the present, prevention tactics are not a fool-proof solution as evident in the varied approaches taken by different countries to offset the pandemic's spread, with some faring better than others.

In short, it is, in hindsight, clear that had more attention been given to the risk of contagious Diseases much earlier, possible outcomes could have been dramatically different. Accordingly, the world is now learning from the fallout of the recent pandemic with many countries taking steps to build a better medical infrastructure and increase access to medical resources. With any luck, this will lead to a pandemic-resistant future, one that could have been achieved had important lessons been taken to heart many years ago.

The sheer irony of it all is almost too much to bear; instead of having conversations and spreading awareness about how health, exercise and immune-boosting superfoods and dietary supplements can be powerful tools in helping people battle against COVID-19, we are essentially forcing people indoors in the hopes of containing the virus. This, however, is in fact only exacerbating the problem. I truly believe that we will face very similar obstacles related to influenza in the years ahead. People's immune systems will be facing more strain than ever before due to lack of fresh air, sunlight and Exercise, not to mention the fact that food shortages and rising prices for essential health-foods are pushing people towards processed, junk food. It's a recipe for disaster.

Sadly, it looks like Big Pharma's coffers are going to be overflowing for the foreseeable future, given their preference for treating symptoms rather than curing underlying illnesses. The idea of curing entire diseases would put an end to their endless stream of profits, so as long as people keep getting sick, Big Pharma will remain in business. It's an unfair situation, and it's up to us to make sure that our collective health is put first in this complex time.

The Big Reset – They Didn't See That Coming.

It is incredibly disheartening to look at our current government and be reminded of the frightening shifts in society that occurred in Nazi Germany in the early days of Hitler's reign. By systematically and consistently blaming the Jewish population for Germany's economic woes through a powerful and effective propaganda campaign, an entire nation was hardened against this population. This gradual manipulation of opinion and blatant manipulation of facts led to the Jewsh people being denied basic humanity, by their fellow countrymen. We all know the horror of what followed in the wake of this mass scapegoating and dehumanization. It stands as a lasting warning to us all of the dangers of unchecked scapegoating and should cause us to re-examine the implications of today's government tactics when fingering groups as the source of our nation's issues.

Today, Prime Minister Justin Trudeau has repeatedly made it clear that he holds the unvaccinated population responsible for impeding the economy's recovery. His stance reflects that of Adolf Hitler, who similarly urged his associates to repeat his orders and make them widely known. This parallels numerous accounts which detail the atrocities of the Nazi regime, depicting a disturbed ruler who escaped responsibility and placed the blame on those of contrasting beliefs. In modern times, this reeks of eerily similar behavior, as Prime Minister Trudeau places blame on the unvaccinated population in order to prevent himself from facing the brunt of responsibility for the economic crisis. Ultimately, such tactics reflect a leadership approach riddled with manipulation and maltreatment of its people.

The efficacy of the COVID-19 vaccine has been widely discussed in recent times, but one question that has yet to be definitively answered is whether it is actually effective at immunizing people who have received the inoculation. If it is as effective as it is purported to be, then those who have been vaccinated should theoretically be able to go about their daily lives without fear of contracting the virus. However, there remains the question of what the impact would be if a significant portion of the population is not vaccinated. It is possible that a large unvaccinated population could create a pathway for the virus to persist through a number of different carriers and perpetuate the pandemic. This could be especially hazardous for vulnerable populations such as the elderly or immunocompromised, who are at risk of enduring serious complications from contracting COVID-19. Fortunately, there are now a number of different treatment options available if someone becomes infected, which could reduce the mortality rate of the virus and ultimately help to bring an end to the pandemic.

It is difficult to understand how government logic works when it comes to protecting our healthcare system. Unfortunately, the protective measures that are taken are often seen as too drastic by citizens. I find myself asking "why are these steps needed?" Many members of the public would like a more in-depth explanation as to why they should accept the expansive regulations on healthcare that are being proposed by the government.

The truth is, many governments enforce stringent rules and regulations on health care services in order to ensure that citizens are receiving the best possible care. By doing this, it helps to control costs, improve healthcare quality, and expand access to medical attention. To illustrate, some countries implement price ceilings on medical treatments. This prevents healthcare providers from overcharging consumers on medical services and keeps people from being priced out of medical care altogether. By imposing such governamental measures, people receive necessary healthcare without being financially burdened.

Therefore, it is essential that people recognize the value of such regulations and accept the government's logic. It is not for the purpose of limiting health care rights, but rather to ensure that citizens receive the absolute best healthcare possible. This does not mean that the government's logic is perfect or free of criticism, but it should be acknowledged and respected nonetheless.

As we grapple with these tumultuous times, we often forget that our current situation may seem small in comparison to some of the more pressing issues of our world. It is easy to get wrapped up in petty bickering and fighting, but we mustn't get so lost in the chaos that we forget about the much more significant matters that could be at stake. Though the current pandemic has presented its own unique set of challenges, there are much bigger issues that exist beyond our current tribulations. For example, climate change continues to pose vast threats to our world; from rising sea levels to increased periods of drought and desertification, it could be argued that its consequences could eclipse even the impact of COVID-19. Another pressing matter is the growing international tensions between nuclear-equipped powers, such as the United States and North Korea, that could potentially spell disaster for our home planet if a conflict were to arise. By keeping these dangers in our minds, we are provided a more sobering perspective of the intricate world we live in, in which COVID-19 may not even register on the scale of human tragedy.

Words to remember

We are currently traveling on a path marked by the rallying cry of 'Build Back Better' – a rallying cry that has appeared to gain political momentum within the framework of what has been dubbed 'The Great Reset'. This rather euphemistic term is often used by influential globalists such as Canada's own Prime Minister Justin Trudeau, and it has become the banner for his ideological movement designed to redefine the modern capitalist system, further entrench an international green agenda, and ultimately reshape the established global order. While this may seem far-fetched, it isn't beyond the realm of possibility when taking into consideration the numerous policy announcements from the Prime Minister over the past few months. For example, Trudeau has recently announced the introduction of initiatives such as the Canada Infra Bank – an independent entity intended to oversee the investment of billions of dollars into infrastructure projects, as well as the Green the Economy Fund – a government fund established to provide capital for environmental sustainability projects. These are just some of the examples signaling that the 'Great Reset' is underway, and it would go beyond the scope of this piece to cover them all. It's clear, however, that this ambitious project to redefine the world as we know it is an unmistakable reality.

It is my sincere wish that my apprehensions are unfounded, but the recent spate of large-scale transformations in the socioeconomic sphere have been occurring with such frequency and magnitude that it is becoming increasingly difficult to reconcile the likelihood of these upheavals as mere happenstance, when they could quite possibly be the result of a carefully orchestrated campaign seeking to create a 'Great Reset' of the global economic order. After all, the implications of drastic shifts in the global political and economic landscape are simply too monumental to ignore. Take the recent introduction of massive fiscal stimulus packages in many of the world's leading economies as an example; the long-term effect such a move could have on prices, incomes and job security is something that should not be taken lightly. We are seeing the redistribution of resources from those who can less afford it to those who can. This profound change in the balance of global economic power is no mere coincidence and is indicative of a greater agenda to shift the traditional locus of economic power from the west to the east.

The signs of this Great Reset are scattered far and wide, from the emergence of a global digital currency to increased trade barriers and regulations that have effectively reduced the mobility of labour and capital – no doubt part of a larger, deliberate plan to produce a new economic reality. And this gravity-defying global economic rearrangement is not exclusive to the financial side of things; the current focus on green initiatives, ethical investing, privacy and digital transformation are not isolated phenomena, but a visible channeling of collective resources towards a pre-defined goal. The goal is nothing short of a total re-orientation of the global economy, with the power and responsibility for creating an equitable, sustainable and economically just future resting largely on the shoulders of those who are in a position to effect it. Only then can we look to a more equitable global economic future.

As the global pandemic stretches on, the fear that society will experience an economic crash as severe as the 1930s Depression is becoming a very real possibility. The threat of COVID-19 has already inflicted catastrophic damage to many industries, most notably air travel and hospitality, as booking cancellations and business closures become increasingly common. Additionally, the halt of global workforces, trade restrictions, and a drop in key commodity prices are all leading to a decrease in global GDP; according to the International Monetary Fund, the world's GDP is set to shrink by nearly 10% this year.

Not only will this downturn have a dramatic physical effect, but it is expected to spur a mental health crisis, as well. With unemployment levels reaching record highs, financial insecurity will become a pressing issue for many individuals and families. The effects of this can be seen in people of all ages and backgrounds, from youth to seniors, with a particularly marked impact on those who had been living paycheck to paycheck prior to the pandemic. The psychological toll of the economic tumult will assuredly be immense, causing distress for those who are uncertain about their financial prospects in the coming months and years.

Given such a dire situation, it is unsurprising that the public is becoming increasingly anxious about the world's social and economic future. This sentiment is especially strong among major economies such as the United States, which is on the brink of its worst downturn since the Great Depression. It has become clear that, if greater preventative measures are not taken, the world could face a prolonged period of economic hardship; a period of despair that is already impacting vast swathes of the population.

The Big Reset is sure to be a seismic, transformative event in our world, and one of its most crucial and far-reaching components involves the close coordination of the private sector with government regulation. The reality is that profiting at the expense of long-term societal well being cannot be allowed - governments and regulators therefore have a duty to intervene, to ensure that companies are internalizing the cost of environmental and social harm. This raises all manner of questions about the nature of the state we may find ourselves living in once all of this has been implemented. How much power, exactly, will the government have over corporations? What form will this regulation take? Will companies have to adhere scrupulously to a set of rigid and inflexible guidelines, or will they have the ability to use discretion and innovate? All of these are pertinent questions, and only through thorough consideration of the specific details and an in depth analysis of the implications of such measures can we truly understand the implications of these big shifts in the current state of affairs.

For decades, some people have maintained that powerful individuals and organizations are conspiring to control and manipulate large populations of citizens. This so-called 'conspiracy theory' has now been confirmed by Prime Minister Trudeau, as we see the elites using the current COVID-19 crisis as a pretext to consolidate power and shape our lives in accordance with their own agenda.

This agenda includes adopting and strengthening measures that allow for greater monitoring and surveillance of citizens. In Canada, for instance, the arrival of the "digital infrastructure law" will soon expand the authority of the federal and provincial governments over internet circuitry, opening the door for increased online police presence and monitoring of citizens. And of course, governments are now encouraging the use of cashless and digital payment systems, while government drones circle overhead to monitor population density and movement.

At the same time, we have to consider how this power consolidation and reshaping of life can have lasting economic, social, and political ramifications. All of these sweeping changes can have a deep and lasting effect on our daily lives. We can already see the increased risk of cyberattacks, privacy violations, and a new wave of unemployment, poverty, and inequality in our society.

What was once dismissed as fantasy is now all too real and is something we must prepare for. The COVID-19 crisis has allowed for a new level of intervention in our lives by the elites, and it should be seen as a wake-up call for all of us to confront the growing power struggles that affect us.

If Prime Minister Trudeau were interested in transforming Canada into a European-style green socialist utopia, it would be necessary for him to gain a mandate from the population. It is only fair, after all, for all Canadians to have an opportunity to vote on the country's recovery plan - a plan which could have a huge impact on future generations, spanning our children, our grandchildren and beyond.This is an issue that would significantly change the direction of our nation; therefore, it would not be responsible for a leader to impose this plan on the people without their consent.

Whatever the outcome of a popular mandate may be, it is essential that all Canadian citizens have a chance to express their opinion on the direction of their country. This includes members of all backgrounds - urban, suburban, and rural; First Nations, Metis and Inuit communities; people of all ages and genders, and economically disadvantaged individuals. By giving all members of society the chance to have their say, we can ensure that any alterations to Canada's path will be handled in a unified, collective manner.

The recovery plan that Prime Minister Trudeau may decide to put before the people could have a drastic effect on the nation's character. It could alter taxation, healthcare, education, and environmental policies - among a multitude of other potential changes - and thus the decision must not be taken lightly. Canadians must be well-informed of the implications of the plan and given adequate time to consider it before making their choice in the ballot box.

Ultimately, it is the Canadian public who will decide whether or not to accept the Prime Minister's vision for the future of the country. But whatever their preference, all citizens need to have a voice in the matter. Only then can we be confident that the country's transformation will be a decision that is representative of the people and in the best interests of its citizens.

The Flu Season vs COVID Season

The phenomena presented by the two graphs is quite striking. The data has made it apparent that the spike and decline of COVID-19 infections are in line with eight flu seasons identified by the layered line graphs. This showcases the incredible similarities in their genetic makeup and demonstrates that COVID-19 has inherited the same seasonality of the flu. This may be seen as the ravaging pandemic of 2020 replacing the flu as which has existed for over a hundred years.

Studies have shown that like the flu, COVID-19 follows the prevalence of certain climatic weather patterns that are directly related to transmission of the virus. Subsequently, it appears that similar to the flu, COVID-19 too is driven by seasonal factors. Yet, based on the data presented, it appears the duration and severity of the seasonal stages of the virus are both greater and longer than those of the flu.

As the world continues to grapple with the raging pandemic, looking back on the data presented by these two graphs provides us with insightful information. It has become increasingly clear that dealing with COVID-19 requires an understanding of the seasonal patterns of flu-like viruses and an adaptation to the same. Though more research is needed in order to truly understand the dynamics of this virus, the comparison illustrated by these graphs is certainly a start in the right direction.

It is sadly quite clear that the fears surrounding COVID-19 were not unfounded, for it is a real threat that has since caused irreparable damage and profound grief. But its potential for destruction has been overstated and sensationalized in an egregiously irresponsible manner. The massive public health response to the virus has been nothing short of disastrous, ripening the effects and threats of COVID-19 with an unnecessary sense of urgency and pandemonium. How did we, as a society, let this happen? How did we squander the facts and latch onto lies concerning a virus that is essentially no more serious than the average flu?

Rather than taking the opportunity to inform the public of the relative dangers of the virus and highlight preventative measures to guard against its spread, many chose instead to stir the pot with exaggerated stories and alarming claims meant to induce fear and compliance. Furthermore, the ramifications of the sweeping dragnets and decrees ushered in by governments and reinforced by overzealous media outlets were often overlooked or even deliberately concealed.

It is absolutely essential that we take actions to educate the populace and implore our government to be judicious in their response to future medical events. This is the only way to stop the barrage of exploitation and misinformation campaigns that have the power to divide us and leave us defenseless against the forces that spell our demise.

We dramatically overreacted to the current Corona Virus pandemic, and it has caused irreparable damage to many lives. The panic we experienced has led to immense changes in how we detect viruses, as well as how we record deaths due to these viruses. This has had a devastating effect both literally - with medical and civil systems restructuring to accommodate the changing threat - and metaphorically - with the fear resulting in hundreds of thousands of lives changed forever.

For instance, the number of hospital beds, medical supplies and protective equipment available have all been drastically altered in the face of the panic, leading to reduced healthcare and availability for those with existing conditions and pandemics. On the other hand, the emotional and mental distress caused by the mismanagement of reports and the resulting wave of fear has impacted businesses and families alike. We are now seeing millions of people being laid off from their places of employment, losing their businesses or investments, and facing homelessness.

It is time we assess the consequences of our panicked reaction to this virus, and turn the tragedy into a beautiful opportunity to redefine our healthcare and safety protocols, ensuring that the same mistakes are never repeated again. Making sure that people's lives and livelihoods remain intact and that our reactions are based on proof and facts, not fear.

The COVID-19 pandemic has presented an unprecedented challenge to health care services around the world, and it is vital that we take seriously our collective responsibility to monitor and protect against the spread of the virus. However, it is equally important to recognize that this virus should not be viewed in isolation from other diseases. Many of the mortality rates and other related statistics associated with COVID-19 are very high in the grand scheme of things and in comparison to other infectious disease, showing it can and does cause serious ramifications. However, it is critical to note that similar mortality assessments for other afflictions and risks – such as seasonal influenza, measles, heart attacks, or auto accidents – are rarely conducted and tracked on a cumulative basis across multiple years. In other words, while any single year's influenza death count might not seem so high, if one instead looked back and tallied the toll over an extended period of time, it would likely skyrocket to may hundreds of millions if not more.

Each year, society is faced with a certain number of seasonal flu-related fatalities that provide an indication of how much suffering a community can tolerate before drastic measures such as the lockdown, masking, and the closure of schools and businesses, become necessary. Thus the current focus of government concerns comes from the introduction of COVID-19, and the consequences of this pandemic, which have brought about a vast change within our mortality data collection and interpretation. Before, a COVID-19 diagnosis was only obtained if it was the primary cause of death, and a tangible and immediate result of the condition was present. However, since the findings of this virus, and the subsequent reworking of the mortality standards, when COVID-19 is merely a contributory illness, it can now be catalogued as a COVID-19 death. This new reality has now been in place for over fifteen months, and has led to more confusion and public fear than before.

What this ultimately means is that a person diagnosed with Alzheimer's disease nearing the end of their life, who has tested positive for COVID-19, despite not exhibiting any of the characteristic symptoms of the virus, can now be classified as having died due to COVID-19. This is markedly different to how we treat the flu annually; statistics for flu-related deaths are not tracked as vigilantly as for COVID-19, and flu-related deaths are not necessarily classified as a flu-related cause of death unless the flu was particularly severe at the time of death. This shift in record-keeping has enabled more of a sustained sense of panic in relation to the COVID-19 pandemic. As our ability to detect COVID-19 improves, we have become more likely to attribute a COVID-19-related cause of death to a given patient, even in cases where the virus may have only played a fairly minimal role in the patient's death. It is unfortunate that not more people are engaging in further research on this matter, given that were they to do so, they may come to the same conclusions as many public health professionals who have spoken out on the matter. Some believe that the hype surrounding COVID-19 has much more to do with advancing certain political agendas - such as the 'Great Reset' - than it does with any genuine concerns for public health.

Show Me Your Papers

If you assumed that when walking into a restaurant or other businesses that require you to prove your vaccination status, this only means having to display your proof of immunization, then you are unfortunately mistaken. Aside from proof of vaccine, these establishments also demand your proof of identification. You might believe that this is related to COVID-19, however, by providing this evidence, you are being tracked. If you do not think this is a substantial matter, then I regret to inform you that you are burying your head in the sand, and it is time that you begin your research.

Tracking of individuals' information is a very real and impending situation, with big companies and government entities taking advantage of the data they already possess, such as the food you order, the places you visit, and purchases you make. Companies such as Amazon and Uber can use this to gain insights into individual's preferences, while governments, such as the United States, can use it to draw conclusions about your political standpoints and beliefs. Furthermore, this practice could lead to the estimation of health outlooks, being used to determine rates of insurance, among other things.

It is imperative that we take this issue seriously and do our part to stay informed. To deny the issue of tracking is to ignore the severity of it and its effects on our lives. We should all aim to be conscious and educated on the realities of these technological advancements, in order to make sure that our rights and liberties are respected.

The Chinese government is swiftly advancing the implementation of their digital ID plan, using smartphone data to verify the validity and expiry dates of your Covid-19 vaccination records. Little do people understand that this is the first step towards a nationwide social credit system. Were this to take effect and be accepted, the Chinese government would have access to every detail of every individual's life. Every purchase, every tweet, every post, every text message, and every photo - they will all be visible to them. Even worse, they would be able to turn on your camera whenever they wanted - and you would never know.

This level of social surveillance would amount to an attack on basic freedom of expression. Individuals would be forced to stifle themselves, their opinions, and their activities out of fear they'd be tracked, monitored, and judged by the government. Ultimately, this system would further destabilize the already sensitive trust between citizens and the government, as Chinese people would soon be living under the illusion that their every move is on a credit score. Such a blatant disregard for individuals' privacy goes against all accepted norms in a democracy, making the world stand back and question the decisions of the Chinese government.

Will We See a Viable Vaccine?

Misinformation spread by the Centers for Disease Control (CDC), the Food and Drug Administration (FDA), and other government and health advisors can have devastating consequences, particularly if it is not addressed or corrected. In some cases, the information provided can be misleading, and it could lead to a misrepresentation of what may actually be happening in the world. This can have a corrosive cycle, where distrust festers and people become increasingly divided and pitted against one another.

A perfect example of this is the recent pandemic caused by COVID-19. The misinformation disseminated throughout the outbreak stunted the efforts of governors and health professionals alike to create effective responses and strategies for stemming the virus's spread. Many people became frustrated with the lack of clear guidance and information from the federal government, fuelling the spread of hoaxes and conspiracy theories about the virus. This type of behaviour only served to sow further discontent and divide people on every level.

In order to address these issues, it is crucial that the government and health experts look to create more effective communication policies and protocols to ensure accurate information is shared with the public. This could involve dedicating additional resources to verifying the facts behind a story, creating more trusted media channels through which to share updates, or organising regular virtual health classes in tandem with local media outlets.

Ultimately, the spread of misinformation should not be taken lightly, as it can cause irreparable damage to our social and political fabric when it goes unchecked. Therefore, it is essential that we work together to create an environment that encourages transparency, truth-seeking, and constructive disputes to ensure accuracy when it comes to health-related content and advice.

The National Institute of Health (NIH) and the World Health Organisation (WHO), along with our government, have made the deliberate choice to suppress information about the efficacy of options for treating COVID-19 in addition to the often-cited alternative of "mass vaccination". This unscientific decision breaks with the established practices of the past 70 years.

We are not supposed to know that the data increasingly suggests that vaccination is only sensible for those over the age of 30, particularly in light of the fact that the delta variant is estimated to be seven times less deadly than its alpha counterpart. The data simply does not support the idea of mandating vaccination as the only option.

In fact, scientific evidence provides evidence that alternative treatments, such as Ivermectin, could provide a higher risk reduction and come with a better safety profile than the current selection of vaccines. Moreover, not only is early treatment a superior alternative, but it is becoming increasingly clear that very early interventions like prophylaxis may be even more effective at controlling the virus and reducing its severity.

It is time for our governments to show respect for science and to make this information public, so that individuals may make educated choices about their own health.

Recent findings have startled the medical community as new statistics suggest that, contrary to popular belief, vaccinations may actually be making us more vulnerable to dying from COVID-19, instead of less vulnerable. Scientists are now beginning to suspect this phenomenon is caused by antibody-dependent enhancement, which could be determined with ease through autopsies. Unfortunately, the Food and Drug Administration (FDA) has thus far failed to provide an adequate explanation as to why autopsies are not yet being conducted to confirm or refute this concerning possibility. This lack of action from the FDA is perplexing, and leaves the medical field wondering why the collection of this invaluable data is not taking place.

Certain infectious diseases, such as smallpox, rarely, if ever, present asymptomatic carriers—that is, people who are infected but do not show any signs or symptoms of the contagious disease. The challenge posed by asymptomatic carriers is that it makes tracking down the microbiological source particularly difficult, as it is difficult to pinpoint the pathogen's origin when it may be concealed within certain individuals. Perhaps the most renowned example of an asymptomatic carrier is "Typhoid Mary"—who was chronically infected with Salmonella typhoid without having any discernible symptoms. As a result of the latter, the woman unknowingly spread the pathogen far and wide, to the extent that wherever she went, she would unwittingly spread the infection, and with potentially tragic results. This episode in history serves to underline the dangers associated with asymptomatic carriers, and serves as a potent reminder of why it is so important to isolate and eliminate deadly microorganisms.

The same is true of SARS-CoV-2. Far too many people can be asymptomatic carriers, making the virus tough to track down.

Vaccines are often suspended or withdrawn from the market once the number of fatalities related to the vaccine rises to over fifty. An example of this can be seen in the case of the 1976 swine flu vaccine, which was stopped in nine states after only three deaths had been recorded. In total, it is known that up to thirty-two people passed away due to the swine flu vaccine, and as many as 450 individuals were found to be suffering from Guillain-Barre Syndrome. This prompted the suspension of the vaccine across the entire country. It's heartbreaking to think of the number of lives negatively impacted by this event. This underscores the importance of taking the time to review the safety and efficacy of a vaccine's effects prior to releasing it to the broader population. The development of a scientifically sound approach to testing a vaccine and its possible side effects is critical in order to protect innocent people from harm.

In modern times, the rules around clinical trial testing have been completely abandoned. This means that there is no longer any ceiling on how many people could potentially be affected by the trial. Alarmingly, no public official has yet to step forward and put an end to it.

The most recent numbers related to this trial are particularly unsettling. 438 people have been diagnosed with Guillain-Barre Syndrome as a direct result. Additionally, more than 9,000 documented deaths have been reported directly as an outcome of the trial. Lastly, the trial has been connected to 438,440 adverse events.

Unfortunately, this set of unparalleled, catastrophic statistics shows us that the clinical trial in question is no longer supported by any type of stopping condition. This lack of a safeguard illustrates a dire lack of responsibility and accountability on the part of those involved in the trial's planning and execution.

Looking at the graph below, it's clear that there has been a dramatic shift in reported injuries through the VAERS system since the rollout of mass immunization. While it could potentially be attributed to the new vaccine, our focus should not be placed exclusively on the cause of this change, but also on how we can ensure that any future vaccines are even safer. Achieving a safety rating of 99.99999 is much more desirable than something like 99.97 and I recently wrote a detailed blog post explaining why this is the case.

The safety and efficacy of vaccines is understandably one of the primary concerns for public health experts. Every scenario should be addressed thoroughly and a plan for dealing with any potential issues must be in place. The analysis of the VAERS data and subsequent conversations will help to that end, and hopefully lead to the development and distribution of a safe and effective vaccine.

In short, while further study is needed, these trends uncovered within the VAERS data are concerning, and suggests an urgent need to ensure the safety of all upcoming vaccines to the highest possible standard. We need to emphasize the importance of this goal, and strive to get as close to the 99.99999 standard as possible.

Since the potential adverse effects of the vaccine are yet to be approved by the FDA, it is up to the manufacturers of the drug to provide evidence proving that the reported side effects were not caused by their own vaccine. Unfortunately, they have not been successful in providing any death or adverse events that were not either caused or worsened by the vaccine. Furthermore, no one else who was consulted was able to find an event which did not appear to be caused by the vaccine either. This is why it is essential for the public to assume, until it is otherwise proven, that all of these side effects are in fact linked to the experimental vaccines. If that is the case, the appropriate reaction would be for the government to stop and terminate the clinical trial out of safety concerns as soon as possible.

The mainstream media outlets have rarely, if ever, taken it upon themselves to question the research conducted by the Centers for Disease Control (CDC), the National Institutes of Health (NIH), or other public health and governmental entities when it comes to claims made regarding the safety and efficacy of vaccinations in recent times. This neglect has become even worse in recent months, as the truth behind the death and injury rates reported in relation to vaccines has become increasingly clear, despite the lack of media attention devoted to the topic.

The hard evidence that has come to light in recent months has demonstrated, beyond any doubt, that the mainstream media's negligence of the vaccination topic has enabled the CDC, NIH, and other governmental authorities to cover-up their lies regarding the dangerous side effects associated with the vaccine program since its inception. Had the media done its job by questioning and properly analyzing the death data made available from public health agencies, it would have become quickly apparent that the overwhelming majority of deaths that were being reported can only be attributed to the vaccine, with no other plausible explanation being available.

It has taken the brave efforts of independent watchdogs and interest groups to release this data in the face of governmental objections, allowing the Canadian people to finally understand the truth that had been hidden right under their noses all along. This proves once again the importance of an independent and vigilant media that is willing to ask the hard questions and investigate the truth behind the lies that have been sold to the public by our government and public health agencies.

The mainstream media outlets have failed to inquire the Canadian Centre for Disease Control for their comprehensive assessment of the death figures. Had they done so, it would have been glaringly evident that the Canadian Centre for Disease Control, National Institutes of Health, and Food and Drug Administration have been deceiving the people of Canada since the initiation of the inoculation program. It is without a doubt that these national healthcare organizations have had the knowledge that the majority of the demises lack an acceptable rationale other than the newly-introduced vaccine.

The gravity of this situation is unfathomable, as the lives of countless Canadians have been needlessly lost. By intentionally concealing the data of these fatalities, they have jeopardised the public's ability to make informed decisions when debating the curative effect of the vaccine. The truth is that a substantial number of individuals have been subjected to adverse effects due to this drug, and this issue should not be taken lightly by any means.

In order to protect the safety of the citizens of Canada, it is imperative that a full and independent evaluation is done of the vaccines. While the CDC, NIH, and FDA have promised to provide reliable statistics, their past records have evoked considerable doubt into the verity of their pledges. Ultimately, the media outlets should not remain in complacency and should take the necessary steps to expose such corruption as soon as possible.

It has been a hot topic of discussion lately among numerous medical experts and wide-spread public opinion alike, this vaccine. While the majority of us might not possess the same level of expertise as scientists, virologists, and other exceptionally intelligent individuals, the dialogue surrounding the efficacy of this potential vaccine has been both passionate and comprehensive, as discourse of this magnitude has been thoroughly discussed.

For starters, hundreds, if not thousands of individuals have dedicated their careers to studying the science behind the vaccine. Ranging from infectious diseases to immunology, the collective knowledge of these experts have been pooled together to form the foundation of new avenues of research. Within these research frameworks have come insights and breakthroughs that have been then tested in the laboratory and observed in clinical trials, giving us both a greater understanding of the vaccine and it's potential applications.

At the same time, the discussion around this vaccine has become a global talking point due to the unprecedented nature of the pandemic. Citizens from all walks of life have been exposed to the notion of a vaccine, and many have become more informed about the science and technology that goes into it's development, manufacturing, and distribution. This has further fuelled the conversation and brought us closer to an informed consensus and debate with regards to the viability of the vaccine.

To conclude, while it is quite evident that most of us don't have the same depth of knowledge or understanding as the aforementioned experts, the conversations and dialogues around the vaccine have become increasingly in-depth, multi-faceted, and thought-provoking. As a result, the scientific and academic research, as well as the informed opinions from the public, all combined, are coming together to form a much larger and more complete picture of what the future may ultimately hold for us in terms of this vaccine.

Asymptomatic carriers represent a significant challenge when it comes to controlling the spread of infectious diseases as they have the ability to spread the disease without displaying any symptoms, making them practically 'invisible.' Smallpox is one such example of an illness that few, if any, asymptomatic carriers are known to exist. This makes eradication of the virus extremely difficult as it is hard to pinpoint who is carrying an infection. Moreover, asymptomatic carriers have been responsible for pushing through massive epidemics in the past, ultimately leading to a heightened number of deaths.

The most seminal example of asymptomatic carriers is "Typhoid Mary" who was symptom-free but chronically infected by Salmonella. Mary served as a cook which led to the unfortunate spread of Salmonella to her family and the wider public alike. This is a prime example of how asymptomatic carriers can so easily spread illness, causing a large public health issue. Due to Mary's negligence and lack of symptoms, the total number of infections and deaths due to her actions was much higher than it would have been if a more typical asymptomatic carrier had been involved.

It is clear that asymptomatic carriers present a unique yet serious challenge to the effective control of infectious diseases. The example of "Typhoid Mary" is a sobering reminder of the destructive nature of asymptomatic carriers, demonstrating that while they may indeed be "invisible" they can have a more than visible effect on public health.

The deadly SARS-CoV-2 virus is unlike many other infectious diseases, in that it can have far too many asymptomatic carriers who show no signs of being infected. This is a major issue, as it makes it difficult to track and hunt down the virus, since it can essentially be "hiding" inside those people. A vivid example of an asymptomatic carrier is the infamous Typhoid Mary, a cook who was unaware of her chronic Salmonella infection in the early 1900s and who, as a result, spread the disease wherever she went, leading to poisoning and death amongst her unknowing victims. As such, it is critically important to be mindful of asymptomatic carriers when it comes to virus control and containment, in order to prevent such harmful situations from occurring again.

The importance of understanding the reality of SARS-CoV-2 must not be overlooked. A worrying fact is that its spread can be extremely difficult to monitor and contain, due to the potential for a high number of asymptomatic carriers. It is essential to be aware of this and take the necessary steps to prevent the further spread of the virus.

This is demonstrated in the case of South Korea, for example, where an array of track-and-trace methods inclusive of cell phone tracking and quarantine surveillance from hospital visits were implemented in order to lower the impact of the virus. It is these proactive steps that enable countries to even attempt to control the progress of the virus through the population.

In stark contrast to South Korea, the United Kingdom has confirmed over 36,300 deaths as of Oct 21, 2020, with an average of 60.1 deaths per day, as stated by the National Health Service. This serves as a timely reminder of how the lack of preventive measures can have a catastrophic impact on lives and the economy.

It is thus crucial to remember that SARS-CoV-2 is a virus that is difficult to track due to the prevalence of asymptomatic carriers - such carriers can unknowingly transmit the virus to other people. This is why it is essential to implement preventive measures and take every possible precaution to protect ourselves and others.

The results of the ultimately-unsuccessful efforts to develop a vaccine for SARS offer insight that could have prevented the current scenario of vaccinated individuals becoming infected. While the timeline pressures of Operation Light Speed may have instigated the disregard of the report's key recommendation to not rush the process, now it is apparent that a crucial design flaw exists in the current vaccine. Had the report's erudite counsel been heeded, this unfavourable result likely would not be occurring.

To explain further, the SARS vaccine report to the then-sitting government put forward an imperative warning not to speed the development of a viable vaccine, as this might lead to an inadequate end product that had unforeseen deficiencies. However, in the pursuit of eradicating the virus, the implications of this warning seemed inconsequential. Consequently, the recommendation was neglected, and the vaccine was hastily designed with fewer rounds of development and testing than are usually required for such a groundbreaking product.

The lack of time invested in the process has come with a heavy price. Numerous people whom have been vaccinated against the virus have still become infected. This illustrates the necessity for thoroughness when creating any medical product, as the consequences for cutting corners can have lasting effects. Furthermore, this point serves to remind us of the value of considering advice from experts, especially when it involves something as expertise-dependent as creating medical products. Had the SARS vaccine report initially been taken more seriously, it is likely that the consequences of this design flaw would have been avoided.

When the novel coronavirus known as SARS-CoV-2 invaded our planet, researchers and medical professionals scrambled to develop an effective mRNA vaccine that could help protect us from infection. Unfortunately, the process of creating such a vaccine took nearly a decade, and though they invested significant amounts of time and resources into the project, they eventually failed.

The results of the study were concerning, as one of the recommendations offered was the need to increase the breadth of the vaccine to be more effective. Specifically, it said that "The breadth of a next-generation SARS-CoV-2 vaccine that incorporates other viral proteins must be increased to have an effective vaccine that would not require multiple boosters..."

Unfortunately, this admonition may have gone unheeded, because we now find ourselves embroiled in a battle against the virus' continuously shifting variants that outpace our ability to create a viable booster. For example, in the United Kingdom alone, three distinct variants of the coronavirus have been identified - each of which threatens to become more lethal and prevalent than the last.

In order to stay ahead of these new variants and effectively immunize our populations against them, it seems clear that we will need to start from the ground up - reworking our existing vaccine protocols to create a single immunization that is broader in scope and requires no booster shots. Doing so would require a collaborative effort between all of the major players involved in the pandemic, from governments to healthcare institutions, so that we can effectively battle the virus and its variants.

The current situation with regards to the development of a viable, effective and safe vaccine is concerningly bleak. It appears that, even with all of the resources and efforts that have been dedicated to this endeavour, we will still be struggling to stay ahead of virus variants with immunization's that offer extremely limited effectiveness. It is disconcerting that the government has not become wise to this fact, and instead continues to persevere in the erroneous belief that simply listening to selected experts is enough to keep us safe. The truth, however, is that science is much more than just a series of individual opinions from distinguished professionals, it is a comprehensive, analytical process in which conclusions are arrived at via a combination of research and peer review. It appears that two distinct camps are forming, which is of huge concern; there are those who are still adhering to the scientific method, and those who are putting their faith in 'science' that supports the current governmental agenda. If we are to be successful in our quest to develop a vaccination that will protect us from this pandemic, then we have no choice but to keep listening to the collective scientific community and utilizing their insights in our endeavours. If we fail to do so, then the outlook is not hopeful, and we may be condemned to living in a world of overlapping variants and underpowered immunizations, indefinitely.

We Are Doing a Horrible Thing

It is extraordinary how long we have been clinging to a disastrous plan, in spite of widespread condemnation from experts and the public alike. As the truth of the numbers emerges, it has become absolutely clear that far from the dangerous pandemic we were originally led to believe, this is a situation that most nations are managing through existing public service infrastructure. Despite mounting evidence to the contrary, many news outlets continue to propagate erroneous information, unwillingly or otherwise.

We must start to pay attention to what our common sense, reason and logic tell us, and begin to set about reforming our approach to this situation. Clearly, the original plan of 'lockdown' was unsuccessful and we must now strive to create alternative solutions. From contact tracing to targeted testing, these more considered approaches have proven to be far more effective in curbing the spread of this disease and are more cohesive with our goals of public health and safety. The United Kingdom, for example, implemented an effective contact tracing system which has made a dramatic difference in their response to the virus. Similarly, other nations such as South Korea and Japan have effectively used targeted testing to track outbreaks, in conjunction with proper social distancing and personal protection measures for their people. Compared to the disastrous results of blanket lockdowns and draconian restrictions, these strategies have proven themselves to be worthwhile and merit further examination.

It is time we cease clinging to the failed methods of the past and start embracing more reasonable and informed solutions. Such approaches have been shown to be successful and could form the basis of a much more constructive worldwide response. We must acknowledge the fact that our original plan was misconceived, and make the decision to implement a different path if we are to progress and defend ourselves against this virus.

The current governmental and media directive to vaccinate children against various diseases and illnesses, though well-intentioned, is posing an unexpected danger and risk to the wellbeing of our most precious and vulnerable population - our children. Vaccination, as a medical health measure, is only recommended and employed in situations where the risk of illness is deemed to be high, yet, inexplicably, the powers that be are attempting to apply this approach to young children. This runs contrary to a rational scientific basis and there is absolutely no evidence to suggest that this behaviour is beneficial or necessary. Rather, it would be far more advisable to target the vaccines, masks, and other protective measures to those who are most at risk: the elderly, chronically ill, or those living in close proximity to others, such as medical or care-giving staff in a clinical or residence setting. These individuals stand to benefit significantly from these interventions and thus, it would be wise to prioritize and focus these essential resources in a strategic and meaningful manner.

It is true: the chances of a young person in Ontario contracting and dying from COVID-19 is an almost non-existent 0.000014 percent, yet we have kids wearing masks in schools and the government is pushing an unapproved treatment on minors. What is going on? The answer lies in the ongoing problem of governments interjecting politics into healthcare.

Instead of looking to real-world evidence as a basis for decision-making, politics has taken over. Doctors and healthcare professionals are being given explicit instructions on what treatments they can, and cannot offer; Ivermectin and Monoclonal Antibodies, for example, are proven and effective treatments for COVID-19 which are not being recommended or available to the general public. With the latest scientific evidence not being taken into account, and instead political bias dominating the conversation, millions of individuals are being put at risk.

It is essential that we take the politics out of healthcare and allow healthcare providers to provide the public with treatments that have been proven effective, instead of policy and rhetoric. Moreover, besides education and raising awareness within the public domain, governments should be engaging in greater scientific collaboration and mutual respect between healthcare experts and politicians. We must move beyond the current state of affairs and start taking science, more than politics, into account, if we aim to make any real progress in our battle with the virus.

Where are the deaths?

As I was browsing the web today, I decided to take a peek at the Statistics Canada website to see what new information has come to light over the last little while. Curiously, it seemed that Stats Canada had gone to extra lengths to ensure the legitimacy of the data they were presenting; I was pleasantly surprised by what I saw! Statistics Canada has delivered a stark snapshot of the immense toll of the pandemic in its first year. From an estimated 1.36 million deaths worldwide, 5,405 were right here in Canada. That staggering figure represents a heartbreaking loss of life, not to mention untold other losses – job security and health complications, to name but a few – caused by this pervasive virus. In addition to this harrowing statistic, Stats Canada reported a total of 413,527 confirmed cases of covid-19 and heart rendingly, 3,726 deaths attributed wholly to the virus. Unfortunately, it looks as though the true toll of the pandemic stretches much further than this, as many have had to say goodbye to a friend or loved one without the luxury of a positive diagnosis.

In its recently published report, Statistics Canada noted the substantial increase in the number of deaths that happened between the beginning of the pandemic in March 2020 and the start of March 2021. This figure, 18,511, was found to be an excess of 6.7%, conveying an excess mortality rate nationwide. It is important to take into account that the excess deaths have occurred as a result of far more than just COVID-19; other contributing factors such as aging and the indirect consequences associated with the pandemic, such as postponed medical procedures, increased substance abuse and a decrease in deaths attributed to different causes, such as influenza, make up a portion of the excess deaths as well. Statistics Canada stated that, "Indeed, beyond deaths attributed to the disease itself, the pandemic could also have indirect consequences that increase or decrease the number of deaths as a result of various factors, including delayed medical procedures, increased substance use, or declines in deaths attributable to other causes, such as influenza." Excess mortality due to COVID-19 and its indirect consequences create a dangerous strain on the health and well-being of many people globally. Unfortunately, in Canada this strain is being experienced on a broad scale, with the nation's total excess death rate steadily climbing due to multiple factors.

It seems clear to me that the process of data qualification and evaluation has been conducted in the same manner as many of us. This involves taking into account comorbidities, as well as determining the number of deaths attributable to COVID-19 as opposed to deaths that occurred in spite of the virus. I anticipate these calculations might put the figure at around a conservative 9,000 deaths or higher. There are obvious discrepancies between the early estimates reported and the data already available, so any effort to try and spot potential inaccuracies is almost too easy for words. I can't help but wonder whether I, together with the numerous other statistical researchers who have taken an interest in Statistics Canada, have been left with seemingly no other option but to report accurate figures. I had already obtained the majority of their data before it was expunged from their website, so I was in a position to cross-check the figures provided in order to verify whether or not there had been any restatements made to the numbers in 2019 and early 2020 as a means of displaying a higher spike in mortality.

In recent years, it has been revealed that the province of Ontario holds highly meticulous records from 1949 up to present, documented with physical evidence such as death certificates. Although not every death in Ontario is reported by official papers like a death certificate, it's been noticed that the paid subscription based data service that has been used to analyze the Ontario data, shows a correlation between those who have official death certificates and those who don't. As depicted in the Figure, the numbers illustrating the deaths have shown a noticeable decrease, as people have mostly stayed home and halted major activities. What's remarkable is that despite the downward trend present in the numbers, the death rate remains consistent contrasting the vast numbers of predicted deaths. Thus, confirming that death certificates are a dependable judging factor when assessing the overall death rate in Ontario.

Given the immense challenges posed by the novel coronavirus pandemic, the idea of recorded zero deaths due to the flu being factually impossible is bewildering. After all, according to figures gathered from more than two decades of data, the prospects for a terrible flu season in 2020 were strong. Now, one must wonder - are flu deaths being incorrectly reported as COVID-related fatalities in order to manipulate the numbers? And if so, then where are the fatalities that we were told would be expected?

These are difficult questions to answer, and make it all the more baffling that such deaths are not reported with the standard protocol of a formal death certificate. This makes it hard to avoid the assumption that the data itself may have been removed. If this is true, then it could have grave implications, not only for our understanding of the numbers, but also for holding those accountable to the valid statistics they provide. It also necessitates that steps are taken to ensure accurate death reporting, allowing us to make decisions based on the most reliable information available.

So could math be prevailing, and fully qualified numbers are being reported instead of inflated speculative numbers?

WILL IT EVER END

As we stand on the edge of an upcoming 'next wave' of this never-ending COVID-19 nightmare, everyone is left wondering what will happen and if there is a light at the end of this tunnel. But this situation is unlike that of other illnesses that have happened in the past because the vaccines found in medical facilities have yet to provide us with the one thing we desire in this time of need: immunity. Though some of the existing vaccines do resemble previous ones, there is still a fear that COVID-19 will not go away. This virus was not created by our Mother Nature, as it was artificially manufactured in a lab in China with the intention to act as a bioweapon. This notion has been justified by its rapid mutation rate and its 'gain of the function', which is considered to be the worsening of an already existing virus that occurs in animals to target humans. Unfortunately, these modifications were done in secrecy, yet somehow there was a leak, leaving us to ponder over what this virus could continue to do in the future.

The grim reality is that COVID-19 is not leaving us anytime soon; the World Health Organization has reported that up to 150 distinct variants of the virus are dynamically circulating, representing a dangerous threat to human health. Each variant has a range of severity, with some, such as the African variant, being far more aggressive and deadly than any other strain seen to date. In this sense, the virus is acting precisely as it is made for, spreading its infectious 'payload' to as many individuals as possible, rapidly adapting and mutating in order to remain resistant to potential treatments or cures. Fortunately, however, the 'package' that SARS-CoV-2 is delivering is not anything as devastating as the Ebola virus, otherwise entire populations would have been wiped out by now. Unfortunately, the highly controversial practice of 'gain-of-function' experimentation, which entailed genetically modifying infectious diseases for medical research, was not globally halted until it was too late, largely due to the incompetence of Dr. Fauci and his colleagues. If it had been stopped sooner, then many more lives could have been saved. This situation has put a spotlight on the need for greater and more responsible oversight of such research activities, as any oversight gaps and unintentional oversight can lead to disastrous consequences.

It seems to me that, unless we make a radical shift in our approach, including enlisting the support of our governments for treatments like Ivermectin, Z Pak, and monoclonal antibodies, it's going to be a long and difficult journey. If I had a magic wand and could make any wish come true, I would implore everyone involved to leave the politics at the door, take greed and global concerns completely out of the equation, and make decisions based on the best available science instead of relying on the output of a few handpicked individuals. We need to be aggressive seekers of truth and action if we are to see real and lasting results.

The fact of the matter is, if we are to see real success in fighting this virus, we must be willing to alter our existing paradigms and embrace different solutions that have been shown to work. For example, recent studies have shown that Ivermectin had a positive effect on the duration and severity of symptoms in COVID-19 patients when administered within 5 days of symptom onset. This is an example of the kind of success that we can achieve if we look beyond conventional treatments and open up to unconventional ideas. The same can be said of other therapies such as Z Pak and monoclonal antibody treatments. We just need the right combination of courage, determination, and vision to make it happen.

Data drives decisions and it is clear that during the pandemic, it is of utmost importance to follow the available data in order to keep our citizens safe. That means protecting those who are most vulnerable while simultaneously ensuring the economy continues to run in a safe and secure fashion. It is also critical to hold the media accountable to responsible standards that uphold the best interests of society. Today's world is full of various expert opinions related to the virus, which is why it is essential to always verify the advice we receive before implementing any sweeping changes. As a society, we must take the urgent steps of removing the burden that the COVID-19 pandemic is placing on our children and education system. We must also hold large pharmaceutical companies such as Pfizer and Moderna accountable in their vaccine contracts, and work together to renegotiate these contracts for the benefit of our citizens. It is also important that big tech companies play an impartial role in our society, without favoring left-leaning socialist agendas or censoring those who oppose them. Ultimately, this may mean that companies such as Facebook, Google, and Twitter become further dismantled and that their separate assets, such as Instagram and WhatsApp, will become their own independent businesses. These changes will ensure that everyone is held to a higher standard and will create an even playing field of competition.

The COVID-19 pandemic has become far more than just a search for a cure. It has become entwined with billions of dollars of economic losses, the wielding of political power and a sense of the need for a 'reset'. The world has become increasingly divided, segregated, and the plague of hate that is quickly taking root has become far worse than the actual disease. When we look beyond the headlines, what we are seeing is a deep disruption of our humanity and civility, and if we do not take a step back, this will all be reduced to a crater of despair where devestation has taken its toll.

It is essential that we remember this when the time comes to cast our voting ballots. We must consider which leaders spoke up, stood up for what is right and who fell into line, blindly accepting whatever came their way without questioning or validating any of it. We must not allow ourselves to be classified as 'conspiracists' or 'anti-vaxxers', we are simply paying attention and the rest of the world needs to wake up soon, or all may be lost.

Did We Ruin the Vaccines? The Cold Chain

Recently, I've made an interesting observation which could be of great importance. A few days ago, I penned a blog discussing the relevance of temperature when it comes to administering the Pfizer COVID-19 vaccine. As I conducted research I found news items from years ago surrounding the Ebola outbreak and how the mRNA vaccine being developed needed to be refrigerated in specific freezers to maintain its 'active' state, meaning it had the capability to immunize properly.

Now in more recent times, the Pfizer break-through is arguably the biggest news in the world right now. A vital issue when it comes to receiving and delivering the vaccines efficiently is to keep them at low temperatures, between -94 to -76 degrees Fahrenheit, otherwise the vaccine is deemed expired and useless. This is why, for instance, medical facilities have had to invest in extra equipment such as ultra-cold freezers to increase the safety and efficiency of their operations.

In the past, a vaccine would be created and delivered using traditional cold storage methods as that was the reliable way of keeping it preserved at the required temperature, however, with the advances made thanks to mRNA technology, new storage solutions have come to light, ones which we may have been unaware of without doing further research.

So, to conclude, it's essential that temperature is taken into account when dealing with different vaccines, especially those that incorporate mRNA technology, to ensure their effectiveness and longevity. Finding out more about previous outbreaks, such as Ebola, helps us to put the current state of affairs into perspective.

A new groundbreaking study released by the University of California, Davis Genome Center, in collaboration with the Chan Zuckerberg Biohub and University of California, San Francisco has recently highlighted an important new finding: there is no major difference in virus load amongst those who have been vaccinated against SARS-CoV-2, and those who had not, but who tested positive for the delta variant. Moreover, the study demonstrated that there is also no considerable variance between infected individuals with and without symptoms. This finding has already generated a considerable amount of media attention, and yet its implications remain largely undiscussed.

At its core, this study underscores the importance of taking preventive action in order to protect oneself and others from catching the virus. Although much of the research around SARS-CoV-2 thus far has mainly focused on the virus' capacity for transmitting itself and causing severe illness, the recent findings from the University of California, Davis Genome Center suggest that an individual's vaccination status does not make much of a difference when it comes to the virus' potential for widespread infectivity. As a result, the need for preventive measures such as social distancing, wearing face-masks, and practicing good hygiene remains as important as ever, regardless of whether an individual has been vaccinated or not.

Moreover, this study's findings could also have far-reaching consequences for future public health policies. By understanding that individuals with and without immunization status are both equally likely to develop the virus and to potentially spread it to others, governments and corporations can take a more comprehensive approach in developing protective measures. This could include establishing stricter control measures for certain populations, or providing more resources to certain communities in order to ensure that everyone has access to vaccinations.

In summation, this recent study from the University of California, Davis, Genome Center, UC San Francisco and the Chan Zuckerberg Biohub is incredibly significant, aiding in our understanding of the ways in which SARS-CoV-2 spreads and the implications of preventive measures. It is important to remember that even those who have been vaccinated are still at risk of contracting the virus, and thus it is essential that our public health policies take into account all possible scenarios in order to better protect our population.

When December of 2020 rolled around, talks of the much-anticipated vaccine became the primary focus, but soon enough people began raising concerns about transport and cold storage. There were fears that without the right provisions, the global rollout of the vaccine could be severely compromised and the costs of specialised freezers and other such facilities to properly store the Pfizer vaccine ran into hundreds of millions of dollars. This posed a huge logistical nightmare and compounded the difficulties associated with this global effort.

Thankfully, when Moderna advanced their vaccine trial results, which did not require the same level of cold storage as Pfizer did, this bitter pill became much easier to swallow. This meant those costly temperature controls were no longer an issue, removing them from the equation and making it that much easier to push forward in ensuring the safe distribution of the vaccine across the globe.

Since posting about my concerns a few days ago, I have come to realize how widespread my worries about the Pfizer vaccine really are. It seems that millions of individuals who have received the vaccine may have been unknowingly administered an inactive or spoiled dosage, and it could potentially explain why there have been so many reported cases of people who need to be placed on a ventilator due to their inability to breathe on their own.

I have read numerous reports detailing instances where this has been the result after receiving the vaccine. One particularly concerning case, for example, is that of a man, in his 40s, who a few days after being administered the Pfizer vaccine, was placed on a ventilator due to the fact that he could no longer breathe on his own. These are the kind of issues that are making many people question the safety of the vaccine and whether it is a viable option, leading many to seek alternative solutions to protect their health and well-being.

Canadians across the country, in both densely populated suburban towns and remote municipalities, have eagerly accepted the millions of doses of mRNA vaccines available in recent months. Unfortunately, not all of the doses have been handled properly due to varying levels of technical competence and varying levels of conscientiousness, leading to the potential risk of spoiled or otherwise inactive doses. This raises a question: what happens if an individual is injected with what is essentially a "dead" mRNA vaccine?

From the research that has been conducted, it appears that the injection of a so-called "dead" mRNA vaccine does not have any adverse consequences. To illustrate, a recent animal study conducted by researchers at the University of Toronto found that the injection of a "dead" mRNA vaccine led to minimal inflammatory response and no discernible long-term health effects. Anecdotal evidence from human trials for other mRNA vaccines have indicated similar findings.

Given this evidence, it is safe to assume that the injection of a "dead" mRNA vaccine into a human will not have any substantial consequences. As such, it is important to remain vigilant and suspicious of the handling of these vaccines, as the issue of spoiled doses is quite concerning. It is also highly recommended to only accept doses of the vaccine from certified medical professionals, to avoid the any potential risk of improperly stored or deactivated vaccines entering the arms of people.

The introduction of the Pfizer vaccine has been a momentous step forward in the fight against SARS-SoV-2. But what if these vaccines are not kept and stored within the temperature ranges necessary for their potency? If this is the case, the mRNA will become compromised, or 'spoiled', and it will no longer be able to effectively stimulate the immunity necessary for immunity against the virus' spike protein. This would be akin to receiving a placebo shot; there would be no vaccine protection. It is feasible, even likely, that there are many people in the world who have been 'fully vaccinated' with these spoilt doses and who, ultimately, are not fully immunized. In fact, with the sheer numbers of people that are receiving the vaccine, it is virtually certain that there will be some people who will receive two spoiled doses, leaving them completely unimmunized despite having been 'fully vaccinated'. This is a deeply concerning revelation, and one that requires significant effort and attention to prevent.

While it is difficult to predict the exact spoilage rate for the mRNA vaccine created to fight the Covid-19 virus, it is fair to assume that anywhere between 1-10% of doses are either partially or fully spoiled. That said, with millions of Americans having received the vaccine so far, this means the potential number of people thinking they are protected when they are, in fact, not, is staggering.

It is possible, even likely, that this percentage of spoiled doses has already been reached. Take, for example, Utah's Tooele county, where when 500 vaccine doses were sent home with seniors, over 1000 were administered due to expired product. There were instances such as this all over the United States, with many individuals believing that their shots were viable and effective, when in fact their doses had either partially or fully spoiled and were no longer viable. This can be a major cause of concern for individuals who believe they are immune from the virus when in actuality, they are not. To ensure the vaccine is administered appropriately, individuals should always verify the vial's expiration date and read the accompanying information to check if the vaccine is still effective.

When it comes to any product, the consumer is faced with an important decision: do they trust that the product has provided them with the desired result? In the case of a vaccine, the consumer must be confident that the vaccine has effectively worked for them to induce a good immune response. In effect, their body must be strong enough to protect itself from illness, particularly those caused by bacteria or viruses. To make this determination, many individuals will call upon the advice of healthcare professionals, such as doctors and pharmacists, who are familiar with the product and can provide informed guidance. Additionally, consumers also need to review the literature associated with the product and make an informed decision regarding whether or not it will provide positive results for them. Ultimately, the consumer is best equipped to make an objective assessment about the efficacy of the vaccine for their own body.

In recent years, the possibility of checking antibody levels after receiving a vaccination has become a topic of much debate. But is it plausible that such a routine practice could save hundreds of thousands or even millions of lives? As much as we'd all like to believe that the current recommendations from the Food and Drug Administration (FDA) are in our best interests, it appears to be part of an overall strategy to reduce what many are referring to as "vaccine hesitancy". FDA experts tell us that if people were to discover that vaccines are incapable of giving them complete immunity, it could risk further diminishing their trust in the efficacy of the vaccine and create even more vaccine reluctance. In November of 2020, IBM posed this query in a thought-provoking article.

In short, it appears that the FDA is attempting to protect public confidence in vaccinations but, ironically, the agency's needs could inadvertently be leading to more vaccine hesitancy and heightening the likelihood of severe disease outbreaks. In order for us to take a proactive approach against potential infections, it's essential that we look at ways to make antibody testing more widely available. This could provide us with invaluable insight into the effectiveness of our preventive efforts, ultimately resulting in a healthier world for us all.

The complexity of the supply chain for vaccines and prescriptions has become all the more apparent as we grapple with the Coronavirus pandemic. Vaccines, which start their journey at production plants, must then be rapidly, consistently and safely transported to the point of sale - a potentially difficult task given the need for temperature control and a reliable shipping process. Each step of the way, from production to sale, middlemen often form the backbone of the supply chain - providing order fulfilment, product tracking and inventory monitoring - but these intermediaries are vulnerable to system failure and can be overwhelmed by unprecedented and unpredictable demand spikes. In times like these, it is essential for all the stakeholders involved in the supply chain, from suppliers to sellers, to invest in an "intelligent" supply chain system that can offer near-real-time transparency over product movements and inventory positions. With large-scale deployment of "smart" technologies, like Artificial Intelligence (AI) and IBM Sterling Supply Chain Suite, companies can be better equipped to anticipate potential disruptions and deploy resources faster and more effectively in the event of a crisis. From proactive stock management, to rapid re-allocations of supplies when needed, an AI-enabled supply chain has the potential to simplify an otherwise complex and demanding process.

And yet, despite the many success stories, it's easy to become jaded and concerned when facing down a pandemic. After all, the Cold Chain had been a matter of contention for some time prior to the onset of COVID-19, and it makes one wonder if the pandemic has merely exposed fears that have gone unaddressed otherwise. It's therefore of the utmost importance to ensure transparency and trust through the entire vaccine supply chain, and to invest in technology that can empower greater control during times of crisis.

The world has seen an increased number of confirmed Covid-19 infections as more nations reach herd immunity through vaccination. A recurring question has been whether these outbreaks are a result of so-called 'vaccine-hesitant' members of the public recklessly disregarding social distancing and mask-wearing protocols, believing themselves to be impervious because they've been partially immunized. This unfortunate phenomenon has been dubbed 'vaccine-hesitancy false-security syndrome' and it appears to be contributing significantly to the present wave of infections in some countries.

As if the current situation wasn't dire enough, reports of Covid-19 vaccine malpractice - unwitting or intentional - are becoming more and more widespread, giving greater cause for concern. This latest slew of vaccine-related mismanagement and disinformation only serves to deepen my conviction that these issues require urgent attention and systemic reform.

Moderna – Flush with FU Money

Back in 2019, Moderna's future was uncertain. Investors had deserted the sinking ship, and the worrying signs of collapse seemed to be all around. Internally, the company was in a state of panic as employees initiated conversations with regards to the need for drastic layoffs and substantial cost-cutting measures. The hand-wringing and tension was thick in the air - all seemed lost and completely hopeless.

Then, like a beacon of light in dark times, came the news of a savior that the ex-employees dubbed a 'Hail Mary.' This brave decision to take a risk and go in a different direction resulted in the revitalization of Moderna, providing a lifeline that afforded the company the opportunity to outlast the difficult period and ultimately emerge, in 2020, as a major player on the world stage.

This turn of events serves as an example to other businesses of the importance of strategic risk-taking and the potential that a brush with disaster can have for shaping the fortunes of an organization. To borrow from the ex-employees of Moderna, sometimes, a Hail Mary is necessary to give yourself a fighting chance.

At the end of 2019, Moderna was seeing considerable growth and gaining momentum. The head of the National Institutes of Health (NIH) had given the firm a subtle nod, indicating that 2020 would be an important year. Very few in the know were privy to the hush-hush news coming out of Wuhan, China, a mysterious yet highly significant event that was unfolding before their eyes. Little did they know that this incident would have far-reaching global implications in the years to come.

Over the coming weeks and months, Moderna was urged to progress with their mRNA vaccine trials, even though the rest of the world had yet to gain insight into the events to soon follow. Moderna was the first to have knowledge of the chaos that the emergence of an infectious disease would bring, with the time of its emergence and even the type of virus being known to them. Their daring decisions to deliver their pioneering research and develop the vaccine would be valorized, as the world soon came to realize the life-saving potential of Moderna's mRNA vaccine. Taking a comprehensive stance on safety and efficacy, not just rushing to get their vaccine on the market, saw Moderna rise to the challenge that the world was facing and ultimately bring an end to the infectious disease. Through their clear insight, strong science, and timely execution of the vaccine trials, Moderna made a difference in the lives of many, leading to a positive health outcome and a sense of joy around the world.

The news of Moderna's early vaccine trials looked to be a promising event in the fight against the COVID-19 pandemic, however, things took a turn for the worse when it was revealed that the mRNA concoction had significant side-effects. In response to their findings, Moderna made the decision to reformulate the vaccine candidate and pause trials on it, forcing delays to the potential availability of a vaccine. While this was certainly a setback for Moderna, the company no doubt understands the importance of safety when it comes to testing prospective vaccines. This is why the pharma giant has made it clear that they are not willing to settle for any formulation that is not proven to be safe and effective in humans. In fact, it was only when Moderna began to conduct small human studies that the side-effects were revealed, demonstrating the importance of these thorough trials. In any case, this latest development is a reminder for us to proceed with caution and understand that, when it comes to researching vaccines and other treatments, we must take our time to ensure that products reaching the market are safe and effective.

The preliminary research and development of the mRNA based COVID-19 vaccine began as a mere showcase project. On the surface, it appeared everyone's primary purpose was to prove the vaccine's feasibility, no matter the cost or consequences. However, below this facade, it was clear the nation was under tremendous pressure to come up with a solution for the pandemic that was sweeping the entire world and of course, there was the promising profit potential of creating an effective vaccine that was tempting investors to pour in resources. The promise of a financial gain from the vaccine's success was enough to fuel the idea that this endeavor was worth the risk, no matter what reservations people had about its potential outcome.

Despite financial difficulties, Moderna had the inside scoop and knowledge to pour all resources into creating a vaccine for the Wuhan coronavirus. Despite the challenges of their precarious financial state, the biotech firm persisted in their innovation, as if their coffers were full of money. Even when faced with the rigorous and lengthy trial and approval process from governmental authorities, with their funding in question, Moderna kept pressing forward with developing the potentially life-saving vaccine, in the hopes of bringing restoration of normal life back to the world. Their tenacity, in this instance, was laudable, and although there were many who discounted the work the firm was doing, the availability of the COVID-19 vaccine today stands as a testament to their dedication and commitment. Even after the company had already invested a great deal of time, effort and resources, they had the courage and determination to continue, despite the numerous obstacles they faced. The willpower demonstrated by Moderna played a major role in the ultimate success of the vaccine and in saving countless lives.

Moderna has been on a remarkable journey in recent times, with the biotech company made a truly inspiring recovery since the pandemic first struck. It has been a triumphant and resounding success reminiscent of classic Hollywood fairytales; it's almost like the plot of the classic Rocky Balboa movie when he goes up against the Soviet Union's best fighter Ivan Drago in a do-or-die showdown. It was inspiring to witness the American icon prevail against such adversity and it appears that Moderna has gone through a similar story of resurgence. The pharmaceutical giant has shifted its focus to Covid-19 vaccine development and is now one of the industry leaders in the fight against the virus. Through the revolutionary mRNA technology and their ingenious vaccine, millions of lives have been saved and their success has inspired countless others in their ongoing battles with the pandemic. It has been a remarkable turn of events and a truly inspiring comeback story.

In late 2019, as the race for an effective vaccine against the novel coronavirus intensified and politics took center stage, Moderna was bestowed with special permission to bypass the traditional processes of research and development as well as quality control, in a bid to fast track the development of the vaccine. This meant that, despite the usual stringent processes, this vaccine was on the fast track to completion and was effectively being used as a political weapon in the fight against the pandemic.

While this seemed like the only viable solution to quickly get a safe and reliable vaccine at the time, it was met with a fair degree of skepticism. After all, sidestepping safety protocols that had been developed over years of research and practice is never a decision easily taken. However, it soon emerged that the vaccine developed under these less-than-ideal circumstances was indeed effective, and the calls of anxiety, hesitation, and criticism were ultimately silenced.

In any case, Moderna's incredible achievement in developing a viable vaccine despite their short-cuts highlighted how much potential had been unlocked when the right incentives were in place. Thanks to this decision, which was at first deemed to be too risky and unorthodox, the world was able to find a viable solution to the pandemic and ensure the safety of millions of lives.

Following a thorough sterility test and numerous other obligatory trials, Moderna Inc. was at long last able to ship out their inaugural batch of the much-anticipated vaccine. Unsurprisingly, their stock value soared in the moments that followed, providing much-needed relief to investors and market analysts who had endured a particularly long drought of optimism. The relief was palpable, and yet they knew that the long road to recovery was only just beginning. An incalculable amount of resources, man-hours, and research had gone into the making of this vaccine, and the outpouring of joy after such a monumental success was understandable. Although this was just the first step to fully releasing the vaccine, it provided a beacon of hope for a more optimistic and healthy future.

During the pandemic, regulatory agencies granted Moderna a drastic exemption from the traditional animal trials which medical experts found highly atypical. This waiver was justified by the urgency of the pandemic, and the immense political pressures from the U.S.

government's initiative, "Operation Warp Speed". Whilst this decision was potentially justified in the immediate concern for public health, it runs rife with ethical dilemmas, raising controversial questions about the safety of the vaccine and potential long-term public health outcomes.

The need for rapid development has been further complicated by questions pertaining to appropriate balance of scientific rigor, cost effectiveness, and even considerations of legal and moral liability, warping the revered ethical traditions which had previously been the basis behind medical research. Consequently, any decision made is bound to evoke contentious debate and public scrutiny. At the same time, exploring new technologies and taking risks may also be critical to advancing public health initiatives. Considering such complicated variables, it is no surprise that finding the right answers is both difficult and delicate.

Rather than relying on the traditional approach of completing each step of the vaccine development process sequentially, the Moderna vaccine was approved to conduct each testing process simultaneously. This move saved time, but it did not allow for the needed data to be collected from animal trials before the process moved to human trials. This decision has proven to be shocking, as it has already presented serious concerns. An array of issues have already been observed, such as reduced fertility in rats that received the vaccine, and a greater proportion of aberrant bone development in the fetuses. In addition, the vaccine has been found to accumulate in the placenta.

Now, we are beginning to see the long-term effects of such a decision - especially concerning the effects of vaccine doses administered during key embryonic development milestones. Unfortunately, this has been linked to the increasing number of stillbirths in local hospitals. Going forward, it is important to prioritize conducting the necessary animal trials and analyzing the data before proceeding with human trials. Such a decision can ensure that these issues don't arise in the future, and that the vaccine produced is truly safe for human use.

It is a shocking reality that the safety testing of the mRNA in pregnant women only started as recent as mid-July 2021, and is projected to not even conclude until early 2024. This timeline is incredible and begs the question, if something wrong was to be found in the tests, is it possible to travel back in time to fix it? Of course, time travel is a futuristic concept that is housed in the realms of science fiction, however, if the magnitude of the consequence was of a massive enough scale then the idea of travelling back in time to 'fix' a problem may become more of a reality. For example, if a pandemic were to occur due to an error as a result of the testing, then a team of scientists may be assembled with the task of travelling back in time to undo the decision that lead to the pandemic. Of course, this is a hypothetical example and is most likely an impossible feat, but the idea that such an extreme course of action is a necessary solution is an unnerving prospect. Still, caution is being taken in the testing of mRNS in pregnant women to ensure its safety and potential harm is prevented. There is no doubt that due to the enormity of the task at hand, the testing process will not be completed until early 2024, however, the measures that are being taken in the testing should be commended for their necessity in ensuring the safety of pregnant women and their babies.

The Centers for Disease Control and Prevention (CDC) is taking an incredible risk in relying on passive reporting to determine the safety and efficacy of the vaccine for pregnant people, breastfeeding mothers, those trying to conceive and those who may become pregnant in the future. This passive reporting approach lacks the rigour and scientific clarity of comprehensive trials. It is an extremely concerning that the CDC would take such an approach to approving and administering a vaccine which could have long-lasting implications for those who take it and their unborn or nursing children.

This lack of thoroughness when it comes to their approach towards the safety of the vaccine particularly raises alarms when considering the large number of individuals who this decision will affect. In the United States alone, in 2019, 3.83 million babies were born, more than half of whom were born to mothers aged 30 and under. This figure proves the enormity of the potential consequences and raises concerns over the lack of scientific trials in this area.

Furthermore, this alarming decision also has huge implications for the international community. The number of individuals who may be affected by this decision is hugely concerning, with an estimated 43.22 million babies due to be born worldwide in 2020. This shocking figure further highlights the risk being taken by the CDC and reinforces the need for scientifically rigorous trials to be conducted before taking any decisions regarding the vaccine's use.

It is clear that the CDC has a responsibility to the public to conduct thorough research and clinical trials to verify the safety and efficacy of the vaccine for pregnant people, breastfeeding mothers, those attempting to conceive, and those who may become pregnant in the future. Without this rigorous scientific research, the CDC is effectively putting millions of people at risk and leaving them subject to the unknown implications of a vaccine that has not been properly studied or tested for safety for these groups of people.

The timing of the announcements of the Moderna deals were especially suspicious, as though they were planned so that the elite investors and senior management had the chance to maximize their potential to gain monetary gain. This is particularly indicative of the circumstances Moderna were in prior to the COVID-19 pandemic, where they were on the brink of financial ruin.

The US government took extreme measures to help by undertaking 'Operation Warp Speed' and providing huge financial aid in the sum of 483-million-dollars, hoping to expedite the production of the vaccine. Within one year, the investment from the US government rose to roughly six billion-dollars, which nearly matched the company's value at their pre-COVID-19 IPO.

This violent shift in the economy created the perfect opportunity for many to become millions and billions in the matter of a few days, while the rest of the world was still eagerly awaiting the completion and distribution of the vaccine. This stands as a testament to the great amount of unevenness in wealth across the world, when the upper echelons are able to take such advantage of a situation that would supposedly benefit all.

The Moderna vaccine has sparked a great deal of discussion and debate; a fundamental issue being the delivery system used to distribute the vaccine. Unfortunately, this raised questions when it was discovered that Moderna's vaccine delivery design was patented by another firm. Subsequently, there have been numerous legal disagreements surrounding this matter, as the government was pressured to simply "trust" Moderna's initiative despite not fully knowing what the vaccine consists of. This raises the controversial question of whether the risk of proceeding with the distribution of this vaccine is worth the potential benefits; and with new advancements in artificial intelligence, it has become easier to evaluate situations such as this. Artificial intelligence systems can detect potentially harmful or dangerous materials in vaccines and provide accurate information on the exact composition of the medical product. Therefore, AI technologies can be incredibly useful when navigating complicated medical matters and legal issues. Still, a great many questions remain unanswered surrounding the Moderna vaccine and its delivery design, leaving a significant amount of doubt and hesitation around its use.

In recent months, the effects of both the Moderna and Pfizer vaccines have been observed around the globe with reports of injuries suffered as a direct result being made. Such cases have cast further suspicion on the efficacy of the vaccines, with experts increasingly acknowledging that they may have been a contributing factor to the emergence of certain variants of Covid-19. This newfound knowledge serves as a compelling reminder of why testing processes and research initiatives concerning immunisation cannot ever be made a political move, with the potential implications of such actions being disastrous.

In the United Kingdom, for example, the overall efficacy of the Pfizer vaccine has been predicted to reduce by as much as 99.68%. This is thought to be largely a result of a new Covid-19 variant first discovered in South Africa, which has proven to be increasingly resistant against many modern vaccine types. Worryingly, this variant is estimated to been the most prominent variant around the world, significantly reducing the efficacy of immunization efforts worldwide.

The experience of the Covid-19 pandemic thus far is a significant indictment of why the development and distribution of vaccines cannot be politicalised. By not taking sufficient precautions and conducting the most comprehensive of trials and testing procedures, the disastrous consequences of investing in an ineffective vaccine have become reality. This sees nations around the world having to face the music with regards to the premature rollout of vaccines, with numerous cases of injury now being reported to healthcare professionals. This is why it is of paramount importance to get such matters absolutely right, without any element of political motivation or malpractice, as the ramifications of failure could be cataclysmic.

The horror that we find ourselves in today is the consequence of an intricate mix of greed, politics and incompetence, which lies at the heart of what can only be described as a complete and utter clusterfuck. We are in a situation where we are left with no other option than to remain tied to the big, fat and deeply unfavourable Pharma. This could be seen as us being 'pot-committed' and 'married' to this distasteful situation for years and years to come.

The problems of greed and incompetence are becoming increasingly well-known in modern society due to the glaring inequality in the distribution of wealth and resources. Take, for example, the disproportionate taxation which applies to the highest earners, allowing them to escape paying their fair share of taxes. Additionally, when it comes to incompetence, one only needs to consider the litany of major catastrophes which have been caused by ill-thought out policies, negligence and a sheer lack of foresight.

It is time for us to call for a radical and constructive re-evaluation of what a successful society should look like in order to move away from the nightmare of dependency on Big Pharma. Only by doing so can we take the necessary steps to tackle the issues of inequality and incompetence, root out the greed and politics that lay at the heart of this clusterfuck, and develop a more equitable and harmonious world.

Our governments are locked into regrettable contracts that have been shrouded in secrecy and purposefully lack transparent oversight. Many of these contracts are so one-sided that the public never had an opportunity to scrutinize them in a normal and open process. To make matters worse, these contracts are typically accompanied by iron-clad nondisclosure agreements, implying that the contents of the contracts will not be divulged in public. The documents that have been made available for viewing are often heavily redacted, supercharging the suspicious that their contents may be of a dubious nature.

It is this lack of transparency that led to the outcry against the controversial Canada-China, Foreign Investment Protection and Promotion Agreement (FIPA) of 2014. This agreement was negotiated and signed in secret, and it was only when details of the contract were made available through Freedom of Information requests that the public became aware of its one-sided nature and its implications for Canadian sovereignty. As a result, it has become a case study for just how far governments can go in concealing information from their constituents, and for the lack of accountability and oversight that can occur when these opaque contracts are put in place.

As the extent of the coronavirus pandemic continues to remain a top news story, we have witnessed a dramatic surge in reports related to potential scandals involving mRNA and Covid-19. A simple search engine query for the terms 'mRNA' and 'scandal' generates a sizable list of hits on the surface web, and while some results may be questionable, there are also those from established and reliable sources. This data points to a potentially larger problem that could emerge in the coming years and require legal action.

What's become clear is that the body of existing research and investigatory evidence has yet to be fully untangled. As CNN recently reported, allegations of racism, sexism, and ageism exist in the scientific community related to the development of the Covid-19 vaccines. Part of the challenge is that a lot of the critical material is buried in complex, undisclosed proprietary networks and it may take further scrutiny to uncover any potential wrong-doing.

Suffice to say, there is much still to be examined in this case. While it may feel like there is no further news to report on the topic, we believe that, in the future, we may just be scratching the surface on this matter in terms of uncovering potential things that have been overlooked and of possible legal significance.

I just for this messaged to me, check out their stock prices.

Increases in COVID-19 are unrelated to levels of vaccination across 68 countries and 2947 counties in the United States

For months now, we've been voicing our concerns about the concerning rise in the number of COVID-19 cases, both in breakout locations and throughout the rest of the world. We've all seen the tweets, the posts, and even heard the whispers surrounding the unvaccinated and their supposed role in all of this, but in truth, the actual numbers of cases tell a very different story. After meticulously analyzing no less than millions of rows of extensive case data, I began to recognize that the amount of unreached people as well as the odds of catching COVID-19 made it seemingly impossible for the unvaccinated to be responsible for all of this. And now, it appears that the world of academia has done their research and are coming back with the same findings, confirming what I discovered back in August. It is only now becoming public knowledge that the unvaccinated simply can't be the sole source of blame here.

The findings of the Harvard report on the efficacy of vaccines in combating COVID-19 have been unequivocal and the data presented in the report irrefutable. The report observed a strong correlation between the number of fully vaccinated individuals in a given population and the observed level of new COVID cases. Areas with low rates of vaccination experienced a disproportionate surge of infection, despite the significant progress that had been made in the rollout of the vaccine in those regions.

Recent weeks have seen a significant resurgence of the virus in countries such as Germany, the United Kingdom, and Israel,which have seen extensive and successful coronavirus vaccination campaigns. Although the reasons for the continued spread of the virus in these countries are varied, one possible cause of the uptick in cases lies in disparities in vaccine access and effectiveness. For example, while the United Kingdom was reportedly vaccinating more than 600,000 individuals per day, vaccine uptake in developing countries was significantly lower, leaving them more vulnerable to the virus.

Improving vaccine access and efficacy has thus become a top priority for governments around the world in order to effectively combat the spread of the virus. Nations must continue to invest in researching, developing, and deploying effective vaccines to those populations that need them most in order to ensure a more equitable and accelerated fight against the pandemic.

As the United States continues to grapple with the ongoing impact of the COVID-19 pandemic, new research has explored the connection between county-level vaccination rates and the prevalence of COVID cases. Published in the journal 'Nature,' the study assessed the correlation between the number and percentage of counties that experienced an uptick in COVID-19 cases, and the rate of people in those counties who were fully vaccinated.

The research team calculated the increase in COVID cases by differentiating the total cases from the last seven days and the seven days prior. After analyzing this data, it became evident that counties with higher rates of full-vaccination tended to see lower numbers of COVID-19 cases. Notably, the study found that counties where 60-65% of individuals were fully vaccinated had, on average, an 11% decrease in COVID cases. Meanwhile, counties with 75% or more of their population vaccinated saw on average a 35% decrease in COVID cases.

While this research provides further evidence for the effectiveness of the vaccine, it also speaks to the importance of community-wide vaccination efforts. To achieve the most effective protection from COVID-19, it is imperative that individuals within a given community are vaccinated to the highest degree possible. This type of research serves to reinforce the imperative of widespread vaccination and demonstrate how, collectively, this public health measure can contribute towards curbing the spread of COVID-19.

The findings were as follows:

"At the country-level, there recent research conducted indicates that there is no obvious relationship between the percentage of the population which is fully vaccinated and the number of reported new COVID-19 cases from the last seven days. Interestingly, the trend line implies a slight positive connection, hinting that countries which have a larger percentage of their population fully immunized actually have more COVID-19 cases per 1 million people. This finding is demonstrated through the example of Israel, where 60% of the nation has been inoculated and yet still experienced the highest number of

COVID-19 cases per 1 million people over the last seven days. This paradoxical outcome proves that there isn't a meaningful correlation between full vaccinations and the severity of the virus.

Given how long it typically takes for individuals to gain full immunity after completion of their two-dose vaccine regimen, experts conducted a series of in-depth, sensitivity analyses that examined the impact of a one-month lag on the percentage of the fully vaccinated population in different countries. To elaborate, they took the percentage of the population that had received their completed vaccine series a full month ago, and compared it with the total percentage of the population that had been fully vaccinated prior to that period. This kind of analysis can be extremely useful in predicting the efficacy of inoculation programs—especially those that are still ongoing—depending on the overall rate of vaccine uptake throughout a given community. For instance, if the percentage of the population that is fully vaccinated with a given vaccine drops significantly in one month, it could indicate a trend of waning enthusiasm for inoculation, potentially necessitating greater public education efforts regarding the importance of vaccination.

The researchers of this study argue that the sole reliance on vaccination as a primary means of managing the spread of COVID-19 and its associated ramifications should be re-evaluated in light of the prevalence of the Delta (B.1.617.2) variant and the potential of further variants coming to fruition. A more comprehensive strategy that combines both pharmacological and non-pharmacological approaches alongside upping the vaccination rate should be taken into account. Given the observed findings from this study, to simply advocate for a rise in current vaccine distribution may be an inadequate response. Instead, a focus should be put on developing more preventative strategies and potentially, newer vaccines that are better able to keep the increasingly dangerous strains of coronavirus in check.

This will no doubt come as a sobering conclusion for politicians and other decision makers, and illustrates the continued complexity of tackling the health crisis posed by COVID-19. Allowing for such a multifaceted approach and actively seeking out innovative interventions may well be key in helping to stem the rising tide of this ongoing pandemic.

Are We simply tracking the realities of Life Expectancy?

The headlines boast an alarming number of COVID-19 deaths in the United States, but upon closer inspection, what I have seen in the statistics raises some questions. Something I remembered hearing months ago was that the calculated chance of dying from the virus was actually lower than life expectancy. So, I decided to do some research and look for data to download and analyze. What I found was quite eye-opening. The overwhelming majority of those categorized as 'COVID deaths' were people over 65 years of age. This is a critical detail to bear in mind if we are going to make sense of what is happening, especially since the media reports these numbers as pure 'COVID deaths' - but in actual fact, the majority of these deaths were associated with the virus, not necessarily caused by it. This means that there are a range of other factors at play here that could have also contributed to the cause of mortality, making it all the more important for us to take a step back, look at the numbers and make sure that we are getting the whole picture.

The documented impact of coronavirus-related deaths in the United States and Canada suggests that those individuals who had three or more comorbidities were more likely to be classified as succumbing to the virus. For example, an individual presenting to the hospital with heart issues can be given a COVID panel to determine the underlying cause of their condition. However, due to the PCR cycle threshold (CT) used for PCR tests currently being set higher than recommended, the accuracy rate of the test has been reported to be just 65%. This has been found to lead to an abundance of false positives, with many asymptomatic individuals testing positive, even if it is only dead coronavirus particulates that are detected.

This problem is very concerning, and leads to preventable misdiagnosis, resulting in undue stress and incorrect treatment plans. For example, the individual described in the first paragraph who is coming to the hospital with heart issues may find themselves added to the official death toll and treated as if they were COVID positive, when in fact their true condition is something unrelated and less serious. There is a real danger that the misdiagnosis of positive cases based on this inaccurate testing can be contributing to the high mortality rate we are currently seeing across the US and Canada. It is clear that further steps should be taken to accurately identify and treat those with the virus, and to ensure that incorrect readings are prevented from tainting the data being used to track the spread of the virus.

As COVID-19 has left an indelible mark on the world, there have been numerous reports of 'COVID Deaths' with a focus on the elderly. This has led to an interesting question: Are we merely documenting expected mortality, or are COVID-related deaths truly increasing? From a statistical point of view, the confidence level of the reported data is a little shallow - it is important to evaluate the confidence interval (CI) to get a better understanding of the accuracy of the reported data. Confidence interval is a statistic measure which reflects the reliability of the data and a range of probability that the result could be true. Typically, a confidence level of 95% is quoted in data research reports but anything lower than 80% can be deemed unreliable.

To illustrate, consider a study on the mortality rate of elderly in a particular region. After running the sample set, the data analyst might find that the CI is 80%. While this implies a high degree of certainty, it has to be noted that it is lower than the standard measure of 95%. In such cases, one must consider long-term trends to ensure that the results are reliable and to conclude whether COVID-related deaths have indeed increased. This could be further verified by looking at historical mortality figures of the elderly in the same region and it would be beneficial to take into consideration other factors such as other medical conditions, age and population. Only then can we be sure if the 'COVID Deaths' are actually higher than the mortality rate of old age.

When it comes to the report of COVID deaths among the elderly, it is important to have an accurate count of those deaths. While some have argued that the possibility of death in such elderly populations would have been high even with or without the presence of the virus, it is important not to discount the very impact that COVID is having on the elderly population. For example, in Alberta, 93% of deaths due to COVID were among those over the age of sixty, and of those, 75% had three or more comorbidities. While it is not entirely possible to have an accurate estimation of the exact number of COVID-related deaths, a more conservative estimate may put the total deaths of those sixty and over from COVID at 809, when taking into account life expectancy.

On the other hand, there are some estimates that even this number may be too high, and that the number of COVID-related deaths may be closer to 162. However, there is simply no denying the fact that a large majority of COVID-related deaths are from among those fifty years or older. According to the Centers for Disease Control and Prevention (CDC), approximately 80% of fatalities caused by the virus have been people aged sixty-five or over, a statistic that simply cannot be ignored. It is clear that COVID has had a significant, disproportionate effect on the elderly population, and efforts must be taken to ensure their safety and wellbeing in the future.

The devastating effects of the coronavirus pandemic have been felt across the U.S., with nearly 9 million Americans now infected and over 229,000 deaths reported. However, when accounting for adjustments, such as factoring in the lack of flu cases due to the onset of the pandemic, the number of reported COVID-related fatalities drops significantly to around 13,740. This tally, which can also be seen on various online discussion boards and reports, suggest that the total number of deaths has been slightly overstated.

In the months leading up to the onset of the pandemic, millions of Americans were annually afflicted by the flu, resulting in numerous fatalities each year. While we cannot accurately determine how many of those cases were mistakenly categorized as COVID-related deaths, we can observe the drastic effect the pandemic has had on flu cases. Since the start of the pandemic, there have been very few cases of the flu due to the diligent efforts of health care workers and everyday citizens to take precautions and adhere to social distancing guidelines, aided by the development of vaccines and efficient testing procedures. Subtracting the deaths that could have been attributed to the flu had it not been eradicated due to the pandemic, we can gain a more accurate assessment of the true death toll of COVID.

While it may take years for us to truly understand the numbers related to COVID-19 in Canada, it is clear that things are not what they seem. In Ontario, for example, provincial authorities have taken extreme measures to suppress data from 2020, likely due to the desire to avoid comparisons between COVID-19 and other illnesses. Although the official figure for COVID-related deaths in the province is around 25,000, I believe that this number is grossly overestimated and the true toll of the virus is much lower, likely somewhere between 2600 - 4100 people. This is more in line with a bad flu season, and provides an indication of the misreported nature of the pandemic. It might take some time for an accurate representation of the losses sustained to be determined, however I feel confident that the figures I presented in this statement will be the closest to the truth. Learning the true numbers of this virus will be a major milestone in eventually accepting and moving on from this unfortunate event.

As we remember the days when the world was not so concerned with the threat of COVID-19, we can only imagine that those feelings of security and comfort will eventually come back. Unfortunately, the events of the past year have only shed more light on coercion and payoffs, displaying just how vulnerable many governmental processes are to corruption. Though corruption is always an issue with public procurement, the healthcare sector's focus on pharmaceuticals and medical devices makes its potential even more troubling.

Various attempts to mitigate the risk of corruption throughout the healthcare sector, particularly in the procurement of medical goods, have been made throughout the years. Unfortunately, big pharmaceutical companies have proven to be adept at devising lengthy legal processes to delay settlement proceedings- meaning that even the most diligent efforts can be undermined by those wielding power and capital. It will not be until the judicial inquiries come to a closure that we can finally learn the full extent of the corruption throughout this sector and in government processes in general. Hopefully, with the right policy changes in place, we can create an setting where these answers will come to us not as a result of such a dismal debacle, but through careful discussion and ethical action.

COVID Vaccine – The Deal with The Devil

The Pfizer Corporation, headquartered in Manhattan, New York, has tightly guarded the details of their negotiations with governments around the world. These contracts can have a life or death impact on populations, yet the details of the agreements placed Pfizer's corporate interests ahead of public health necessities. An investigation by the Associated Press in 2009 uncovered a situation where the pharmaceutical giant had a consolidated partnership with five nations in Asia and Europe as part of their "patent sharing deals." By doing so, they blocked generic competitors which in turn increased prices of their drugs, preventing lifesaving treatments from reaching smaller, more impoverished populations. Other examples of Pfizer's prioritization of profits over public health have been evidenced through the years, proving that it often takes a major scandal to make them waver from their drive for financial gain.

When discussing common themes, one that can often be overlooked is the fact that many disputes are eventually settled in secret arbitration courts. This is an intriguing arrangement that can be seen in business operations across a wide range of industries. For example, pharmaceutical giant Pfizer has been known to utilize this type of arrangement, where they are able to change the terms of important decisions, such as delivery dates, and may even demand public assets from their partners as collateral in the event of potential non-performance.

Naturally, these types of proceedings can be very beneficial to publicly traded companies like Pfizer, as they are often able to enter into a resolution without creating public knowledge or scrutiny of the case. While these secretive courts can provide an invaluable service, it is essential that standards remain in place and monitored to ensure fairness and reasonable and equitable dealings.

Pfizer and Moderna, two major American companies who have both harnessed revolutionary mRNA technology to develop life-saving vaccines, have come under criticism for potentially creating a "duopoly," a situation where two major market players dominate both competition and pricing within a sector. Many view this as unfair, as it limits the ability of other companies to compete successfully and thus potentially restricts consumer choice or drives up prices. For example, if two large mobile network companies dominated a local market, they could independently agree to price services far above what competing providers could offer. Such a situation would leave the small players unable to keep up, and the market becomes an oligopoly where limited products are available at inflated prices. By achieving a duopoly within the vaccine sector, Pfizer and Moderna may be able to introduce similar pricing structures and dominate the market for their products entirely.

Despite not tapping into the government's Operation Warp Speed funding, Pfizer has found success by selling a substantial number of orders to the United States government. Historically, companies have found an economic advantage to getting early support from the government during product development. Janssen Pharmaceuticals, a Johnson & Johnson subsidiary, accepted Operation Warp Speed funding and in return got an agreement to supply the U.S. with 100 million doses of their Covid-19 vaccine. Similarly, Merck & Co. received funding from the government and had purchased orders they expected to supply by the end of 2021.

Pfizer however chose to go a different route and not accept the government funding but instead focus on pre-selling the vaccine. These vaccines first became available for sale after being reviewed and approved by the Food and Drug Administration. Through an accelerated approval process and over 200,000 studies, the FDA has stated that these vaccines are safe and highly effective. By pre-selling these vaccines ahead of time, not only did Pfizer get a significant amount of capital to continue their development but they were also able to secure their place in the vaccine market ahead of their competitors. Ultimately their decision has resulted in the company providing more than 300 million doses of vaccine for the U.S. government by the end of the year.

Contracts between governments, especially with foreign nations, can often involve extreme provisions or demands. Countries around the world, such as Brazil, Chile, Colombia and the Dominican Republic, are discovering increasingly stringent clauses when they sign on the dotted line. A particularly notable example is the immunity from "precautionary seizure of assets". This means that even in regions where legal and economic stability is not a given, a state is allowed to protect and preserve its assets in spite of the situation.

This kind of contract provision isn't unheard of in state-to-state relations. In cases such as trade agreements, one government often seeks to ensure it has financial security and that its assets are not harmed by the volatility of its partner in the agreement.

Contracts of this kind require excellent legal counsel, as well as effective and open negotiations between two or more nations. This is because the terms of such agreements may be vast and the consequences of not adhering to them, substantial. Consequently, it is critical that governments not sufficiently familiar with the rules of engagement that govern their negotiations are supported by legal professionals well versed in international law to ensure they are not taken advantage of.

Ultimately, contracts between nations tend to be complex and often involve unusual clauses. As a result, they must always be treated with the utmost attention and foresight, as the implications of not adhering to their terms can have far-reaching consequences.

Many financial agreements involve the concept of collateral, where a party agrees to provide a certain asset as a guarantee against the payment of a debt or other obligation. For example, a company might pledge a specific real property as collateral for a loan from the United States. After all, the Grand Canyon could certainly be seen as an asset of considerable value and the company could use it to secure the loan, though it would be a highly unusual situation. Furthermore, the pledged collateral can be used by the creditor to economize their risk if the debtor fails to fulfill the obligations of the contract. In the case of the aforementioned hypothetical agreement, any failure by the company to make debt payments would allow the US to seize control of Grand Canyon and use it to clear the debt. As such, collateral is a fundamental part of loan agreements and other contracts and can vary dramatically in terms of the type of asset used.

Since the start of the COVID-19 pandemic, Pfizer has finalized 73 agreements to distribute its highly anticipated coronavirus vaccine. Of these 73 deals, only five have been publicly published by the respective governments.

This lack of transparency within some of these negotiations has caused serious concerns among the public, especially given the multi-million dollar contracts involved. It has also raised questions among the Transparency International, a leading global nonprofit organization which works to fight against corruption. Based in London, this advocacy group called these contracts out publicly, noting that they had been subjected to severe redactions.

The issue of access to the Pfizer vaccine has been the subject of numerous debates across the world. As governments have been hesitant to provide extensive information about these deals, many have argued that decisions have been guided only by political considerations, rather than the public health needs. Furthermore, there have been various reports of wealthier countries hoarding stockpiles of the vaccine while poorer nations are left with insufficient resources.

It is imperative that governments provide greater disclosure with regards to their agreements with pharmaceutical companies, particularly in times like these. In order to create a more fair, equitable and open system of access to lifesaving treatments, governments must be willing to share pertinent details to ensure public trust and confidence.

The utter lack of transparency when it comes to the procurement of vital vaccines means that everyday citizens and members of the public are kept in a state of uncertainty and confusion. The shroud of secrecy surrounding contracts with drug companies and the resulting documents which are often filled with redacted text is a major cause for concern. We are being kept in the dark over such vital issues as when the vaccines will arrive, what will happen if things go wrong, and what level of financial risk the buyer is taking on; this information is solely held by the government and the drug companies, greatly hindering any sense of transparency from the government.

Such a lack of openness leaves us wondering what kind of agenda is being hidden from us. It becomes even more concerning when we consider the fact that these vaccines are being procured to fend off a health pandemic that is adversely impacting us all. In a situation such as this, transparency and openness should be paramount but instead we are left in a state of confusion, exacerbated by the lack of forthrightness from the government. A situation like this makes one truly ponder what is being kept under lock and key and why.

The Swiss-based pharmaceutical conglomerate, Pfizer, has recently come to an agreement with the Brazilian government, in which both parties have agreed that the Brazil will not issue any official public disclosure or commentary about the Agreement and its associated subject matter or stipulations. This is a mutual agreement for both entities, as it ensures Pfizer's rights as an individual business entity. Not only does this agreement protect the privacy of their chosen agreement and its terms, but it also allows Pfizer to uphold their social and ethical standards when associated with Brazil's government activities.

The contract between Brazil and Pfizer is a very important document, as it ensures that neither party takes advantage of the other. Because Pfizer is a large global business, they have to be careful when dealing with governments to ensure that their practice are ethical and meet their internal standards. By having written documentation of their agreement with Brazil, both parties can rest assured that no matter what is published in the media, their agreement will remain an exclusive matter between them. It also gives Pfizer the right to review any public information that is released by Brazil to ensure that it is in accordance to the Agreement they have in place.

In an increasingly globalized world, it is essential that businesses are aware of their potential international business partners, and the mutual responsibility each has to the other. When signing contracts like this, companies like Pfizer must take into account both their own interests, as well as those of the other party. By negotiating a legally binding contract with Brazil, it allows Pfizer to protect their own interests and also to ensure that government activities adhere to their own standards and criteria.

The signing of contracts with multiple Latin American countries, including Brazil, Chile, Colombia, the Dominican Republic and Peru, poses a legal responsibility that goes beyond a typical agreement. Each of these governments must sign a document in which they expressly and irrevocably waive any right of immunity they normally have. Moreover, they must acknowledge in the contract that either they or their assets may be acquired down the road. Four of the five countries have also granted permission for the precautionary seizure of their assets, if and when deemed necessary.

This kind of signing of waivers is somewhat unique, as it often goes beyond the usual scope of contractual requirements. Furthermore, it is an example of governments cooperating and agreeing in the interest of financial and legal issues, though these kinds of agreements can be controversial. Due to their permanent nature and irrevocable agreement, these contracts are not something to take lightly. Nonetheless, they are necessary for maintaining fair and clear legal structure between countries when trading and acquiring assets.

The signing government is obligated to follow the terms of the contract, which include guarantees of indemnification, defense, and holding harmless of Pfizer from any lawsuits, claims, actions, or damages resulting from vaccine intellectual property rights. In other words, the signing government should be responsible for any legal costs, fees, and expenses that are incurred as a result of an intellectual property dispute regarding the vaccine. Examples of such disputes may include, but are not limited to, infringement of copyrights, trademarks, patents, or trade secrets. It is essential that the signing government bear the monetary costs and damages to ensure Pfizer's legal rights are properly maintained and protected should any legal issue arise.

Whilst Moderna has faced its fair share of public scrutiny in recent months, with accusations of price-gouging and deliberate delays to vaccine delivery, Pfizer has seemingly been able to avoid the same level of attention. Nevertheless, the success of Pfizer's trials and distribution of their vaccine means that the pharmaceutical company is predicted to generate a staggering $54.5 billion worth of sales from their COVID-19 related products next year, almost double the value of Moderna's predicted sales.

This enormous surge of revenue, once unfathomable prior to the pandemic, comes as a result of the fact that Pfizer have been able to vaccinate the largest amount of people during the early stages of the pandemic, thanks to a well-developed plan to distribute the doses quickly, alongside their collaborative work with BioNTech. Furthermore, the Pfizer vaccine has become the most popular vaccine which has become available to the public.

Despite the controversy of cost, the success and impact of the Pfizer vaccine in terms of stemming the tide of the pandemic has been undeniable. It is estimated that in 2021 alone, Pfizer's coronavirus related vaccine sales are expected to exceed all other industries profits for the year, demonstrating just how impressive their accomplishments truly are. It could, quite fittingly, be seen as the symbol of a hope of a brighter future, and a reminder that through dedication and commitment, even the most difficult journeys can be achieved.

Nestled among the Middle Eastern countries, Israel is home to 9.3 million inhabitants and one of the first to sign a deal with Pfizer to administer their COVID-19 vaccine. As of June 2021, a staggering 55% of Israel's population has been vaccinated, with over 11 million doses distributed thus far.

However, on the 30th June, a concerning announcement was made: the number of fully-vaccinated individuals being admitted to hospital with COVID-19 in Israel had doubled in just over a month. This escalated worryingly fast, mirroring similar spikes in the UK, Sweden, Italy and other countries around the world. The exponential growth of this statistic poses a troubling reminder of how much further there still is to go in order to protect our citizens from the virus.

In recent months, an intrepid coalition of investigative journalists in South America has uncovered a troubling situation involving the pharmaceutical giant, Pfizer. Through thorough research and interviews with key figures, they were able to reveal that Pfizer had been in negotiations with countries such as Brazil and Argentina, among others, and that during the course of the negotiations, the company had requested that in exchange for the medical treatments they were providing, the countries should give up some of their state assets such as diplomatic buildings, military bases, and other key structures as a form of security against potential future legal costs.

This news, upon being made public, raised serious questions about the ethical behavior of transnational corporations, as not only did it put the countries at a financial disadvantage, but it was seen to be an egregious power grab. After all, these state assets often provided revenue for the countries in question, revenue that Pfizer sought to usurp for its own benefit. Not only this, but the act of pledging such assets as a guarantee indicated that the company was expecting something of a dispute in the future, which is worrisome given the nature of their business.

The voices of the intrepid investigative journalists set off a wave of public outrage, and the implications of this scenario have seemingly been heard loud and clear by other corporations, who have since displayed more caution in similar activities. Should the powers that be fail to prevent the abuse of corporate power in such glaring instances, then the violations are only liable to continue unabated. One thing is clear, however: the power of investigative journalism is alive and well and continues to serve as a vital force for good in the realm of public knowledge.

Pfizer, the multinational pharmaceutical company, has engaged in aggressive and controversial negotiations with more than a hundred nations around the world. Consequently, the company has struck accords with a dozen Latin American countries that have ultimately been compelled to meet the untenable demands made by the pharmaceutical giant. For instance, it was reported that Colombia had to agree to a pricing scheme proposed by Pfizer that greatly increased the cost of a variety of essential medicines. According to some public health officials, the arrangement placed an undue burden on low- and middle-income countries who rely on these medicines to combat a host of illnesses. Additionally, this has also prompted some Latin American countries to pursue legal action against Pfizer due to the company's alleged excessive pricing, questionable business dealings and activities often deemed to be anti-competitive. Despite the criticism, the company has managed to extract a series of advantageous agreements from Latin American countries that have significant implications for their respective health care systems.

The egregious avarice and maltreatment of the vaccine industry is plain for all to see. Pharmaceutical giant Pfizer, as well as many other vaccine formulators have not only accepted, but have profited extensively from hefty grants ranging from hundreds of millions of dollars furnished liberally by a number of governments. Germany, for one, dolled out an impressive $445 million in funding to vaccine firm BioNTech, in tandem with Pfizer. This unconscionable largesse of taxpayers' money characterizes the egregious wastefulness and blatant profiteering of the vaccine industry, sickening to behold. Many governments throughout the world have indulged vaccine firms with a generosity that could only be deemed improper and unjustifiable. If left unchecked, such practices of misuse and exploitation of public funds could well ruin the global healthcare system.

Sweden recently acquiesced to the EU's latest agreement for increased vaccine prices with Pfizer/BioNTech and Moderna. In what many are considering an unfortunate inevitability, there has been a rise of SEK 160 to SEK 200 per dose for the Pfizer/BioNTech vaccine, and a further increase from SEK 190 to SEK 220 for Moderna, according to the Financial Times.

This is a financially challenging issue, as the Swedish state has been expected to already secure and fund large quantities of the vaccines, due to their critical importance during this global pandemic. As the prices per dose rise, the overall financial burden is likely to have a significant impact on the Swedish economy, while the demand and need for the most effective vaccines is still paramount.

While professionals in the field have estimated the costs of administering the vaccines, the reality of the situation is that the investment Sweden is willing to commit to combating the coronavirus pandemic and purchasing large quantities of the various vaccines is much higher than expected. In some cases, this can even mean sacrificing other necessary investments, such as health care infrastructure, to afford the quantity and quality of vaccines that the government has committed to acquiring.

In conclusion, as Sweden continues to comply with the terms of the EU's agreement with Pfizer/BioNTech and Moderna, it is likely to result in increased financial uncertainty and likely detrimental effects on the Swedish state's budget. The importance and necessity of the successful administration of the vaccines, however, still remain.

When it comes to purchasing a vaccine, there is no "out clause" permitting the buyer to walk away if it turns out to be harmful. If significant death or serious side effects occur in the population, the buyer must still complete the Agreement and pay the amount in full, as a way to mitigate the buyer's risk of liability.

For instance, imagine a company purchases a vaccine from a manufacturer and discovers after the rollout that the vaccine is ineffective and potentially dangerous. If this was to happen and there were consequences for the people affected, the buyer would still be responsible for paying the full amount of the Agreement even if the vaccine was found to be ineffective or dangerous. This type of clause may appear unfair, but it serves an important purpose- to protect the buyer from being sued if the vaccine is linked to any resulting harm incurred.

The takeaway is that buyers must carefully consider the risks associated with any agreement they are signing, and if they have any doubts they should always err on the side of caution. That way, they can avoid entering into a contract that could leave them exposed to liabilities if something goes wrong.

SOME SAMPLES FROM THE CONFIDENTIAL AGREEMENTS

The buyer of the vaccine is fully cognizant of the fact that the long-term efficacy and consequences of the vaccine are unforeseeable and that possible adverse effects that are presently unknown may emerge. For instance, there may be an associated rise in certain allergies or, in a more severe case, unforeseen anaphylactic reactions. Furthermore, it is also possible that the vaccine could produce adverse impacts in the months and years following the injection, such as an alteration in fertility or the emergence of new conditions. These types of outcomes are impossible to predict and, in view of this, the buyer is aware of the precariousness of utilizing any vaccine.

It is essential for the buyer to fulfill their financial obligation to Pfizer for the doses ordered, regardless of situation. This holds true even if Pfizer has not yet had the preparation approved by the appropriate governing body. This pre-dates the FDA's emergency approval of the treatments, dubbed medications by some, most commonly known as vaccines. Even at this time, when proverbial "unknowns" linger, it is the responsibility of the buyer to still make payment for the ordered doses. Various scenarios could play out, yet the buyer's obligation remains intact and unmoving. It is a commitment no matter the amount of doses actually used or the approval status. This is an example of the immutable nature of contracts and unquestionable intent to go through with the deal made.

The buyer hereby agrees to bear the responsibility of indemnifying, protecting, and shielding Pfizer/BioNTech and their subsidiaries from any and all potential claims, documentation, legal claims, losses, damages, settlements, penalties, fines, costs and expenses incurred by them. Additionally, the buyer will protect and defend any lawsuits or accusations brought against Pfizer/BioNTech and their subsidiaries for any of the aforementioned and bear the overall financial burden of resolving any such disputes. As an example, if a customer sues Pfizer/BioNTech for a defective product, the buyer will be responsible for all costs related to defending that case in court, along with the cost of any settlement or award the customer may receive. Similarly, the buyer would be responsible for covering the costs of any debts, fines, or penalties associated with any incident.

By agreeing to this indemnification clause, the buyer agrees to financially compensate Pfizer/BioNTech and their subsidiaries for any potential costs, expenses, or liabilities that may occur as a result of any claim brought against them, as outlined hereinabove.

The buyer must pay all losses, including and without limitation costs for legal fees and other legal costs.

Buyer must indemnify Pfizer for claims and all losses and must implement this through statutory or regulatory requirements.

Pfizer retains the right to adjust the total amount of contracted doses and the delivery schedule in accordance with their internal principles of conduct. This may require the buyer to accept any changes that Pfizer decide to implement. As an example, should there be an abrupt demand spike or a significant unavailability of raw materials, Pfizer may have to adapt their production plan in order to be able to fulfil the obligations of the buyer. This could include delaying delivery, reducing the number of doses, or even rejecting an order altogether. In such a case, the buyer is expected to accept the proposed changes without hesitation. Similarly, any other unforeseen circumstances, such as an equipment breakdown, may also require the two parties to renegotiate the obligations in order to ensure an agreeable outcome. In all such cases, the buyer is expected to concede to Pfizer's adjustments due to their status as the primary party of the agreement.

The Agreement must be kept secret for ten years.

Despite the fact that the Israeli government is renowned for their transparency with going public with their healthcare agreements and legislation, in the case of the state of Israel's arrangement with the pharmaceutical giant Pfizer, a full thirty year secrecy clause is remarkably in place. This poses the question: why?

While details of the agreement remain surprisingly opaque, recent gossip indicates that the government of Israel wanted to secure a minimum of one million doses as quickly as possible - and did whatever necessary to ensure that these supplies would be available for their citizens with the uttermost haste. Shortened timelines for the acquisition of research related details and production information may have been traded for the guarantee of ample supplies in record time - hence the secrecy agreement.

Whatever the reason, the state of Israel's arrangement with Pfizer is a unique one and has stirred up plenty of debate and speculation, thus making it a particularly notable agreement due to its degree of secrecy. Reports suggest that the agreement was signed and sealed on January 6th. With the adherence to such a lengthy and unknowable timeline for disclosure of information, one can only conjecture as to the contents - and true reasoning - behind the signing of such an agreement.

Despite being underreported, a remarkable global initiative was recently launched that saw many of the world's biggest media organisations coming together to support the rapid and equitable rollout of vaccines around the world, and to combat the spread of false and damaging information and disinformation about vaccines.

The initiative brought leading media companies such as The New York Times, BBC, Japan's Kyodo News, Germany's DW and India's NDTV, together with the WHO, GPEI, the Bill & Melinda Gates Foundation, the UNDP and a range of other global partners.

Each of these media partners has committed to greater collaboration to promote the vaccine rollout, such as providing dedicated reporting on the efficiency and reach of the rollout, and raising awareness of the efficacy and safety of appropriate vaccines.

In addition, they have also agreed to work together to place more emphasis on fighting the spread of misinformation, false narratives, and 'fake news' that can lead to confusion, mistrust, and contribute to the vaccine hesitancy seen in some countries.

This global collaboration is a recognition of the need to promote the positive aspects of the vaccine rollout, while ensuring that the public is provided with truthful and accurate information that allows individuals to make an informed decision about whether or not to receive the vaccine.

Ultimately, this agreement between the world's biggest media outlets will help ensure people around the world have access to the accurate, balanced and credible information they need to make the right decisions about taking up the vaccine, and help to make sure the world is a healthier and safer place.

For years, media consumers have observed the pro-vaccine bias exhibited by the world's most powerful and influential media organisations, who appear to act as megaphones for the interests of the vaccine industry and state pharmaceutical executives. This observation has been substantiated by recent agreements which confirm the media's singular and striking dedication to propounding the benefits of vaccines. Unfortunately, this one-sided description drastically changes the narrative and often neglects to explain to the public the potential risks of vaccination.

Such a focal point on vaccines, unfortunately, takes resources away from other pressing matters that should receive the same level of attention. For example, despite technology companies' ability to swiftly detect and respond to trillions of Facebook posts, tweets, or other communications, they have often failed to detect and act upon cases of child pornography. Considering the number of malicious individuals and organizations worldwide utilizing the internet to facilitate exploitation, this dereliction of duty on behalf of big tech companies is concerning and should be addressed appropriately.

By now, it is clear to the general public that many media leaders and tech giants prioritize advertising for the vaccination industry rather than providing the public with a comprehensive understanding of the nuances associated with vaccinating oneself or their children. It is time that media companies and tech organizations become more selective with their resources and redirect some of the time and money dedicated to promote the vaccine industry and instead focus on protecting vulnerable members of society.

The world of international media is dominated by a powerful and influential selection of global news outlets. Of them, a few of the preeminent organisations include the Associated Press (AP) and Agence France-Presse (AFP), Reuters, the British Broadcasting Corporation (BBC), Canadian Broadcasting Corporation (CBC Canada), European Broadcast Union (EBU), Facebook, Financial Times, First Draft, Google, YouTube, The Hindu, Microsoft, Twitter and the Washington Post.

These globally recognised media houses and news outlets are renowned for offering comprehensive journalism of the highest calibre, with expertly curated content and information from around the world. Individually, each has its own strengths; the AP focuses its coverage on politics, health and business, Reuters is a self-described 'authoritative source' for financial news and events, the BBC is renowned for its public service radio, television and online output, while CBC Canada has a strong record of providing high-quality Canadian coverage.

Meanwhile, the breadth and diversity offered by the rest of the names on the list is hard to ignore. Google and YouTube are leading examples of tech-driven media outlets, and Microsoft and Twitter are unwavering in its commitment to deliver content and news in an innovative and modern way. The Financial Times and Washington Post, meanwhile, are equal match for their long history of delivering in-depth and comprehensive coverage on regional, national, and international news.

Regardless of its form and origin, however, the news and information disseminated by these influential outlets remains a vital part of today's news ecosystem, informing and engaging audiences both domestically and overseas – a responsibility they take seriously.

In recent years, millions of users have felt the brunt of a disturbing tendency among several of the world's most profitable privately owned online platforms - Facebook, YouTube, Twitter, and Google - to collude with the established media in order to restrict freedom of expression. While it is likely that this practice was initially implemented as a means of enhancing profitability, it has led to a variety of negative outcomes for users far and wide.

For example, many users have found that posts of a political nature have often been flagged as containing 'offensive content' and subsequently blocked or deleted without explanation. Meanwhile, the number of searches with politically-charged terms returning results that could be considered to be of an impartial nature can often be frustratingly low. This is particularly alarming when viewed in context of the large volume of evidence suggesting that this practice has been employed as a means to suppress certain points of view.

Ultimately, such behaviour is highly detrimental to the concept of freedom of expression. It is of paramount importance that the aforementioned online platforms and the established media be held accountable for their part in helping to suppress information and opinions, as such practices are in direct conflict with the values of a free and democratic society.

In Sweden, the mainstream media has been seemingly lobotomised and paralysed, operating more like a megaphone for the country's Public Health Agency and vaccine industry, promoting and pushing only one-sided vaccine propaganda. The same faces appear over and over again on television and on various media platforms, without any form of criticism, opposing views or even questions being voiced which would provoke a more healthy public discussion.

A stark example of this was recently seen with many reports vilifying those who had their doubts about the efficacy and safety of the Covid-19 vaccine, when the European Medicines Agency released an adverse reaction report stating that two people had died after receiving the first jab. The media largely failed to report on the report or discuss how grave it was, instead encouraging people to take the vaccine anyway, creating a sense of mistrust with the public.

This type of dismissive attitude from the Swedish media further reinforces the idea of a lobotomised and paralysed press that fails to perform its traditional journalistic quirk of presenting different sides of the story. The press should also be impartial and objective in its reporting, something that's evidently missing in Sweden when discussing the issue of vaccines.

It is particularly alarming, yet not altogether unexpected, that one and the same individual, James Smith, has both a directorship of Reuters news agency and of the pharmaceutical company Pfizer - two entities that represent two very incompatible roles. This is profoundly troubling when you consider the global impact of both institutions, and their influence over the dissemination of accurate and factual news and of drugs.

It sets a troubling precedent and raises questions about the ultimate objectives of the individual in question, particularly when considering the longstanding journalistic integrity that Reuters News Agency has been known for its entire existence, in contrast to the objectives of a corporate pharmaceutical organization whose primary commitment is to maximize profits.

Does James Smith intend to influence the former in the interests of the latter? Or, to use the informally obtained resources of the one to advantage or further the mission of the other? The implications here are far-reaching, as the outcomes of decisions made by those involved in organizations like Reuters, and Pfizer, have a long-term effect on the lives of individuals the world over.

We must confront the existing conflict of interest, and ensure that public safety and trust are not compromised by this situation. This is both a repomial obligation, and an ethical responsibility.

If you've ever considered why Ivermectin's potential use to treat COVID-19 has been so widely suppressed, it's due to the agreements countries have with pharmaceutical giant Pfizer. These agreements are ironclad, dictating that any discoveries made regarding the drug's efficacy in fighting against the virus cannot be included in the contract or used to break it. This constraint has been shared amongst numerous sources, consistently being posted and discussed in detail.

This limitation highlights how corporate contracts can restrict governments from making decisions in favor of public health, even during a global health pandemic. Take for example the arrangement between the United States and Pfizer. This agreement divested any authority from Washington to order huge quantities of the drug for public-sector use and the most vulnerable of society unless it was done on a local level; due to the government having already agreed to this clause, it eliminated their options to manipulate the agreement. A similar circumstances happened in South Africa where the fines for breaking the contract with Pfizer were so drastic, that it was deemed just to pay for the drugs in full as opposed to finding cheaper alternatives.

The constraints resulting from the agreements countries have with Pfizer and other relevant parties around Ivermectin illustrates how corporate contracts can sporadically overpower and impede governments from making decisions in favor of public health and welfare, even in the midst of a worldwide pandemic.

As of July 15 in the United Kingdom, a staggering 87.5% of adults have already received at least one dose of the COVID-19 vaccine, with 67.1% already having had the two doses necessary to be fully vaccinated. Despite these overwhelming statistics, new infections of the virus are on the increase. The daily average of 15,537 new symptomatic cases being detected is a worrying 40% rise in comparison to the numbers recorded during the previous week.

The implications of these findings are far-reaching and of great concern to government and medical professionals. Despite a high portion of the population being immunised, the virus is continuing to spread rapidly, suggesting that partial vaccinations may not be enough to adequately protect against the virus.

Experts advise that the vaccine is still the best method of protection from becoming infected and from contagion to others. While evaluating the effectiveness of the vaccine, it is necessary to take into account the virus' ability to mutate, and the fact that several new variants, such as the

'Delta strain' or B.1.617.2, are not as easily protected against by current vaccines.

However, alongside the continued importance of the vaccine, it is also important to consider the impact of abiding to social distancing and contact tracing regulations. The importance of reducing the number of occasions during which individuals may come into contact with COVID-19 is of the utmost importance and will help to lessen the spread of infection, thus reducing the effectiveness of the virus' mutation and spread.

Ultimately, it is essential that individuals continue to abide by the restrictions and improve their personal level of protection through the regular reception of vaccinations while remaining mindful of the risks of contagion. This will serve to reduce the number of new cases being detected, mitigate the development of new variants, and ultimately protect the population against this global pandemic.

A new report from the Centers for Disease Control and Prevention (CDC) on July 19th revealed an alarming finding: the number of reported deaths following a COVID-19 vaccination had more than doubled since the prior week. The Vaccine Adverse Event Reporting System (VAERS) now shows 12,313 deaths associated with the vaccine, up from the 6,079 previously reported.

The VAERS is a tracking system maintained by the CDC and the U.S. Food and Drug Administration (FDA) that monitors the safety of COVID-19 vaccines and other immunizations. Any reports of adverse events following vaccination — including serious events, such as death — are recorded in the system to alert the public and healthcare providers to the dangers posed by the vaccine.

VAERS data can help public health officials and scientists better understand the safety of the vaccines, making sure that people receiving the vaccine are being given the right information and that any potential serious reactions are identified quickly and appropriate action is taken. The recent doubling of reported deaths in the VAERS data underscores the importance of continuing to monitor the vaccine's safety and ensuring that individuals are given accurate information before receiving a COVID-19 vaccine.

Shortly following the release of the report, health officials were swift to revert the number of reported vaccine-related fatalities back to 6,079, the same tally as the week prior, without offering any explanation as to why the figure had been altered in the first place.34 This shift, by default, supposes that no deaths were attributed to the negative effects of the vaccine during that specific week, prompting concerns over the transparency of the process and the safety of the vaccine.

These worries have been to the fore in recent months following numerous reports of troubling side-effects, from nausea and headaches to severe medical conditions for some, in line with what experts suggest may be the result of an allergic reaction. This has led to a sense of skepticism and distrust among medical professionals and the general public, who now believe that health authorities, and the pharmaceutical companies at the helm of administering these vaccines, are not always forthright with all the facts.

In a further attempt at mitigating any further public unrest, it is essential that all parties involved act with integrity and transparency when entering into, or discussing, matters related to the safety of the vaccine and the effects of administering it. Disclosing the relevant information and communicating the potential opportunities and risks associated with this new treatment could give citizens not just a greater sense of control but also the freedom to make their own informed decisions.

Throughout the world, the rollout of the SARS-CoV-2 vaccine has caused many people to raise their eyebrows in concern due to the emergence of numerous adverse events. These reactions range from risks associated with the biologically active spike protein used in the vaccine, to worrying long-term effects such as blood clots, reproductive toxicity and inflammation of the heart. Unofficially, the issue is causing alarm amongst many members of the scientific and medical communities, particularly in light of the contents of the confidential indemnification agreements which are making their way to the public domain. As these documents show, in the event of the vaccine proving to be a dismal failure and a risk to short and long-term health, countries have no legal way to take any form of recourse, with individuals receiving the injection facing the same dilemma.

The matter has become even more worrying as more and more premature deaths have been linked to the vaccine, with India and Bangladesh being among the worst-hit countries in recent weeks. The Sunday Times reported on the 9th May, 2021, that a 19-year-old in Bangladesh died two days after receiving his second Covid-19 shot, becoming the nation's youngest Covid-19 death, and raising serious questions about the safety of the vaccine.

The lack of legal measures to protect against the aforementioned risks illustrate the dangerous and irresponsible nature of introducing such experimental bioinoculants without adequate clinical trials. With cases of vaccination-produced adverse event continuing to increase, it may be yet to be seen the full damage of the rollout of this experimental inoculant.

The question that is lingering on all our minds is this: if the vaccines for COVID-19 are truly safe and effective, then why are the manufacturers seeking such significant indemnification? To understand the importance of this issue, we must delve deeper into the implications of immunity and indemnification.

Due to the unprecedented level of research and development being expended towards these new vaccines, it is understandable that the manufacturers would want to take precautions to ensure they are not facing a financial burden in case of an unforeseen issue. In the process of creating and distributing vaccines to a population, a range of potential liabilities can arise – from potential adverse reactions, wrong dosage levels, or manufacturing defects.

This is why the issue of indemnity should not be taken lightly – the right level of coverage could help the manufacturer protect against financial losses due to any unforeseen issues related to the vaccines. At the same time, it could also help protect the public against any faulty product or mistakes made during the production of these vaccines. With so much riding on the safety and efficacy of the new vaccines, it is vital that all stakeholders involved make sure that the appropriate level of indemnity is secured.

Ultimately, it is paramount that individuals seek clarity on these important issues before assuming any level of liability or responsibility. Consumers of the new COVID-19 vaccine should be asking questions about the question of indemnity and making sure that their interests are adequately protected. We should all make sure that the proper tests have been conducted and the right protocols are in place to guarantee the safety and effectiveness of the new vaccines. Without the proper assurances in place, we could be facing a difficult situation further down the line.

Vaccine Math – Stupidity!

Through rigorous examination, I have researched the issue of differences in reporting between vaccinated and unvaccinated individuals and the possible positive cases of Covid-19. Upon closer inspection, there appears to exist certain key deficiencies in the reporting of such figures, giving an inflated impression of how effective the vaccine is as an initiative to control the spread of the virus.

A recent study conducted in the United States revealed that although there have been marked improvements in terms of the number of vaccinated individuals, their efforts to compare the number of positive cases between those who have been vaccinated and those who have not been inoculated appear to be incomplete. Praiseworthy attempts to bolster the approval of the vaccine have been quashed by the inclusion of sporadic data in forming a comprehensive summary. This is particularly pertinent when considering the need for a uniform set of rules to be adhered to across the breadth of the United States, in all sectors of the industry, in order to ensure a conscience effort towards accuracy and reliability.

My findings have, arguably, highlighted the imbalance in reporting which can lead to false conclusions and an underestimation of the impact of the Covid-19 virus. More attention needs to be paid to this problem to ensure that the government, public health organizations and the general populous are all on the same page when it comes to analyzing the data and making decisions based on the results. This should be done through the right process and tools in order to assess the pros and cons of vaccination as well as to properly manage public health that is not intended to be misconstrued. Moreover, more detailed follow-up questionnaires could also be used to gain further insights in order to assist stakeholders in making an informed decision.

The Polymerase Chain Reaction (PCR) test and its various shortcomings have been an ongoing problem for health professionals and policy makers for years. Originally, the test was not designed for the purpose of infectious disease diagnosis. When the Centers for Disease

Control and Prevention (CDC) obtained the tests, they were informed that it was not designed with this intent. Furthermore, the use of the test has been marred by wrong testing methods and has caused numerous false positives. In particular, the issue lies in the cycle threshold used and the way it is often presented as something of great importance, making headlines seemingly whenever possible.

One example of the aftermath of this wrong use of the PCR test is the misclassification of patients as having active coronavirus infections in countries all over the world, resulting in an alarming surge in presumed cases which may not have been accurate. The problem here is that the cycle threshold used was not properly adjusted, resulting in false positives. The same has occurred in other tests as well, such as the lung capacity test studied by the New England Journal of Medicine, wherein 85% of the positive results were false positives due to faulty testing procedures.

There is a clear need for more research on the PCR test and its uses, as it is evident that wrong use and/or incorrect setting of the cycle threshold can lead to disastrous and misleading results. Yet instead of implementing more rigorous protocols and more extensive research on the underlying issues, some seem to view the test as a solution in and of itself, relying over-zealously on its results. However, for the PCR test to be of any valid use, it is of the utmost importance that authorities implement stringent procedures and adhere to the proper setting of the cycle threshold. Otherwise, the whole exercise of using the test can be rendered useless.

According to the Centers for Disease Control and Prevention (CDC), accurately identifying the presence of any live virus in a sample when the threshold is set at 33 cycles is exceptionally challenging. This is based on their own statistical calculations. Unfortunately, the CDC has not been forthcoming in accurately describing the data, which suggests that if the cycle threshold is lowered to 40 or even 37, there could potentially be a marked increase in positive diagnoses that would otherwise be negative.

It is essential for organizations such as the CDC to accurately communicate the findings of their data to the general population. This is especially true during times of global health crisis such as the current pandemic. Without the correct information, lives can be put at risk and costly damages could result due to misinterpretations and misguided decisions. For example, without an accurate understanding of the data regarding cycle thresholds, organizations may be more likely to wrongly quarantine people or entire communities, preventing people from engaging in their regular daily activities such as work or school, resulting in financial and emotional distress to everyone involved.

Therefore, it is vitally important that the CDC accurately communicate the findings of its data so that organizations and people around the world can make informed decisions in response to the pandemic. Anything less than complete transparency in the communication of data can lead to disastrous outcomes and have hidden costs that far outweigh the pain and suffering caused by the virus itself.

As I delved deeper, I discovered that the Centers for Disease Control (CDC) were not only having, as they had called it, an "extreme difficulty" locating any viable virus in samples whose cycle thresholds (CT) were over 33. The truth was, they were utterly unable to find any at all. Consequently, the CDC issued best practice guidelines for COVID-19 testing that, based on their research, were likely to lead to many individuals that were not infected by the virus receiving erroneous test results stating they were.

This illustrates a potential problem with COVID-19 testing that mandates use of a significantly high cycle threshold to determine results. This overly conservative approach to testing can lead to false negative results, meaning that people may be falsely informed that they do not have the virus, when in fact, they may have been infected. This could have tragic consequences as those misdiagnosed could unknowingly continue to spread the virus.

It is thus essential for laboratories to judiciously adhere to the most up-to-date testing protocols and standards, striving to ensure the accuracy of results. Only then can healthcare providers and scientists gain the insights into this lethal virus needed to fight the pandemic and protect public safety.

Results of recent studies have indicated that an astonishingly high percentage, as much as 80%, of reported positive cases have been false positives. The sheer number of such cases makes it apparent that the majority showing symptoms were not necessarily suffering from the illness, which could explain the impressively high rate of recovery amongst those reported positive. More importantly, this suggests that a greater number of individuals may have been affected than originally estimated.

False positivity of such rate indicates a flaw in the testing system, and further research must be done to ensure the accuracy of results. This is of particular relevance due to the potentially serious nature of COVID-19, since in accurate reporting of the illness will afford officials and medical personnel the opportunity provide adequate treatment, and potential protective measures, to the individuals and communities affected by the virus.

Studies have also speculated that the high rate of false positivity can be attributed in part to the use of straightforward laboratory tests, which do not take into account the range of levels of viral activity, or asymptomatic patients. Advanced testing systems, such as those involving the use of antibody tests or antigen tests, must be employed in order to ensure an accurate diagnosis and reduce the number of false positives, providing for a more consistent and reliable system.

By gathering more accurate and comprehensive data, medical professionals and scientists can develop a better understanding of the virus, allowing for the more successful treatment of patients, and the implementation of effective preventative measures. Only with accurate positive reporting, will the true spread of COVID-19 be known, allowing for stellar outcomes in terms of containment and treatment of those affected.

Governments, in their infinite wisdom, have decided to use a somewhat questionable method to investigate whether individuals who have not been vaccinated show a higher rate of testing positive for the virus than those who have received the vaccine. This has proven to be a most intriguing area of inquiry, one whose implications - both positive and negative - are far-reaching.

For example, if it is found that individuals without the vaccine have a higher rate of infection, then this could indicate several things. For starters, it might be evidence that the vaccine is in fact effective in providing protection from the virus. It might also point to a need for expanded education and outreach efforts regarding the need for prompt vaccinations of those at risk. On the flipside, it could lead to increased scrutiny by governments to ensure that safety measures such as masks, social distancing, and other precautionary measures are respected and adhered to.

Moreover, the findings of such research could also impact the public's perception of vaccines and vaccination programs. If it is found that there is an increased rate of positive tests amongst those who are unvaccinated, it could cause widespread fear and mistrust of these programs, or potentially even inspire public backlash against those who refuse to receive the vaccine. In the same vein, it could potentially lead to a reduction in voluntary uptake of the vaccine, making it difficult for these programs to be effective in the fight against the virus.

All in all, governments are right to be utilizing this method to try and determine whether there is any disparity between the unvaccinated and the vaccinated in terms of testing positive. Even though the results may seem inconclusive or perhaps uncertain at times, they nonetheless remain an invaluable tool in helping to gain insight into the efficacy of both the vaccine and the current safety measures, ensuring that we remain well-informed and can move forward in the most responsible and effective manner possible.

In recent months, multiple reports from journalists and brave medical professionals suggest that an alarming breach of trust may be taking place. Daniel Horowitz, writing for the Blaze Media, outlined one particularly disconcerting occurrence in relation to the new guidelines for the COVID-19 vaccine breakthrough case investigation. This investigation is used to examine cases in which individuals test positive for the virus soon after being inoculated. According to Horowitz, the Centers for Disease Control and Prevention (CDC) has instructed PCR tests to be set at a maximum of 28 C.T. or lower, apparently with the goal of stifling the potential for false positives in vaccinated persons. This would consequently dissuade people from getting the vaccine, a notion that the public should be made aware of and greatly oppose.

It is essential that we are made aware of any potential violations of trust. As people around the world harbor an already-existing hesitancy when it comes to vaccinations, there is a delicate balance between trustworthiness and disinformation. The goal now should be to make sure that the public's best interests are being looked after and that there is transparency and truth being communicated by those in the healthcare industry. If people suspect they are being misled or not being provided with accurate information, then the public's confidence in the healthcare industry and vaccine efficacy will likely suffer a great deal. Therefore, it is important to continue the stern investigation into the CDC's guidelines and the possible discrepancies, whilst simultaneously emphasizing the importance of staying vigilant and to take preventative measures for everyone's safety.

Polymerase Chain Reaction (PCR) tests are the most popular and widely used tool for diagnosing COVID-19. This technique uses cycles to amplify tiny strands of the virus and make them visible to scientists. In order to understand this process, it is important to explain the concept of Cycle Threshold (CT). The number of cycles needed to detect levels of the virus is inversely proportional to its contagiousness. Therefore, a lower cycle threshold implies a higher concentration of the virus and a higher frequency of contagion, while a higher cycle threshold indicates a lower viral load and thus sites a lower likelihood of transmission. This trend has been experimentally verified, with many samples of the virus exhibiting higher levels of contagion when smaller numbers of cycles are necessary to detect the virus. For example, in the testing of an active SARS-CoV-2 infection recently conducted at a clinic in the United States, a CT of 34 was obtained, meaning a high level of the virus was present in the samples, and a high likelihood of infecting others.

A 28 C.T. means the tested material is amplified or doubled 28 times.

This recent example of selectively applying scientific best practices is indicative of a troubling trend. The Centers for Disease Control and Prevention (CDC) unfortunately has not chosen to lower the threshold for determining a positive Covid-19 result for anyone tested. Such an approach is essential in order to reduce the risk of false positives, particularly with regard to vaccination programmes. However, the same is not applicable when it comes to the utilisation of such tests to keep children out of school or to implement other forms of government-imposed restrictions. In effect, the CDC is using science to reinforce political objectives and rigid regulations.

What is particularly disturbing is that such practices could have a dire impact on the public's trust in science. After all, if scientific advice is only heeded when it lines up with what those in positions of power find politically convenient, then why should the average citizen have faith in the scientific process and medical professionals who work within it?

Unfortunately, the pandemic has only exacerbated this issue. As the situation has evolved, so too has the public's confusion and distrust of science-based policy. In order to regain the public's trust, it is essential that we begin to apply scientific best practices consistently and fairly, across the board. Only then can we hope to rebuild a strong level of trust and confidence in science and the real-world solutions it can offer.

Last summer, the New York Times reported on the efficacy of C.T. tests - the cycle threshold tests most widely used to detect viruses. Although the test typically detects dead nucleotides that are not contagious, there is an inherent risk with C.T.s set below 34, as there may still be a potential for positive detection of a materalistically contagious virus. For example, here in Canada, many provinces have raised the threshold of C.T.s to levels of 38, 40, and even higher, in order to more accurately report virus cases. This has allowed Canada to be more proactive in addressing the potential spread of illnesses such as COVID-19, and to have better strategies in place for controlling and ultimately eliminating the pandemic.

The spread of the virus throughout the world has had a devastating and far-reaching impact, especially for those not directly infected with the virus. As a result, entire families or small businesses have found themselves in the difficult position of having to self-quarantine due to a single individual testing positive for COVID-19, despite not having any symptoms associated with the virus. This can not only be emotionally devastating, but also economically crippling, as many people and companies are forced to drastically alter their lives or stop business operations in order to prevent the further spread of the virus. Unfortunately, this often occurs without warning, and without any form of due process or evidentiary standards that prove somebody is actually carrying the virus. This often forces the effected people to have to incur significant financial losses, damage their personal and professional reputations, and generally disrupt the safety and security of their families, all without a single shred of evidence that somebody ever actually had the virus in the first place. If a family is quarantined for two weeks due to a single family member testing 'positive' for the virus, only to find out later on that there was no evidence to even indicate the family member had been infected, that family's trust and confidence in their government and its policies could be greatly damaged, while the psychological effects of the quarantine could take a long time to repair. It is no wonder why so many people, especially those in small businesses, remain uneasy and uncertain about the lasting consequences of this virus.

The recent discovery of the power of cycle thresholds on PCR among those at the helm of this pandemic situation has been nothing less than extraordinary. It is as though these inner circles, shrouded in secrecy and deception, have created a greater problem than the virus itself. It is unsettling to speculate, for I cannot be sure, yet I am forced to confront the possibility based on the evidence I have gathered over the last several months. It is like the cliché of the pin-drop tinfoil hats; not to be dismissed lightly, however strange the ideas may seem at first. These days it is wise to consider all theories, as oftentimes they turn out to be accurate. There can be no underestimation of the significance of such an occurrence when considering the range of implications it has. For instance, if cycle thresholds are indeed reliable, it would mean an immense shift in the way we identify and treat those exposed to the virus. Furthermore, the increasing sense of distrust and paranoia could make it difficult to accept the results of such tests, adding to the complexity of the situation. Consequently, a thorough examination of this dynamic is warranted to gain a better understanding of its potential ramifications.

In the context of PCR tests, the vaccine status of a person is assessed first before the test is conducted. Depending on whether they have been vaccinated or not, the cycle threshold (C.T.) used to run the test can vary widely. Vaccinated people typically have PCR tests conducted at a lower C.T., i.e. the threshold at which the test turns positive. On the other hand, unvaccinated individuals are commonly subjected to a much higher C.T. when their PCR tests are performed, significantly increasing the likelihood of a positive result. It goes without saying that the difference in these two thresholds can have a tremendous influence on the results of PCR testing in the long run.

To illustrate this point, take the example of a person in the United States who has been vaccinated against SARS-CoV-2, the virus that causes COVID-19. In this instance, routine PCR testing on this individual may be conducted at a C.T. value of 20 or lower, meaning that even low levels of the virus will be detected. However, if this same individual had not been vaccinated, the C.T. used in the PCR test might be pushed as high as 40 or even higher, greatly decreasing the precision of the test and the odds of it yielding a positive result. This demonstrates the profound impact the vaccination status of a person can have on the effectiveness of PCR testing.

As we come to the end of 2020 and beyond, there has been a constant discussion among the health and safety experts regarding the reliability of the PCR test when it comes to detecting COVID-19 infections. While some argued that the test wasn't reliable enough to be trusted and scrapped, others contended the PCR test could actually be an effective tool towards combating the virus. It was then that this discussion started to be swept under the rug, which may have led some to believe this was all part of a secretive conspiracy. However, as the events of the pandemic have unfolded, it has been found that some of the most outlandish claims and information has been true.

The Centres for Disease Control and Prevention (CDC) has further endorsed the use of the PCR testing, although with a bit more caution and rigour than before, in order to cater to those who have yet to be vaccinated. This issue of secrecy and half-truths have lingered in the air ever since the onset of the pandemic, and when it comes to the talks about the collaboration of big Pharma, politicians, and big tech, it is unsurprising that some sense an aura of mystery and opaque motives. The notion of fear-mongering as a form of control and manipulation is something that has been suggested, as well as officials being told what to say so it pleases those in more powerful positions.

I always suspected something like this, but I never expected it to be so real. It's almost as if the Flu has been completely eradicated and replaced completely by COVID-19, at least that's how the data reports seem to be pointing. We have never been this aggressive or even this focused in tracking the Flu, so if we were to start doing the same type of assessments and testing for the Flu as we do now with COVID, I'm certain that we would be encountering near-identical outcomes. It makes me wonder whether this is some sort of new phenomena or perhaps a case of déjà vu, where we are simply tracking variants of the Flu, just in a much more intricate and intensive manner.

To fully understand the truth we must delve further, and this is a mission I am fully committed to accepting. I will perform my rigorous investigations and report back my findings, though one thing I can tell you for certain is that nothing could ever surprise me again, not after this.

Just some other notes to think about

The U.S. Centers for Disease Control and Prevention (CDC) states that a person won't be accounted for as completely vaccinated until fourteen days after their subsequent infusion, in the case of the Pfizer and Moderna vaccines, or fourteen days after their one dose of the Janssen vaccine, even though more than 80 percent of deaths supposedly caused by the vaccine happen during this time. The fact that this critical window is conveniently ignored by the CDC appears to be rather specious.

Unfortunately, this has become a concerning pattern for the CDC in regards to the handling of the Covid-19 pandemic. It has been frequently observed that the agency is slow to respond to new developments in the virus and fails to update safeguards accordingly. For example, the CDC's guidelines for using masks did not reflect the fact that wearing a cloth mask can protect asymptomatic people from transmission until early 2020. It took months for the agency to advise healthcare workers to wear N-95 masks even though evidence of how contagious the virus is was already available.

It appears that by neglecting to address the potential complications of the vaccine within a two-week window of its administration, the CDC is setting the public up for unnecessary risks. Going forward, it will be essential for the agency to more thoroughly examine the potential risks of each vaccine, and factor these considerations into their guidance. Technology and research have allowed us to better prepare for health emergencies, and it is the CDC's responsibility to keep up with the times and alter their protocols as the status quo shifts.

The Centers for Disease Control and Prevention (CDC) have issued two distinct sets of guidelines regarding coronavirus testing – one for individuals who have been vaccinated and the other for individuals who have not received the vaccine. For those who remain unvaccinated, the CDC guidelines advise using a cycle threshold (CT) of 40, which has been known to produce false positives. On the contrary, for individuals who have been vaccinated, the CDC recommends using a CT of 28 or lower, which greatly reduces the chances of false positives.

It is important to understand the distinction between these testing guidelines. False positives can lead to unnecessary hardship for any individual – from additional time off from school or work to being subjected to unneeded medical procedures. For example, if a vaccinated citizen receives a false positive outcome, they may be advised by a medical professional to enter self-isolation and miss vital family gatherings or job obligations. On the other hand, if the CT value was 28 or lower, the individual is much less likely to receive a false positive, thus avoiding any potential disruption to their work and family schedules.

Ultimately, these testing guidelines issued by the Centers for Disease Control and Prevention play an important role in helping to keep individuals protected from the spread of coronavirus. It is imperative that we follow CDC best practices and use the correct cycle threshold value so that we can accurately identify any potential cases of Covid-19 in accordance with the latest scientific evidence.

The Centers for Disease Control and Prevention (CDC) has been accused of concealing evidence of failures from COVID-19 vaccines, downplaying the seriousness of when vaccines don't work. This is done by only counting cases of breakthrough infections resulting in hospitalization or death, leading to the widespread narrative that unvaccinated people pose the largest risk to public health. While it is true that those who are not vaccinated are more vulnerable to catching and spreading COVID-19, the CDC's studies should focus on other factors, such as the strain of the virus, the length of time it takes to build immunity after getting the vaccine, and how genetically diverse the virus is. By focusing solely on hospitalization and death rates among the unvaccinated, the CDC has created a one-sided view of the implications of not getting the vaccine.

By overlooking other data points, the CDC is missing an opportunity to help inform public policy decisions and promote preventative health measures. For instance, the CDC's decision to only highlight hospitalization and death rates among unvaccinated individuals fails to account for lower forms of infection that can still be detrimental to public health, such as milder forms of the virus that can still be spread to others. Without taking into account this data, we cannot make fully informed decisions on how best to protect our communities. To ensure the proper dissemination of information, the CDC should expand its analyses of vaccine success rates to account for all symptoms and immunogenicity, not just hospitalization or death. In doing so, the CDC will be able to provide a more balanced and accurate picture of COVID-19 and improve vaccine confidence among the general public.

Hospitals are still also reporting non-COVID related illnesses as COVID-19

Recently, agencies have been marking people who passed away within two weeks post-vaccination as unvaccinated to artificially pump up the numbers of unvaccinated fatalities. This practice also hides the potential health risks of getting immunised, as a majority of deaths after vaccination happen within the first 14 days. This means that these deaths are not classified as vaccine-related resulting in the real danger of these shots being under-reported. In other words, not only are these inaccurate statistics hiding the real death toll caused by the vaccine, they are also misleading public health officials to believe that unvaccinated people are passing away at a higher rate than those who have got the jab, when in reality, it may be the other way around. This is why it is important to accurately report on the dangers of such medical interventions in order to preserve public health and safety.

It appears that there is an incessant stream of scandal and misdirection that bombards me on a regular basis, making it virtually impossible for me to ignore; however, this is largely swept under the rug because most people have been conditioned to put faith in the government and the medical profession. Despite this, the greed and selfishness that one sees from private businesses has been an issue for many years, going so far as to create some of the most significant economic catastrophes in recent memory. Whether it be the Great Depression of the 1930s or the more recent Global Financial Crisis of 2008, the fallout from these examples of corporate greed and manipulation has affected countless individuals. As a result, it makes it compulsory to be ever vigilant when it comes to the actions of both public and private interests, ensuring that any underhandedness or corruption can be called out and the associated parties held accountable.

This event has the potential to be one of the most debated and controversial political scandals of our times. As more details emerge and time passes, there will undoubtably be an abundance of public inquires, law suits and judicial inquests. Harsh charges of bribery and conspiracy will no doubt be at the forefront due to the massive amounts of money and political power that are involved in this story.

Evidence and testimonies provided through these investigations will likely reveal even more corruption that goes beyond the realm of bribery and conspiracy, such as cover-ups, embezzlement and more. They may also lead for calls for additional investigations, with countries around the world ordering probes and criminal investigations into the affairs of all those involved.

The consequences of this scandal are set to be incredibly long-lasting and no one will be surprised when it results in prison sentences and substantial fines for those who have broken the rules. This scandal has the power to do serious damage to the public consciousness and to public trust, and it will certainly serve as a reminder of the significant dangers that come with such immense power and wealth.

It is an unfortunate yet highly significant event, and as time goes on, the true scale of this political scandal will be revealed, with even bigger and more damaging revelations yet to come.

If you haven't already, you're missing out! Be sure to check out the Uncensored Podcast, which can be found on a variety of platforms like iTunes, Amazon, Android, Google and Spotify. With so many different outlets providing access to the show, you'll never miss a single episode! Plus, you can keep up with the latest and greatest content by subscribing to the podcast. Just click the subscribe button below and you're all set.

The Uncensored Podcast is the brainchild of some of the top minds in the business. They deliver an unrivalled mix of edgy content for people looking for sharp, engaging material. Expect intimate interviews with world-class leaders, up-and-coming Entrepreneurs, and more - covering a variety of topics from the latest technology trends to the most pressing cultural & political issues. Each episode promises to inform, empower, and entertain. So what are you waiting for? Hit that subscribe button and get listening - you won't regret it!

The 14 Day Lie

The 14 day wait period following your dose of the vaccine has been widely discussed in media outlets, with debate surrounding why this 14-day rule is necessary. The truth of the matter lies within the injuries that have been recorded in the Vaccine Adverse Event Reporting System (VAERS) database, which has surpassed 700,000 in the US alone.

While the initial dose of the vaccine is judged to be effective much earlier than those 14 days, taking the suggested waiting period is important because it allows authorities to accurately track any reactions or side effects that may be encountered. This provides extremely valuable data to researchers who are continually aiming to minimise the risks of adverse reactions to all doses of the vaccine, as well as ensuring that information is recorded with accuracy.

On the 30th April 2021, the Centres for Disease Control and Prevention (CDC) issued a statement on the importance of the 14-day wait period. As the statement stated "“Vaccine providers should recommend a 14 day waiting period after receiving a COVID-19 vaccine and before an individual has contact with someone who is at an increased risk for severe illness from COVID-19, such as an immunocompromised household contact".

This rule not only highlights the significant rise in injuries seen in the VAERS database after administering the vaccine, but also ensures the safety of vulnerable members of the population who could be at a higher risk of severe harm if they were to encounter any unexpected side effects.

Overall, it is clear to see why this 14-day post-vaccination waiting period is a necessary step in ensuring the safety of those who have received the jab. This is why it is so important to adhere to the guidelines and recommendations of health practitioners and medical professionals to guarantee that the data gathered remains accurate and reliable.

After scrupulously crunching the numbers, I have arrived at a rather disheartening conclusion: the majority of recorded vaccine injuries and fatalities happen within the initial 14 days following administration. This stunningly high statistic is unfortunately rarely given any attention by mainstream media outlets like MSNBC, CNN and CBC which is why I've included a chart below to illustrate this data.

This overly high rate of vaccine-induced harm is particularly concerning when one considers the sheer number of people receiving the vaccines worldwide and the speed in which they are being administered. Indeed, according to the latest stats released by the World Health Organization, as of October 2020 there have been over 321 million vaccine doses administered around the world since the beginning of the pandemic, with a disproportionate number of these occurring within the first 14 days.

This staggering amount of injections in combination with the fact that the majority of these occur within the aforementioned timeframe is a cause for concern and one which should not be taken lightly, hence why raising awareness about this statistic is critical for public health purposes. Ultimately, the public should have access to full and accurate information about the associated risks of new medical treatments and procedures and should not have to rely on potentially biased media outlets.

The U.S. Centers for Disease Control and Prevention (CDC) has set forth a stringent standard for determining full vaccination status - you must wait at least 14 days after your initial injection for Pfizer or Moderna, or your single dose of Janssen, before you can be officially marked as 'fully vaccinated.' Even though research has indicated that over 80% of deaths after receiving a vaccine occur within the 14-day window, the CDC has not modified their policy. This could be seen as rather convenient for responsible public health institutions and convenient for those who are seeking to vaccinate as many people as possible in a timely manner.

However, this is potentially highly concerning to many who are considering taking the vaccines, as these same nefarious death statistics imply that there is a legitimate, inherent risk in the vaccine that needs to be taken more seriously, especially by those in charge of vaccine distribution. Furthermore, the issue should not be overlooked, as these risks might be too great a risk for some and could easily warrant stronger pre-vaccination health assessments or changes to the current guidelines.

The sheer amount of misinformation and data manipulation surrounding the COVID-19 vaccine is both alarming and unacceptable. A prime example of this is the way the CDC handles death tolls associated with the vaccine. According to the official protocols, if one dies within the first fourteen days of receiving a COVID-19 vaccination, the death is attributed to being unvaccinated, resulting in an inaccurate representation of the true death toll due to the vaccine. This false stat means that the dangers of the shots are obfuscated, as most fatalities associated with the vaccine have occurred within the first two weeks following inoculation.

Adding to the confusion, the CDC employs two distinct protocols for testing, depending on the person's immunization status - one for those who have been vaccinated and another for the unvaccinated. For those who have not been vaccinated, the CDC recommends using a cycle threshold (CT) of 40, a known quantity that could result in false positives. Those who have been receiving the vaccine are simply asked to use a CT of 28 or lower, a measure that minimizes the risk of false positives. Both of these protocols, while seemingly benign, merely add to the ongoing misinformation surrounding the vaccine's death toll.

It is evident that the Centers for Disease Control and Prevention (CDC) choose to hide the fact that not all vaccines are entirely successful in providing proper immunization against disease. The CDC instead work to perpetuate the "pandemic of the unvaccinated" myth-by claiming that those who have not been vaccinated are the cause of an ongoing epidemic. To do this, they only report on cases of breakthrough infections, meaning those that occur even after a person has been vaccinated, if those cases end in hospitalization or death. This spin denial of the ineffectiveness of some vaccines ignores the more common outcome of minor illness and discomfort that could still be caused.

Take for example Tetanus, a vaccine which is widely accepted to be extremely effective in preventing a dangerous and often fatal disease, even with just one or two shots. However, the CDC have documented that even immunized individuals can and do get sick, with potentially painful and uncomfortable side effects. As such, the CDC's decision to focus only on the more serious cases where hospitalization or death is involved ultimately paints a skewed picture of vaccine success, as such outcomes are rare.

As a result, by glossing over the less severe, but still common, side-effects of vaccines, the CDC continues to enable the propagation of the "pandemic of the unvaccinated" narrative, dangerously confusing and misinforming the public on the real-world efficacy of vaccinations.

Hospitals are still also reporting non-COVID related illnesses as COVID-19

The Vaccine is a complete Failure

A groundbreaking study conducted by researchers from the University of California, Davis Genome Center, UC San Francisco, and the Chan Zuckerberg Biohub, has revealed that the new SARS-CoV-2 Delta variant does not seem to result in higher viral loads for vaccinated or unvaccinated individuals who have tested positive for the virus. Moreover, the study found no significant difference between those infected with the virus who presented symptoms, and those who did not.

The breakthrough study, which utilized sophisticated genomic approaches to study SARS-CoV-2 evolution and its impact on vaccine performance, has provided researchers and the public a valuable resource in addressing the efficacy of vaccine protection. For example, their work showed that the Delta variant reacted similarly to the mRNA vaccines despite its structural differences. This insight can help scientists better understand how the virus is evolving, and allow medical professionals to adapt vaccines to further protect vaccinated individuals.

Ultimately, this research provides insight on how the SARS-CoV-2 Delta variant has impacted the vaccine's effectiveness. With the knowledge that varying viral loads between vaccinated and unvaccinated individuals were not present, public health officials can deploy strategies for curbing circulation and addressing the risk of spread. Through such measures, we can work together to bring an end to this pandemic.

I have been seeing more news stories like this that I think we are missing a link to why this is happening.

New findings published by U.S. health agencies on Monday revealed that the efficacy rate of the Pfizer Inc./BioNTech SE vaccine in protecting individuals from contracting the coronavirus had fallen from a formidable 88% to 47% just six months after receiving the second dose. This data has been taken into account by the U.S. health agencies when evaluating whether additional booster doses of the vaccine are necessary.

The impact of the Pfizer Inc./BioNTech SE vaccine in blocking COVID-19 infections had initially been considered to be extremely high. As early as December 2020, the percentage of people who would be protected against the virus was estimated to be around 95%. However, due to the emergence of variants, this rate has suffered a significant decrease.

The efficacy of the vaccines aside, it is of the utmost importance for the population to continue prioritizing safe practices such as wearing face masks in public areas, avoiding large gatherings where social distancing isn't possible, and regularly washing hands. Doing so could counteract potential issues related to the diminishing protection rate of the vaccine.

Health agencies are also actively studying the possibility of replacing the second dose of the Pfizer Inc./BioNTech SE vaccine with that of another pharmaceutical company's, or else creating a booster shot tailored to protect against the new variants should additional immunization be necessary.

Experts suggest that boosting immunity could help slow the further spread of the virus, however the effectiveness of such an approach has not yet been determined. Nonetheless, the findings on Monday provide a good starting point to assess the current effectiveness of the vaccine, and an understanding of the steps needed to be taken going forward to reduce the impact of the virus.

The data, published in the Lancet medical journal, had been previously released in August ahead of peer review.

In the wake of sequencing samples diagnosed with various infections, data has revealed that the effectiveness of vaccinating against diseases influenced by the Delta variant is notably strong in the initial month following full vaccination, reporting a success rate of 93%. Sadly, after only four months this efficacy has decreased substantially to 53%. This decrease in protective efficacy emphasizes the importance of administering regular booster shots. For example, the measles vaccine requires a person to be inoculated three times throughout their life in order to confer lifelong protection. Failure to do so might result in the MMR immunity becoming incomplete and more susceptible to contracting the virus. However, the initial months following full vaccination demonstrates a heightened capacity for immunity, significantly reducing the chances of contagion among the population.

I wrote about this a little while back about how we could be giving people injections of dead or spoiled vaccines.

When discussions began in December 2020 regarding the rollout of a vaccine, there was great cause for concern. The issue of safeguarding the transportation 'cold chain' posed a major logistical challenge and threatened the global distribution efforts. In addition to this, the estimated costs of millions of dollars of specialized freezers necessary to ensure the safe distribution of the Pfizer vaccine only added to the burden. Fortunately, upon Moderna's coronavirus vaccine, which did not mandate the same level of temperature control, the issues concerning the Pfizer vaccine's safe shipment appeared to dissipate and were not raised again. The Moderna vaccine allowed for a solution that took into account monetary and temperature requirements of transporting the vaccine, making the challenge of distributing said vaccine around the world became far more efficient. A prime example of this was with the shipment of low-cost and temperature controlled shipping containers to countries such as India, allowing for a much simplier shipping process. It is safe to say that the lack of an alternate shipping option presented by Moderna's vaccine allowed for the entire global rollout of the Pfizer vaccine to be successful.

Canadians across the country were eager recipients of the millions of doses of mRNA vaccines delivered to countless municipalities and towns, both urban and rural, but it's essential to be vigilant and remain wary of the doses. After all, administering a vaccine requires a certain degree of skill and conscientiousness. This can be seen in the differences between cities that are well-equipped with hospitals and healthcare staff and small towns or remote municipalities that may not have the same resources available. And while Canadians generally take pride and responsibility when it comes to the handling of medicines and vaccines, the sheer amount of doses distributed nationwide means that there is still potential for mistakes and unprofessional proceedings.

It is understandable and admirable that Canadians have placed their trust in healthcare professionals, yet it is equally important to be mindful and responsible with taking whatever measures possible to ensure that the correct protocols and safety standards are being kept. Regrettably, it is true that there will always be occasions where such expectations are not met and doubt is unfortunately inescapable. Nevertheless, it is essential to remain optimistic and trust in the healthcare procedures, professional judgement and implementation of proper administrative processes that Canadians are entitled to.

It would not be overly generous to suggest that a considerable proportion of mRNA vaccine doses were not sufficiently taken care of, resulting in them becoming inactivated or destroyed before reaching the people for whom they had been intended. This carelessness contributed to the staggering numbers of vaccine doses that have been wasted due to the failure to observe vaccine protocol. For example, a major issue that has arisen in some countries is the lack of refrigeration or cool storage for the vaccine doses. This can cause a substantial decrease in its strength and efficacy as well as its potential to induce an immune response. As a consequence, when reaching the people, the particular set of vaccine doses is rendered ineffective. Another common occurrence is the mismanagement of vaccine expiration dates and the vaccination of doses that have been kept beyond their expiration date which can also lead to lacking results.

To strive for an efficient deployment of the available vaccine doses and to ensure their potency, it is thus paramount to have clear systems of organization and monitoring in place in order to prevent any unnecessary waste through measures such as avoiding the separation of vaccine vials, maintaining proper temperatures of the dosage, and keeping track of expiration dates.

The significance of maintaining an optimal temperature range for Pfizer vaccine shipments cannot be understated; if not stored in the temperatures reported in the medical community, the mRNA in the vaccine will be rendered inactive, making the shot ineffective in the fight against SARS-CoV-2. This would be the equivalent of a placebo, with the recipient receiving no substantial protection.

Different immunizations must be stored at specific temperature ranges in order to preserve their effectiveness; without proper storage, they are rendered useless. For instance, vaccines containing the measles, mumps, and rubella virus (MMR) must be kept at a temperature range of 35 to 46 degrees Fahrenheit. Variations outside this range will cause the vaccine's effectiveness to diminish or be completely inactivated.

The Pfizer vaccine is no different; it must also be kept at an ideal temperature range. Different cold-storage technologies, such as the dry ice box, are often relied upon in order to ensure that temperature-sensitive drugs and vaccines remain in an optimal condition for efficient transport and distribution.

However, when equipment failure occurs, or when transportation delays cause a vaccine's temperature to deviate from the approved range, the vaccine becomes useless as it cannot stimulate an effective immune response against SARS-CoV-2. In essence, the recipient is no better off than if they had received a placebo injection.

Due to the clear importance of proper storage, careful monitoring strategies and detailed contingency plans must be in place for the successful distribution of temperature-sensitive drugs and vaccines. The Pfizer vaccine is no exception, and preserving its effectiveness is essential for providing safe and effective protection against SARS-CoV-2.

The authors of this compelling study suggest that the findings illustrate the continued necessity for wearing protective masks and regular testing and vaccinations, especially in heavily populated regions. According to the research, which is currently accessible on the preprint platform MedRxiv, there appears to be an increasing amount of proof regarding virus levels in those who have had breakthrough infections despite having been vaccinated in two environments with persistent high levels of the delta variant of the virus. This indicates a need for increased safety protocols, such as mask-wearing and testing, in these kinds of heavily populated locations. For example, in a city like New York, where cases of the delta variant are rising, it is of utmost priority to ensure people wear masks, practice social distancing, and get tested regularly.

After extensive data collecting and analysis, the research team discovered considerable differences in the amount of virus within both the vaccinated and unvaccinated participants. Interestingly, the scientists failed to note any significant variation between the two groups when it came to the viral load, which has prompted a reconsideration of the vaccine's effectiveness. To better understand these results, it is important to look at the research findings in greater detail.

Firstly, the study evaluated the amount of a specific virus present within each patient by measuring their viral load. This involves running tests to detect the presence, or absence, of the virus and quantifying it on a molecular level. When analysed, it became clear that even amongst the vaccinated participants, the viral load varied greatly. In some cases, patients had extremely high levels of the virus and others barely showed any signs at all.

Nonetheless, the same could be said for the unvaccinated. While there seemed to be an overall disparity between the two groups viral load levels as a whole, the team was unable to ascertain any significant difference when comparing an individual from the vaccinated group with someone from the unvaccinated. This suggests that the vaccine itself is not as effective as anticipated.

To conclude, the research findings demonstrate wide variations in the amount of virus present amongst vaccinated and unvaccinated participants. Even with the vaccine, some individuals were still highly susceptible to infection. Unfortunately, the team was unable to detect any statistically significant difference in viral load between the two groups, which weakens the potency of the vaccine and is cause for concern.

Breakthrough infections where vaccinated people do become sick can occur, especially in high virus prevalence.

Recent scientific studies have demonstrated that vaccinated individuals who develop a breakthrough infection may be as infectious and prone to spreading the virus as their unvaccinated counterparts. These findings are particularly disconcerting in light of the fact that these individuals may be significantly less likely to experience severe symptoms than the unvaccinated population. For instance, in a study with 676 healthcare workers who tested positive for SARS-CoV-2 following full vaccination, only 9% of these individuals had to be hospitalized due to the infection, while a much higher rate was seen in an unvaccinated group of the same population. This indicates that although these individuals may not show symptoms or be as fragile to the virus, they can still transmit it to others. As such, it is essential that people continue to practice preventative measures even after receiving the vaccine in order to mitigate against any impact they may have on the spread of the virus.

I am absolutely convinced that the evidence of wrongdoing is right out in the open, plain as day, and yet many people are trying to complicate or confuse the issue. Did the decision to allow reduced standards of cold storage and transportation capability really result in millions of doses of the vaccine being rendered ineffective? It's almost as though it was a deliberate attempt to see the extent to which we would comply with these rules and regulations, preparing us for something yet to come. All of the data indicates that in its current state the vaccine is far from efficacious both in terms of its intended purpose to provide immunity and as an effective therapeutic tool. Instances of people receiving a full course of the vaccine and still getting sick afterwards show just how unreliable this vaccine is in practice. Even more troubling are the anecdotal reports of reliable and unavoidable short-term side-effects such as nausea, fever, and lethargy. The sheer unfairness of this situation can't be overstated. Millions of people are now at risk of developing more serious symptoms of the virus and, worse still, of passing it on to the elders and other high-risk individuals in their lives who weren't able to receive their vaccine. The failure to meet with even the minimal quality assurance standards necessary to ensure that the vaccine remains viable had a catastrophic and avoidable effect on global health.

I have a deep-seated feeling that the official vaccine from Pfizer that is slated for release in 2024 without being subjected to the Emergency Use Authorization decree and full FDA approval is a pipe dream that will never come to fruition. My belief is that this vaccine has been a complete and utter failure from the start, and the company knows this to be true; this lends credence to the rumor that only a few people actually received vaccines that were distributed in accordance with the required protocols.

The very idea of a successful vaccine being prescribed, let alone received, without being subjected to the strict rules and regulations put in place by the government is unfathomable. With a majority of the population perceiving pharmaceutical companies as having their own agenda, it only makes sense that they would install additional guidelines to ensure the efficacy of a product is up to par. Case in point, we've seen examples of unethical practices in the past - such as the controversy that surrounded the generic versions of insulin sold by the U.S. company Eli Lilly.

To put it simply, if Pfizer's vaccine doesn't pass the FDA's stringent standards, there's a good chance it may never make its way out of the lab and into the public. Unfortunately, this means that the company may have contracted a bridge too far and this alleged miracle drug might end up as nothing more than a dud. With the vaccine's future now in doubt, Pfizer might find themselves facing yet another dreaded loss of time, money, and reputation.

The fiasco of Pfizer's soon-to-fail vaccine appears to have taken a turn for the worst; the pharmaceutical giant has recently announced the release of their own version of the much-talked about Ivermectine, an incredibly effective drug that has thus far proven successful in treating the adverse symptoms of the VID. However, in spite of its effectiveness, doctors are continually threatened with reprimands should they dare prescribe it to patients. Despite the risk, Canada has already entered into a long-term contract with Pfizer in order to purchase a staggering four hundred million doses of the vaccine over the next four years. Unsurprisingly, this news has received mixed reception; while some may deem this move to be nothing more than a cleverly-crafted money-making scheme, there is no denying that it has certainly added further insult to the injury already caused by Pfizer's failed vaccine.

No matter the outcome of the coronavirus vaccine, it is inevitable that we are obligated to purchase them. Even if the vaccine is declared ineffective in 2023, we are still going to need to purchase these products, causing a massive uproar among the public. Questions will arise about why so much money was spent on something that has become useless, and whose decision it was to sign these contracts. It all starts with the contracts that were signed with the firm Pfizer, but it's for the whole population that the consequences are dire.

The issue of spending an enormous amount of money on a product that fails to stand the test of time is an important one. We can liken it to investing in a business only to find out that it is no longer profitable, rendering your money spent meaningless. Just like any other major purchase, it is essential to weigh the pros and cons of acquiring a product that may not be functional over a few years. Furthermore, when hundreds of millions of dollars are at stake, the need to be prudent and considerate becomes even more paramount.

It is our responsibility to be prudent and responsible when agreeing to large contracts. Not only is it our duty to protect the public's money, but it is also our job to ensure that any commitment made will serve its purpose in the long run. In this case, the public is counting on us to guarantee that the vaccine purchased will serve its purpose, that being to vaccinate against the novel coronavirus, for the foreseeable future.

Therefore, it is clear that any decision that involves spending a large amount of money must be thoroughly investigated and heavily scrutinized. Otherwise, we may be faced with the same outcome as before; massive spending leading to a useless product. The potential consequences are dire and the public will be holding us accountable.

Given the recent news of government purchased Covid-19 vaccines from manufacturers such as Pfizer and Moderna, one would assume that the public should have knowledge of the said purchases. This would be prudent not just because of the sheer amount of money that the government used of taxpayers' money to make these purchases, but because they are major investments with significant public health implications.

Not only are these purchases public matters, but public documents should be available to the public before being signed by the relevant authorities. This would enable citizens not only to be aware of the specifics of the purchases,but also to review them, ask questions and/or take other measures if necessary.In some countries, for example, citizens have the right to view government accounts and expenditures, regardless of whether the purchase is related to something as vital as Covid-19 vaccine procurement or something more mundane, but still equally important, such as public infrastructure development.

Given the gravity of the situation, not having access to such documents would be a serious breach of the public's right to transparency and accountability.Therefore, it is only reasonable to assume that such documents should be made available to the public before they are signed, or at the very least, as soon as possible thereafter. Anything less would be a violation of the public's rights and freedoms.

Lately, it's really hard to shake the feeling of unease and skepticism that something far more sinister than meets the eye is at play amongst the political and corporate elites. Questions are beginning to surface if this 'game' is really all about power, politics, and a concentrated effort to make the already wealthy even more wealthy while simultaneously turning big-tech into something akin to an Oligarchy. It's as if an invisible agenda is at work behind the scenes and all the pieces just don't quite add up. This is causing more and more of us to take a step back and critically assess exactly what is transpiring on the world stage.

The corporate elite have grown to an unprecedented size in recent years. The number of billionaires has dramatically increased during the pandemic, when the majority of everyday workers have felt the pinch financially. Meanwhile, recently unearthed documents have detailed exactly how big-tech companies like Google, Amazon, and Facebook have been manipulating the public discourse and selectively controlling the narrative in order to further their own interests. These activities are raising worrying questions as to how far into a cyber-controlled future we're heading and how our civil liberties will be safeguarded.

Some reports and studies have indicated that these same big-tech companies, along with the richest in society, are using lobbyists to create self-serving regulations that are detrimental to the public, suggesting that something is truly afoot. The concerning extent of this 'power-grab' is something which we cannot ignore and simply turn a blind eye to. Now more than ever is the time to wake up and take action while we still can. This is a reality that cannot be ignored and we need to act pragmatically and with resolve if we are to ever stand a chance of taking back control of our lives and our society.

Rochelle Walensky, the director of the Centers for Disease Control and Prevention (CDC), has given seemingly conflicting messages recently on the prospect that the delta variant of the coronavirus could infect individuals who were already vaccinated against COVID-19. On one hand, she has characterized the chance of such a "breakthrough infection" as very rare, emphasizing that the risk of infection is far lower for vaccinated people. On the other hand, Dr. Walensky has also suggested that being vaccinated could actually increase a person's chances of being infected.

This hesitation on whether the vaccine can protect against delta variants is fully understandable, given the uncertainty surrounding this particular mutation. According to research conducted by Public Health England, the delta variant is 57 percent more transmissible than the original strain of the virus. It is also believed to be more contagious for children, and is believed to be more resistant to the protection offered by traditional vaccines.

With this in mind, the CDC has taken measures to improve the efficacy of the vaccine against the delta variant. Photos from the agency have been released showing that the Centers is reading and analyzing additional trials, involving the booster shots of the Pfizer and Moderna COVID-19 vaccines to improve protection against the new strain's spread. In the meantime, the CDC suggests that by continuing to practice safe behaviors such as wearing masks, washing hands, and social distancing, people of all ages, including those already vaccinated, can still protect themselves and their loved ones against the virus. It is unclear exactly how much protection a vaccination will confer when a person is exposed to the delta variant, yet one thing is for certain: adhering to CDC-recommended safety guidelines is the best course of action.

The Centers for Disease Control and Prevention have emphasized that vaccination is the most effective way of achieving control over the rising COVID-19 pandemic. Since early January 2021, the numbers of COVID-19 cases and deaths in the United States have significantly plummeted, and experts attribute this in part to wider vaccination coverage. It is well-known that the development of effective vaccines is one of the only ways to decrease the severity of the pandemic, and the data clearly demonstrates the efficacy of the approach.

The importance of rapidly scaling up vaccine distribution and administration efforts has been a focal point in public health conversations across the world. Doing so can not only save lives, but has been proven to be a successful strategy in curbing the spread of the virus and protecting vulnerable members of society. Countries that have adopted high vaccination rates, such as the United Kingdom, have seen higher rates of successful vaccination campaigns, and their case and death numbers reflect that.

The truth remains that until the majority of people are vaccinated, those vulnerable to the virus are still in danger. Although the United States has seen a decline in case and death numbers since January, it is essential to continue vaccinating individuals in order to keep the numbers low and protect those at risk in our communities. It is through continuous efforts to increase vaccination coverage that the pandemic can eventually be brought under control.

In June and July of 2021, an alarming trend began to emerge related to the novel coronavirus, or COVID-19, across the United States. The number of reported cases increased dramatically during this period, with some regions seeing an increase of as much as 300%. This surge was soon followed by an increase in the number of hospitalizations and deaths.

At the center of this concerning surge was the highly contagious B.1.617.2 (Delta) variant of the SARS-CoV-2 virus. The severity of the effects of this new strain was clearly demonstrated by the rapid influx of new cases and the continued strain on hospitals and healthcare systems.

Though it is too early to fully understand the ways in which this virus differ from the original COVID-19 strain, it appears to not only spread more easily amongst individuals but also more likely to cause more severe symptoms and more rapid hospitalizations. This is evidenced in regions like the South and Midwest, where cases began to skyrocket after the emergence of the Delta variant despite having previously been successful at keeping infections numbers low.

The importance of taking extra caution and adhering to government health and safety guidelines in the face of this new strain is imperative. Without proper attention, we could find ourselves in the midst of a new surge in cases, hospitalizations, and deaths.

Why some will not let go of COVID

Recently, I have become very intrigued by the psychology of people and have devoted a considerable amount of my time to exploring this extensively. My interest can be traced back several years, having not just a personal interest in mental wellbeing but also as a general fascination. Over time, through reading and observation, I have encountered examples of human behavior that I have encountered in everyday life, in recent months in particular. Admittedly, I am by no means a psychologist and, while I have done my utmost to become informed on this subject, I could not conclusively provide psychological advice. Instead, I prefer to focus on passing on my observations and the knowledge I have acquired, so as to benefit the reader or listener.

For instance, when studying the psychology of people, I discovered the concept of empathy, which is the ability to perceive, understand and insofar as possible share the feelings of another. I noticed how strongly it influences our interactions with other people, observing how it has enabled me to offer assistance more efficiently to those who need it. This has aided me greatly in times of difficult, where generosity and kind-heartedness can make a real difference. Ultimately, understanding people's psyches allows us to develop greater appreciation of our own minds and identities, as well as that of others, and so I believe it is an incredibly important area of study.

The idea that it is more virtuous to perform good deeds behind closed doors is an important one. The New Testament teaches us that if we truly want to make the world a better place and help those around us, we should do so in a quiet and humble way. By praying in private rather than in public, we are not trying to gain recognition for our actions nor are we attempting to show the world how 'good' we are; rather, we are being sincere in our efforts to connect with something greater and foster genuine change in the world.

This adage has implications that go far beyond prayer. All too often, altruistic actions are performed not for the benefit of those involved, but instead for the purpose of gaining recognition or opportunity. In this way, the old dictum serves as a reminder to stay true to what is truly important, and to reduce the focus on what others may think and believe. Doing good, after all, is its own reward.

This principle also applies to our relationships with others. Struggling to bridge divides between family members, friends, and even enemies is difficult, especially when they appear to be insurmountable. As a result, it's easy to become discouraged and give up. But by standing firm and attempting to mend relationships out of personal compassion rather than a desire for advancement or social recognition, we can truly start to make a difference.

When we take a moment to reflect according to this ancient directive, it helps us to reflect on what is truly valuable: taking action internally and intrinsic motivators. Thus, we can see that practicing this saying of old encourages us to take genuine, meaningful action to better ourselves and the world. And, in the end, that is a cause for celebration.

Contrary to many people's beliefs, externalizing our problems or crises often does more harm than good. This is mainly due to the fact that our issues are rarely found outside of ourselves. Evidence suggests that the only true way to overcome any difficulty is to confront and manage it ourselves. Consider a friend who is constantly worried and suspicious of the world. They might accuse and blame those around them for environmental disasters, poverty, and other injustices. Despite their outward display of 'caring' for the world, we can usually see that, in reality, they are only trying to hide their own issues. By pointing the finger of blame elsewhere, they are inadvertently attempting to distract themselves from the real crisis they are facing.

This may explain why they often jump from one problem to the next. Once the intensity of the current situation dies down, and their life becomes quieter and more peaceful, these individuals are left with more time to think and reflect on the issues that they do not want to face. Furthermore, these individuals would never be able to solve the world's problems until their own life problems are addressed. In conclusion: any attempt to shift stress and pain away from oneself and onto somebody or something else is probably doomed to failure. The only way out of a crisis is through self-examination and resilience-building.

Sadly, even when there is plenty of drama and drama surrounding somebody, the deeper unresolved issues that lurk beneath the surface will ultimately come back to the forefront and require attention. It's almost as if ignoring a problem or avoiding it only delays the inevitable, and when that time comes, the worry and stress of the first glimpse of it can trigger a kind of emotional chaos as a coping mechanism. Paradoxically, many people who are constantly embroiled in chaotic scenarios or who seem to live from one crisis to the next, also happen to be among the most charismatic personalities out there. This is usually because such people tend to be outgoing and extroverted, making them naturally more attractive and charming to those who suffer from a lack of self-confidence and low esteem. Such individuals are essentially drawn to the exciting energy and charisma of their drama-prone counterparts despite their underlying issues.

The drama queen or king often has an outsized influence on their surrounding environment, deliberately exaggerating situations and inciting reactions in otherwise calm people. This can be a result of a variety of clinical issues, generally resulting in long-term difficulties in relationships of all kinds. The person may be highly charismatic but the stress and frustration of their hyperdramatic behaviour quickly become unmanageable for them and for those close to them, leading to further anxiety and tension.

The effects of this type of behaviour can be wide-reaching. Along with creating a sense of unease at the societal level, with more people on guard and less friendly, this toxic state of mind has been perpetuated and exacerbated by the pervasive nature of the media in today's world. The fear and worry that is steadfastly fed to people by the media, a fear of the ever-present and impending threat, only serves to heighten the tensions and add to the fire of this chaotic environment. This has the potential to entrap those living and struggling with mental health issues, leading to a vicious cycle of anxiety and stress.

In this day and age, many of us have noticed that people have become less familiar, less patient, and less cordial than they used to be. Unfortunately, social imbalance is not just observed in humans, but also in wild chimpanzees. Studies have found that when a few group members become unsteady, cortisol levels increased for the whole group. This could lead to aggression, leading to members invading other groups, killing weaker members, and taking over new areas in the forest.

The emotional instability of drama creates serious health problems. The body stresses so much that excess adrenaline and cortisol is produced, and as a result, blood pressure, sleep, and eating are disturbed. People who often try and ignore the real issues they have in life become more prone to depression. This could be anything from mild sadness to more severe clinical depression, which leads to further brain damage. MRI scans have not only shown evidence of such damages, but also shrinkage in the hippocampus area of the brain.

Clearly, dealing with such imbalances will have more than just physical repercussions, but mental ones as well. We cannot ignore the connection between social interaction and physical and mental health. If we want to build strong, healthy relationships, understanding and patience is key.

If your friend or family member is suffering from seemingly endless drama and constant crises in their lives and these issues are starting to strain your relationship with them, it can be helpful to sit back and reflect on the source of their behaviour. It may be that they are in desperate need of assistance and are not getting the help they need. With the constant barrage of news stories, fear-mongering alerts, and negative energy fed to us daily through social media and other outlets, it might be time to take a step back and remove yourself from the bigger issues of the world.

Some people can perfectly handle the barrage of international news, understanding the complexities of various issues, and still retain a healthy level of distance in order to lead their own life relatively unaffected. On the other hand, there are some people who become unable to step away from their involvement with world issues and allow it to take over their lives, as well as those around them. We all know that politics are intertwined with many of these issues, and it can seem that the endless task of trying to tackle them is insurmountable, never ending, and overwhelming. The key is to find the balance between being informed and socially aware, while not letting it overwhelm our day to day lives.

Taking a break from the hustle and bustle of everyday life can be an incredibly beneficial act to take part in for our mental wellbeing. At times, it can feel like the pressure to stay on top of the seemingly never-ending list of the world's problems will become overwhelming, leaving us feeling dissatisfied and unable to make any real change. Yet, by allowing ourselves time to simply 'be' without the feeling of having to fix and mend everything, we can go about addressing the broken areas in our own lives.

For example, in the relationships we have with those closest to us, taking a break from the chaos of this world from time to time can give us the much-needed space to realign and rebuild any fractures that may have occurred due to external pressure or stressors. This can include aspects like apologising for any judgments we may have carelessly passed and strengthening the understanding that we may have with our family and friends and indeed, even ourselves.

All in all, taking a break from feeling that we need to fix the world's issues, however huge they may seem, can free up space in our lives to focus more on the little things in life that are just as important. This mindful approach can bring a fresh perspective to the situations in our own lives, that we may not have noticed before.

It has become increasingly clear that our mental health is taking a toll due to social media use. With the rise of virtual connections, it is becoming increasingly more difficult to effectively give support to those in our lives who truly need it. To counteract the effects of social media, it is important to create space within our communities to periodically disconnect from technology and allow for the exploration of more meaningful connections. This could look like setting aside a few hours of the day to only spend time with your physical, face-to-face relations or engaging in activities that move away from the hustle and bustle of a digital lifestyle such as camping trips, hikes or even painting. When it comes to helping our friends in times of stress, it is important to remember that underlying the situation, there is often a need for companionship and support. Instead of jumping to conclusions on the surface, it is helpful to be mindful and proactive in addressing their needs. This could look like simply lending an ear to listen or providing resources for professional help. Taking these small steps can have an incredibly positive effect on someone who is feeling overwhelmed and can even serve to prevent more drastic outcomes. After all, no one wants a repeat of the horror of the Rwanda genocide, where deep-seated social issues such as lack of trust and communication were ignored and ultimately resulted in one of our darkest chapters in human history.

Follow The Covid Money

It appears that once again, the media and the government have caused a stir in the public domain, this time testing students in school to an unprecedented degree for reasons that don't make an astounding amount of sense. With the reliability of the COVID tests perpetually in question and their value as a tool sometimes debatable, the consequential rise in false-positive readings would create a sense of dread and paranoia in a population yet to adequately prepare themselves for the eventuality of this virus.

This lack of public preparation and readiness for the virus can be so evidently seen in the closure of schools, business and other social functions. With more people misinformed on matters concerning the virus than ever before, they are prone to behave in dubious and sometimes dangerous ways. To put it simply, the occurrence of false positives as a result of over testing will mean that the virus will be further opened up to manipulation and abuse.

It is incumbent upon governments and institutions to provide the tools and information necessary to the public to make informed decisions. Until appropriate and sensible action is taken, the credibility of our social constructs and the implications of this testing will remain in limbo, leaving us even more vulnerable.

The chart provided gives a concrete representation of the delineations I have outlined. My utilization of established, trusted formulas allows us to understand the underlying processes in minute detail. Our calculations are based on factual and historical evidence, giving us assurance that the results are authoritative and precise. By evaluating the data, we are able to come to an accurate qualifying statement of the topic at hand.

For example, in financial markets, the use of these well-defined formulas allows investors to assess whether an equity is a good buy or a bad one. This is based on the stock's price history and other factors such as its dividend yield. Investors are then able to make informed decisions on the stock's performance in the future. This is one example of how the measurement of tangible facts and data can give us concrete information that we can confidently use to predict the future with a high degree of accuracy.

Our CNN and CBC viewers represent a vast community of people who have come to rely on our trusted sources for reliable information. Unfortunately, recent incidents have highlighted the potential for our children to be put at risk by improper handling and transport of vaccines, which can potentially result in spoiled doses. To illustrate what can happen, we can look to the following chart, which shows the results of a well-established, widely used formula based on factual and historical data. This formula can be used to accurately calculate the extent to which the correct transport and handling of vaccines can be critically compromised if not done within the prescribed regulation and guidelines. It is evidence such as this that underlines the importance of taking careful steps to ensure the integrity of vaccine transit and storage, as any failure to uphold the necessary standards can lead to dire consequences.

We can accurately qualify what is taking place as our educational institutions are trying to initiate a swift vaccination campaign which on the surface appears to be for the safety and wellbeing of our children, but if we were to look closer, we could see that it strays dangerously close to an act of child abuse. The way in which it is being marketed is quite questionable: our leaders use 'number fudgery', or the manipulation of figures, to scramble parents and provide the impression of an imminent threat, essentially fabricating an unease that induces parents to consent to the vaccine. Sadly, this practice has become common in our modern-day society, with numerous marketing campaigns creating a false sense of urgency to make consumers choose their products over their competition. However, this type of manipulation when it comes to our children's health and safety is taking it too far, and has to be addressed as such.

Recent data released by authorities regarding the number of students infected with a virus has come under scrutiny, with many calling the figure an 'alarming number.' Let us take a closer look at the evidence, so we can properly assess the gravity of the situation.

To begin with, the figures available to us point to a higher than average rate of infections among school aged children. In many cases, the rate of spread is higher among students than it is within the general population. This can be evidenced in places such as New York City, where at one point in time, nearly 40% of positive coronavirus tests belonged to children under 18. In some places, the number of cases among school-age children is so high that some districts have had to shut down or at least send their students home.

Given these facts, it is understandable why a large number of parents are raising an alarm about the current situation. Furthermore, it is also not surprising that medical professionals, who are witnessing the effects of this outbreak firsthand, are growing increasingly concerned about the virus' affect on younger persons.

Therefore, it is clear that the “alarming number” of cases that is being reported is a legitimate cause for concern, and further calls on the public, the educational system and health care providers to take stronger steps to prevent the virus' spread within the student population.

At the time of compiling this report, the number of cases stood at 6343. It is important to consider the size of the population within the scope of this investigation. According to the data available on Ontario’s official website, approximately 2 million children are currently enrolled in schools across the region. This suggests that the number of cases, while still significant, is relatively small in comparison to the total number of students involved.

In an effort to further contextualize the data, it is important to analyze the progression of the reported cases. From April to June of 2020, the number of reported cases has consistently increased, with a total of 5,455 cases reported by early June. Consequently, this provides an indication that the reported cases relate to the larger population of students, and the current ongoing nature of the situation.

It is essential to consider the steps being taken to mitigate the number of reported cases. Reported efforts by the Ontario government have focused primarily on the implementation of targeted restrictions, such as limited attendance at school and the compulsory wearing of masks. While these measures are yet to be proven effective, they provide an indication of the strategies being implemented to contain the reported cases and prevent the issue from escalating further.

In conclusion, the number of reported cases and the proportion of the student population they represent are both salient components in understanding and addressing the issue. As such, these considerations must be taken into account when assessing the impacts of the reported situation and determining the best methods for containing the reported cases.

When it comes to the current health situation of our children and the potential risks of them contracting COVID-19, hospitalizing, and even death, it is essential to profile the data available to date. With over eighteen months of evidence pointing in the same direction, it is possible to predict and determine a statistically accurate estimate of risk. When the numbers associated with these risks are assessed, the health issues stemming from the virus appear to be significantly less concerning than what's being reported by the government, health officials, and the media. Inaccuracy and exaggeration seem to be the underlying theme as they're pushing their own agenda, using relative risks to overinflate facts and figures to create an illusion of a full-blown crisis. Unfortunately, their accounts are far from the research-backed information, and any such gaps have the potential to create confusion and manipulate the public's opinion in an otherwise preventable situation.

The dangers of COVID-19 with regards to children cannot be overstated as the virus includes certain biological factors that make it exceptionally difficult for them to become infected. Knowing this, it is difficult to determine whether the risks outweigh the potential benefits of administering the still-experimental mRNA COVID-19 vaccine to children.

Children have proved to be notably less vulnerable to the novel coronavirus than adults. Studies of pediatric cohorts have revealed that even the most severely affected children often show a comparatively milder clinical picture than found in adults. Furthermore, adults are found to be far more likely to require hospitalization, or even unfortunately succumb to the virus, than children.

What's more, there are other ways to protect potentially vulnerable children from any risks of COVID-19 without utilization of an experimental vaccine. Children can be shielded from outside exposure through the implementation of safety precautions such as social distancing, masks and limiting exposure to large gatherings. Moreover, the frequent use of disinfectants, hygiene, and other prophylactic measures can protect children even further.

Therefore, the risks of administering the still-experimental mRNA vaccine to children may not be worth the potential benefits. Much research is still needed to fully understand the potential of this experimental vaccine, and its efficacy could determine whether or not its use in children is a viable option going forward. Until then, it may be wiser to err on the side of caution and keep children safe with other proven substitutes.

The question of why schools are encouraging students to get vaccinated for COVID is an important one. It's worth looking at the motivations from the school's point of view. Many schools have received substantial funding as part of the COVID relief effort, with some receiving millions of dollars of money. This funding can be used for myriad school projects. One obvious use is to upgrade HVAC systems, providing fresher air and potentially preventative measures against the spread of viruses. With upgraded Wi-Fi and other technologies, the school experience can be improved in a variety of ways. Additional funding may be used to upgrade the teachers' lounge, utilities like power and water, and other necessary expenses that were not previously available. It is clear why schools are pushing for vaccinations: they are looking to benefit from the money they receive. Whether these benefits are being used for necessary or even justifiable reasons is a discussion for another day. But it is nevertheless clear that funding from the relief effort is incentivising schools to make sure all attending children are vaccinated so that they can access this money and make the improvements that could not have been made prior.

Considering the current climate, investing in a new 85-inch plasma television to show essential educational information about the impressive gains made by popular and reliable vaccines would be an excellent purchase. Visitors to any educational facility, such as schools and universities, should be provided proper information on the efficacy and safety of these life-saving products so that they can make well-informed decisions. With this kind of advanced teaching aid, schools and universities would be able to demonstrate the effects that these vaccines have had on reducing the spread of illnesses across the globe, thoughtfully illustrating this information to their students.

Ensuring that our children understand the science behind the positive impacts of vaccines will help them make smart decisions about their healthy in the future. Once students have been exposed to this type of educational content, seen via the 85-inch plasma television, they may become more aware of the health-care industry, potentially inspiring some students to pursue a career in pharmaceutics.

Once the educational benefits of this television have been fully exhausted, though, the classroom would have the perfect setting for a last-minute viewing of the big game! The new 85-inch plasma television could easily be relocated to the teacher's lounge, to provide a communal area in which faculty can come together to cheer on their favourite sporting teams.

Ontario schools have a wealth of COVID support, with the government providing more than $500 million in COVID cash. This includes $383.6 million to help school boards with staffing and navigating a safe school year, which includes $304 million from the Provincial Priorities Fund and $79.6 million through the Language Grant in the Grants for Student Needs. There is also an additional $20 million to re-engage students and promote reading assessment, as well as $34.9 million for technology funding that includes an annual $14.9 million to support devices for students and $20 million in connectivity supports. Schools can also take advantage of the up to $508 million to access two percent of reserves to cover COVID-related expenses.

The funding is also helping to cover pandemic-related infrastructure upgrades to ensure Ontario schools are safe and secure. This includes improvements to ventilation, installations of hand sanitizer stations, as well as investing in broadband infrastructure to support remote learning. This will allow students to access educational content from the comfort and safety of their own homes. With this increased support, Ontario schools will be well prepared to offer a safe learning environment for students throughout the upcoming school year.

The Honourable Patty Hajdu, our Minister of Health, remarked that vaccine distribution is a key factor if we want to put an end to this pandemic, and young people have a crucial role to play in achieving that goal. By providing funding through the Immunization Partnership Fund, we are getting closer to this goal and bringing people safe and effective vaccinations.

The program going a long way, by involving the youth in promoting, tracking and supporting those in need of vaccines, in summary, it has enabled us to reach more people, especially the most vulnerable members of our Society. For instance, it has allowed us to provide vaccines to elderly people who are unable to go to a regular doctor's office, support Indigenous communities, and many more.

What is more, this program has had a ripple effect in our communities. By making sure vaccines are widely and easily accessible, it has encouraged more people to get vaccinated which has resulted in a decrease of hospitalizations. This has also made it possible for people to safely enjoy activities outside their homes, help the economy bounce back, contribute to the mental health of society, and help children go back to school.

In conclusion, the Immunization Partnership Fund has gone a long way in helping Canada is getting out of this pandemic, thanks to the youths' tireless efforts to ensure that everyone gets the access to the right vaccinations. It is hard to overstate the impact this program has in our lives and we should all be thankful to the Minister of Health and our dedicated young people.

The Pfizer vaccine is a remarkable medical breakthrough and its ability to prevent people from catching the COVID-19 virus is well-established. But many are still questioning how this vaccine, which provides temporary immunity, is able to result in a form of immunization. To truly understand how this works, we need to look at the concept of immunity in more detail.

Immunity is our body's ability to resist disease. It is created through exposure to a virus or other antigen, which stimulates the production of antibodies that enable the body to fight off any subsequent encounter with the same virus or antigen. Vaccines work by exposing your body to a harmless version of a virus or antigen, which stimulates antibody production and helps create a state of natural immunity. This immunity protects us from disease even when we have not otherwise been exposed to the virus.

The Pfizer vaccine is a great example of this process in action. It contains a neutralized version of the coronavirus, which prompts the body to produce antibodies. As a result, those who receive the vaccine are able to build up their immunity to the virus, effectively creating a form of immunization. Evidence of this can be seen in the data that shows vaccination recipients having a reduced risk of catching the virus, compared to those who were not vaccinated.

Therefore, while Pfizer's vaccine does not guarantee absolute immunity to the COVID-19 virus, it is a crucial step in helping to create a state of immunization and reducing risk of infection.

With the world currently facing the unprecedented challenges brought about by the global Covid-19 pandemic, it has become increasingly important for the public to have access to accurate and comprehensive information about the newly developed Covid-19 vaccines. To this end, many organizations and institutions are developing tailored strategies specifically designed to promote evidence-based, trustworthy and reliable information about the Covid-19 vaccines to three important audiences; youth, their parent and caretakers, and teachers.

For instance, public health officials in numerous countries are creating educational material tailored to each age group, providing scientifically-backed information on the Covid-19 vaccine in a manner that is both easily understandable and appealing to young people. This can include videos and interactive resources, such as comic book-style illustrations which illustrate how the vaccines work in a way that is both factual and engaging. Furthermore, to ensure that the information is reaching as wide an audience as possible, many authorities are also sharing this information through online channels, such as social media and websites.

It is also essential to reach out to parents, guardians and caretakers, so as to ensure that they have the information they require to make informed decisions about the vaccinations for both themselves and their children. To this end, organizations and institutions are using various strategies, such as hosting Q&A sessions with experts, to provide parents and caretakers with information and advice on the safety and efficacy of the Covid-19 vaccines so that they can make informed choices.

Finally, organizations are also working to ensure that teachers have access to the most up-to-date information about the Covid-19 vaccines, so that they can accurately inform students about the risks and benefits of the vaccine. This can include providing teachers with detailed presentations, hosting webinars and other online seminars with medical personnel, and providing teachers and educators with accurate and verified resources on the vaccine to further support classrooms and students.

Overall, with the help of tailored strategies to reach out to each of these core audiences, organizations and institutions are taking an important step in helping to spread credible information about the Covid-19 vaccine to the public, this helping to inform and empower people to make informed decisions about their health and safety.

Once again, the Minister of Health must provide concrete evidence that the supposedly-effective vaccine has had a positive impact, and that it is not merely a futile endeavour. From careful monitoring of the figures concerned, it is evident that the success rate of the vaccine to date has been unattainably low – below 40%, in fact. It is also important to realise that the shelf life of the booster shots is already lower than that of the most recent variants of the coronavirus, implying that once again, there will be a disappointing failure in comparison to earlier iterations. Widespread testing, particularly in countries where the virus is now most prevalent, is necessary in order to provide solid evidence of its efficacy and adequate protection against the virus. It is evident that governmental figures and institutes must provide the facts and figures with transparent standards, rather than leaving researchers and the public to speculate in vain. Vaccines are the most efficient defence against the virus, and should remain the primary source of protection on a global scale. Collaborative action should be taken by health organisations across the world to ensure the latest variants and mutations of the virus are adequately understood and effectively combatted. After all, the key to beating Covid-19 is clear: a safe, efficient and widely-adopted vaccine – let's not play WACKAMOL with it.

She continues her message, emphasizing the importance of getting vaccinated against COVID-19, particularly for young people. "Now more than ever, it is critically important that our younger generations get vaccinated in order to protect themselves and their loved ones from the virus. We have seen the devastating effects that the virus can have on the vulnerable, and it is essential that this new generation of adults is educated to arm themselves with the necessary prevention tools to protect from the virus. Getting vaccinated is the single most powerful thing each and every one of us can do to help stop the spread of COVID-19. If you have yet to get vaccinated, I urge you to do so today and help us to create a healthier and safer environment for us all."

At a time when Ontario's statistics are increasingly suggesting that vaccine administration to our children may not be as safe or beneficial as previously thought, more and more worrying reports of unprecedented heart problems in young and previously healthy individuals have been coming in. To make matters even more concerning, there is a burgeoning demand for automated defibrillators to be installed in schools—with the introduction of Pfizer's revolutionary new blood thinner specifically designed to combat blood clots in children, it prompts the question: what exactly are we preparing for?

It is understandable that many parents and guardians are concerned about the long-term implications of vaccine intake during this period of hesitation, especially in light of the advances in health sciences that allow for the early detection and further understanding of any potential side effects or complications. It is for this reason that, for many, the promise of protection offered by vaccines does not adequately offset the ever-growing list of potential risks. In a study conducted by the University of Calgary for example, over 200 children enrolled in a trial to determine the efficacy of the HPV vaccine were discovered to have developed long-term adverse health effects, many of which were deemed severe enough to require prolonged medical treatment and care.

Ultimately, while vaccinating our children may still remain a safe and advisable course of action overall, it is only natural for many to hesitate and have reservations concerning the potential long-term consequences of such an act. Furthermore, with the assurance of appropriate regulatory oversight and a heightened awareness of the possible side effects, it increasingly becomes essential for each individual to make a well-informed, educated decision for themselves or for their children.

FLAWED REPORTING BY THE MEDIA

Recently, it was reported in one of Ontario's most esteemed newspapers that a particular factor has been on the rise in terms of numbers. While at face value this might seem concerning, closer examination reveals that the reporting was potentially sensationalized in order to elicit fear in the general public. As a responsible journalist, one should always be aware of which data elements and indicators are utilized when reporting figures, as the wrong selection can easily alter the story to appear more ominous than it actually is.

Take for example a study of maternal mortality rates that was recently published. The headline of the story proclaimed a "dramatic increase" in deaths, but no one outside of the media took a moment to contextualize the findings. After thorough analysis, it was revealed that the increase in mortality numbers was actually far less significant than originally reported, due to the fact that the journalist had selected the wrong metrics when reporting on the topic. If they had used data that was representative of the entire population, rather than focusing on one subgroup, the victory would have been much more accurate and not so alarming.

It is important that journalists recognize their power and use it ethically. When composing stories, accuracy and careful selection of data pints should be upheld in order to combat misinformed public discourse. Media sources have a duty to present the facts as they are, without relying on sensationalized language to get readership. Oftentimes, the dissemination of misleading information can create confusion and inaccurate impressions, which can be damaging to communities around the world.

All in all, it is essential that journalists practice integrity and thoughtfulness when it comes to reporting on quantitative data. Otherwise, the story simply becomes a form of news that is more "clickbait" than informational, and that spreads fear rather than knowledge.

Alberta's hospitals are struggling to cope with the barrage of COVID-19 patients, and the province's Head of Health Services, Dr. Deena Hinshaw, has laid the grim reality on the line: the sole reason that Intensive Care Units (ICUs) are not completely full is that the death toll is steadily rising. On Wednesday, Alberta tragically logged an all-time high of 34 deaths from COVID-19 in one 24-hour period – or one death caused by the virus every 42 minutes. It is only through the losses of human life that hospital beds remain available.

This unacceptable situation has developed due to the pressure put on health care services by the sheer number of new coronavirus cases contracted; hospitals across the province are simply unable to keep up. In Alberta, some family doctors are reporting waitlists of up to a month and a half for non-en emergency patients, and the province has been outsourcing acute patient care to Calgary and Edmonton hospitals since the beginning of the pandemic.

Sadly, the provincial government's priority has been necessary, but not well-received restrictions on the lives of Albertans. Actions like the confinement of private gatherings and the suspension of voluntary sports activities have caused public outcry, as citizens demand to know why it has come to this. Alberta is a province ripe with resources and professionals, and the inability to cope with the pandemic shows that a lot can go wrong despite the highest of intentions.

As the coronavirus continues to wreak havoc with Alberta's health care system, we must continue to follow public health guidelines and do all we can to lift the burden from our hospitals. The province needs to properly mobilize resources and support its overburdened health care workers to ensure that we don't see anymore of Alberta's beloved citizens perishing from COVID-19 so that ICUs can remain open for those in need.

42 is an intriguing number to examine. It has come to fruition by taking into account two seemingly unrelated metrics - minutes in a day and number of deaths. To elaborate, if we divide the number of minutes in a day (1440) by the number of deaths in a certain timeframe, we end up with the number 42. To put into perspective what we can learn from this number, let's look at an example.

If one were to consider the number of people who perished in the Holocaust and divide that by the 1440 minutes in a day, one would arrive at the figure of 42. This figure, while it obviously cannot represent the enormity of human tragedy, serves to aptly put into perspective the staggering amount of lives affected by the Holocaust.

But these calculations can also be used to put into perspective the goals of the United Nations' Sustainable Development Goals which would be to reduce global mortality. If we take the number of global deaths per day and divide it by the 1440 minutes in a day, the figure of 42 surfaces. This serves to remind us that regardless of landscape or nationality, everyone has access to the same 1440 minutes in a day, minutes which must be used to strive towards making our world a healthier, safer and more sustainable one for everyone.

Whether for the mindful contemplation of the success or failure of past epochs, or for the aspiration of a brighter future, the number 42 is a powerful way to put into perspective the fragility of life and the responsibility we bear towards ensuring collective well-being.

It does not take a genius to figure out this is so seriously flawed.

When we consider the figure of 42 concerning deaths associated with the novel Coronavirus, we should keep in mind that this number does not present a complete, accurate picture of the situation. The province of Alberta has deemed that deaths wherein the deceased was afflicted with three or more pre-existing health issues should not be counted in the tallies. In order to arrive at an accurate representation of COVID-related fatalities, it is necessary to restrict the reported deaths to those in which the virus was deemed the sole cause of death.

To illustrate this further, we can consider the example of an elderly individual who may have suffered from a chronic lung illness coupled with a heart condition. If this person were to contract the coronavirus and pass away, it is likely these two preexisting health conditions played a role in the outcome of death. In this case, their death would no longer be attributed to just one factor, the virus, but multiple and thus it would be excluded from the official death toll.

By taking into account the number of COVID related' deaths that are reported and ensuring that only those individuals who have perished as a direct result of the virus are included, we can arrive at a more accurate, representative figure. Such data can help health officials more precisely evaluate the current threat posed by the virus and act accordingly.

When discussing the risk of death related to coronavirus, it is important to consider the statistics surrounding it. For the purpose of this discussion, we will use the number 42 as a general reference; a more accurate number will be presented later. Taking that number as our starting point, let's look at the population of Alberta, which is estimated to be roughly 4.4 million people. If we divide 42 by 4.4 million, the percentage of the population at risk of dying from the virus is .0008%; this is approximately the same odds of dying from riding a bicycle. By applying this calculation to an entire lifetime, the odds of dying from COVID-19 come to one in 290. This is drastically different from the daily lottery odds that we hear about; these statistics reflect the chances of passing away from coronavirus over one's entire life.

Although the odds of dying in a car crash are alarmingly high, let's take a look at some other life-long odds of dying. According to the Centers for Disease Control and Prevention, the life-long odds of dying from a fall are 1 in 106. That means for every 106 people who live long enough to reach the age of 75, one of them will die from a fall. This statistic is particularly concerning because accidental falls are the leading cause of unintentional injury and death in the United States. In stark contrast, a person's life-long odds of dying from a gunshot wound is 1 in 289. It may seem like a small number, but this number is still concerning when you consider that gunshot wounds continue to be a leading cause of injury in the U.S. In addition, the American Public Health Association has stated that adolescents, particularly young African American children, are particularly vulnerable to gun violence. With these sobering statistics in mind, it is essential to take measures to lower the risk of accidental falls and gun violence.

With the more accurate death count due to the COVID-19 pandemic, the odds of dying entirely from coronavirus on any given day is incredibly small - standing at a whopping 0.00004 percent. If we take into account a lifetime's chance of passing away as a result of this virus, that comes to around 1 in 6,000. To put that into context, you are three times more likely to die following a bicycle accident (1 in 3,800) or as a result of heatstroke (1 in 6,500), as opposed to coronavirus-related death which stands at 1 in 290. In fact, when you compare the death rate of car accidents (1 in 112), heart disease (1 in 6) and cancer (1 in 7) to COVID, the mortality rate from which stands at 1 in 290, it becomes very clear that the mortality associated with the virus is incredibly low.

All of this goes to show that, while COVID-19 is a very real threat to our global population, the actual risk of dying as a result of it is much lower than many believe. We must remember that although there has been extensive and rightly necessary media coverage of this virus, it is not the crisis that some portray it to be. Far too often, scaremongering rhetoric has been used to sensationalise the issue. This can not only lead to unnecessary public panic and unrest, but can also be hugely damaging to the reputations and livelihoods of those on the front-lines, doing their best to contain the spread of the virus.

If you don't take the time to carefully analyze and interpret data, you can end up with inaccurate and misleading conclusions. This can lead to headlines or stories that are distorted or even completely false. Consider, for example, the headline that recently declared that 'one death every minute' is due to COVID-19. If we had examined the facts with more caution, we would have seen that this didn't take into account other variables like the day on which the figures were gathered or how the figures are actually calculated. It's an extreme example of just how wrong the interpretation of data can be.

In reality, there's a good chance you're at a much lower risk of dying from COVID-19 than from other causes. Cycling, for instance, is the most dangerous form of transport in most places. In the US, you are five times more likely to be killed while biking than to be killed in a car accident. That's why it's so important to observe data responsibly and think critically. The data is there, but it's up to each of us to make sure we interpret it responsibly and accurately.

Every day, we take risks in a multitude of ways without thinking twice about it. For example, when our children ride their bicycles, we accept the potential risks because the benefits of exercise far outweigh the potential consequences. We should take this philosophy when it comes to the pandemic and lifted the lockdown restrictions, allowing us to return to life as it was before. This would be beneficial to everyone, as the associated benefits of having a robust economy and access to greater social activities far outweighs the potential health risks of the virus.

It is important to remember that fear is a powerful feeling, and there are many people out there who deliberately seek to prey on fears in order to make headlines. Don't be fooled by these sensational reports and instead look for reliable sources before basing your opinion off of clickbait or false information.

DID WE RUSH THE VACCINE?

When a pharmaceutical product is introduced onto the market, it goes through an exhaustive and laborious series of examinations and trials. Most drugs that experience pre-clinical testing never even reach the point of being submitted to the FDA - U.S. Food and Drug Administration – for human trials. Those that do are subject to the most stringent evaluation by the FDA, analysing all aspects of the drug – from the clinical trial design to the severity of side effects, as well as the type of environment in which it is produced. This rigorous testing process is put in place to ensure that the drug will be completely safe and viable for use when it is made available to the public. As an example of how closely the FDA scrutinizes drug safety information, one recent study found that nearly all commercials screened for drug-Safety warnings were rejected for being too brief. This highlights the thoroughness with which the FDA evaluate the safety of pharmaceuticals and illustrates the importance of its long testing process for new drugs. As a result, the public can have confidence that the drugs that are available are held to the highest standards and are completely safe for consumption.

It is often discussed in the pharmaceutical industry that the Federal Drug Administration (FDA) must review the results of preclinical testing in laboratory animals in order to bring a drug to market. Preclinical trials are typically carried out on cell cultures or organisms that don't include humans, and the results of these tests are used to evaluate whether a drug is safe enough to be tested on humans. At this point, the FDA will decide whether the company can move forward and commence human trials.

For example, in 2002, the FDA designated a cancer drug called Gleevec as a 'breakthrough therapy', meaning that they believed it had potential therapeutic benefits that could provide a meaningful improvement over existing treatments. After reviewing the results of preclinical tests, the FDA allowed Gleevec to undergo further testing in humans, which eventually led to it being approved by the FDA in 2003, becoming the first drug to get approved for a form of chronic myeloid leukemia.

Thus, it is evident that preclinical testing plays an important role in the process of bringing a drug to market, as it provides valuable insight into the safety and efficacy of a drug, allowing the FDA to come to an informed decision on whether it is safe for human testing.

When a new drug is introduced to the market, a rigorous testing process is required by both local and federal regulations, in order to ensure the safety of users and compliance with the law. This testing process typically consists of five distinct phases, but for the purposes of this discussion we will focus on the three initial phases.

The first phase, Preclinical Testing, involves testing the drug in a laboratory setting on drug safety and efficacy using animal models. This is an essential first step because it offers important insight into the drug's effects and helps to identify potential issues early on. During this phase, researchers usually examine toxicity and examine the drug's effectiveness by administering to a range of dosages and modes of administration.

The second phase, Clinical Testing, involves testing the drug on willing human subjects. This follows the preclinical tests, and is designed to monitor the safety of the drug over a period of time and collect data on how well it is tolerated by humans. But importantly, it should also demonstrate the drug's effectiveness as compared to placebo and existing available drugs. Clinically testing a drug requires carefully monitored trials which may involve thousands of volunteers and cost millions of dollars to complete.

The third and final phase that is relevant for this discussion is Regulatory Compliances and Submissions phase. Here, the drug's manufacturing and distribution process is reviewed to ensure it meets all regulations and complies with the relevant laws. The paperwork for regulatory approval must be submitted, and if the drug is approved, the manufacturer is then allowed to market and sell the drug.

All in all, these three testing phases are critical steps that must be accomplished when bringing a new drug to market. Without passing through these three stages, the drug would not be allowed to be available to consumers and thus would pose a great risk to public safety.

Phase 1 clinical trials are typically first conducted in healthy individuals as a way to assess the safety of a new pharmaceutical product. The purpose of this research is to uncover the most common side effects of the drug, gain insight into how it is absorbed, metabolized, and then excreted by the body. The sample size for these studies may range from twenty to eighty participants.

For example, a new anti-inflammatory drug might be administered to healthy volunteers in order to detect any potential side effects and to document its absorption and excretion profile. Or, a new cholesterol medication may be administered to healthy volunteers in order to predict its safety profile, understand its metabolism and determine the most efficient dosage for use in humans.

Ultimately, this initial phase of testing allows researchers to better understand and identify any potential risks associated with the particular pharmaceutical product. Armed with this information, more detailed testing can be conducted in subsequent stages of the drug development process.

If the Phase 1 studies did not reveal any toxicities that are deemed to be unacceptable, then Phase 2 studies begin. In Phase 1, the main focus is on safety whereas with Phase 2, the emphasis is on how effective the drug is in treating individuals suffering from a specific disease or condition. To obtain information on the efficacy of the drug, some clinical trials during this stage may be conducted. For example, the team at Pfizer may look at how the injected compound travels in the body and which tissues or organs it accumulates in. This is highly important data that can provide insight into how the drug performs in the body.

Once the evidence for effectiveness has been identified in Phase 2, Phase 3 studies begin in order to gather more information about safety and effectiveness. It is essential to have a better understanding on diverse groups, different dosage levels, as well as how the drug interacts with other drugs. Such data is essential to validate the effectiveness of the drug. To start, the number of participants usually ranges from several hundred to approximately 3,000 people.

These Phase 3 studies involve an extensive amount of biomedical data capturing. Moreover, there is an array of test formats that vary from study to study, along with an adequate form of statistical analyses to study the efficacy of the particular drug in question. As a part of the protocol, several measures of safety and efficacy of the drug must be taken into account. Participants may be randomly assigned to either receive the drug or a placebo. Subjects are then monitored over a period of time to determine the effectiveness of the drug. Such measures may involve screening for potential side effects and real-world usage of the drug, as well as its impact on mortality rates.

The data collected from Phase 3 studies provides medical professionals with a comprehensive evaluation on the drug's efficacy and safety in order to provide the best possible care to patients. It is because of such studies that groundbreaking drugs are developed to prevent and treat serious diseases, and ultimately save lives.

The development of any medical treatment requires a rigorous series of testing and evaluation at each clinical phase, a crucial set of protocols which are designed to verify the efficacy and safety of the drug or vaccine. During a typical clinical phase it is common for there to be double-blind studies, peer reviewing, and murder-boards, all necessary steps that reduce the likelihood of confirmation bias and ensure that any perceived short-term benefits are not outweighed by the long-term risks.

When it came to the development of treatments for Covid-19, however, these required prerequisites were for the most part bypassed, leading many to suggest that this might have been a contributing factor to the alarming number of deaths and injuries "associated" with the Covid-19 "vaccines." According the Centers for Disease Control and Prevention (CDCP), there have been over 7,000 reported cases of fatalities resulting from the vaccines, with this number likely being only a fraction of the total number of deaths. This lack of proper clinical testing is no doubt a cause for grave concern and must be addressed as soon as possible.

The implications of the findings concerning Pfizer's COVID-19 vaccine development process are clear: they sought to expedite the entire process due to increased pressure from the pandemic itself, as well as the United States government's 'Operation Warp Speed' initiative. The objective of this initiative was to get the vaccine to market quickly, under the United States' 'Emergency Use' Provision.

Given the severity of the virus, Pfizer sought to bring their vaccine to the public as quickly as possible. To do so, they committed hundreds of millions to the project; not only to develop their vaccine, but also to initiate the global search for participants in clinical trials and start production of the vaccine before receiving emergency use authorization from the FDA.

The company's ambition to efficiently develop their vaccine was indeed laudable. However, the risk associated with forgoing the usual, slower but safer protocols was understandable. Nonetheless, it is a testament to the skill and dedication of their scientists, as well as those of other vaccine-development companies, that the pandemic was not even more devastating.

Moving forward, it will be important to determine how such lessons learned during the pandemic can be applied in an effort to develop safe, reliable, and effective vaccines in a timelier fashion in the future.

The challenge is that following strict protocols such as Good Laboratory Practices (GLP) is essential to ensure quality control and patient safety. Without following the proper processes, the outcomes can be far-reaching and detrimental, leading to negative consequences that are far from ideal. For example, in medical research and development, if key steps such as GLP are skipped or hastily done, it can put patients at risk due to inaccurate data or unforeseen results. Furthermore, by taking shortcuts, the integrity of a laboratory's operations is compromised and can have a long-term ripple effect on the efficacy of their products. Ultimately, it is critical to adhere to established quality control measures like GLP in order to safeguard the reliability of the results for both patients and researchers.

The number of reports of post-vaccination injuries and deaths in Canada, the United Kingdom, and the United States has been truly alarming, raising concerns about the potential impact of the vaccine when we consider the much larger population of India. Whilst we can only speculate on the scale of the issue in India, these reports should be deeply concerning to us all, and demonstrate why we must demand more transparency and accountability from the organisations distributing the vaccines.

Considering reports of Canada alone, we have been witness to instances of individuals experiencing intense pain, neurological symptoms, and at worst, hospitalisation and even fatalities. On a smaller scale, news of swollen arms and reactions weeks in the aftermath of vaccination, especially some of those aged 50 and over, have been commonplace. Similarly, in the United Kingdom we have heard of people suffering from severe adverse reactions, and even death, post-vaccination. With the United States and its 330 million plus population, the number of reports of post-vaccination suffering have been similarly widespread.

Given the large scale of the Indian population, tales of post-vaccination injuries and deaths, however alarming and tragic, only represent a fraction of a much bigger picture, and one which must demand all of our attention. Only by actively monitoring the roll out of the vaccine can we ensure that those innoculate are protected, not put at greater risk.

Sandra Kweder, M.D., the deputy director of the Office of New Drugs in the CDER, believes it is the clinical trials that require the most extensive amount of time and patience to complete–typically stretching over a few years. That said, the FDA is relatively prompt when it comes to the review time of drugs and the examination of proposed developmental timelines of sponsors. Dr. Kweder explains, "The emphasis on speed in our department is mainly concerning the timeframes of being able to effectively communicate with sponsors while the drugs are being developed. It's a delicate process, one that must be carried out in a way that meets FDA standards and expectations."

The clinical trial process demonstrates the remarkable amount of diligence and scrutiny the FDA applies to medications. The drug development phases include preclinical, clinical, post-marketing/Phase 4 studies and more, each of which requires compliance with the drug standards set by the FDA.

Equally important, Dr. Kweder points out the significance of sponsors connecting with the FDA in a timely fashion, as the FDA holds sponsors to the pre-agreed upon timeline to go forward with their drug. One example of such communication was the FDA-issued Coronavirus Treatment Acceleration Program (CTAP) in the early stages of the COVID-19 pandemic, in which the FDA indicated their goal to expedite work with Sponsors and streamline the review of certain medications.

It is evident that the FDA takes great care in the development of drugs and ensures that smart and timely decisions are made in order to make the process an efficient and effective one.

Clinical trials, which typically take several years to complete, are one of the main contributors to lengthy FDA product approvals, according to Sandra Kweder, M.D., Deputy Director of the Office of New Drugs in the Center for Drug Evaluation and Research (CDER). Efforts to expedite the FDA process are concentrated primarily on meeting timelines for drug review and sponsoring meetings during drug development.

For example, in response to the COVID-19 pandemic, the FDA expedited emergency use authorization for treatments and diagnostics, allowing patients to get potentially life-saving products and services more quickly. In July 2020, the FDA approved the emergency use of remdesivir, the first antiviral medication for the treatment of COVID-19 infections, reducing the usual six-month review time to just 16 days in order to meetings the medical need of critical patients more quickly.

By focusing on review timelines and sponsoring meetings during drug development, FDA clearly understands the value of speeding up its drug review process, especially in a climate of urgent medical care.

One of the key benefits of modern medicine is the extensive testing, evaluation and regulation of drugs and treatments by government agencies like the Food and Drug Administration (FDA). Through rigorous research and analysis, the FDA has been ensuring the safety and efficacy of drugs, allowing healthcare professionals and patients to trust the medications they prescribe, dispense, and take.

To demonstrate the FDA's efficacy at this role, we can look at three examples of drugs that have been approved and recently released onto the market. Let's use the recently released mRNA vaccines as a comparison.

Our first example is Nulibry, a drug that is used to treat cystic fibrosis. According to the FDA's report, this drug has been found to have little to no fatal side effects associated with it. This is drastically lower than the reported rare instances of fatal complications associated with mRNA vaccines.

Our second example is Breyanzi, a drug used to treat large B-cell lymphoma. This drug is reported to have a side effect profile that is largely mild and manageable, with only a low risk of serious adverse reactions like anaphylaxis. These mild side effects contrast sharply with the potential for serious reactions to the mRNA vaccines.

Finally, there is Primeuvo, a medication to treat adults with active, mild-to-moderate community-acquired bacterial pneumonia. The FDA has concluded that this drug can be taken without increased risk of severe complications or death. Clearly, the risk-benefit analysis is significantly different than what is seen with mRNA vaccines.

In summary, the level of safety, efficacy, and regulation imposed by the FDA has been proven by looking at these drugs. Each of these drugs have a significantly lower rate of fatal side effects than what has been seen with the mRNA vaccines, clearly demonstrating the importance of stringent regulatory guidelines.

When it comes to statistics, it's important to be careful with the numbers – and especially with percentages. High-sounding numbers can be used to create an impressive image, more so when raw figures are toned down. For example, a statement like 'less than 0.1 percent of the population has a negative reaction' may initially seem acceptable, but when you take into account the total population, the consequences can be catastrophic. As an example, if the total population is in the billions, 0.1 percent is still over a million people.

No matter how it's presented, numbers can never be taken lightly. People need to be more aware of what percentages actually mean – particularly when it comes to large populations. Misinterpretation of data can lead to confusion and, ultimately, inaccurate conclusions being drawn. In order to ensure that we don't get fooled by statistics, it is essential that we approach figures with increased understanding and scrutiny. After all, accurate interpretation is key to making educated decisions.

The staggering number of reported deaths in the United States, which has climbed to over seven thousand cases, brings into question whether there has been any consideration of a pause, at least temporarily, on the vaccine distribution. The fear, however, is that this will never happen as even a hint of a pause or a reevaluation would be interpreted as an admission of fault. Therefore, it seems only two scenarios can bring a stop to this situation: an awakening of awareness or a strong public revolt, in response to a tremendous amount of collateral damage.

This fear of damage is a very real one, and we have seen the consequences of it in other situations. Take the Flint water crisis, for instance, where poisoned water was the result of a seemingly nonexistent consideration of a pause and reevaluation. Initially, the public servants had not taken the time to consider the implications of a serious problem, only to be faced with disaster once it was too late. In this case, the collective outcry of the population was strong enough to call for governmental support and lament the damage done to their environment, their health, and their reputation.

It may be that a similar public demand and outcry of this magnitude is what is needed in order to prevent any further incidents from occurring. There may come a time when the public has to make an important choice; to either accept risk, or to demand at least a pause and reevaluation of the situation at hand. With such a high cost in potential human life, it seems only wise that such a measure be taken and that the safety of the population be taken into consideration before any further action is taken.

The illustration below presents an estimate of how long it usually takes to bring a new vaccine to the market. It is quite remarkable that the mRNA Corona vaccines have managed to be developed and brought to market in such a relatively short span of time, prompting us to consider what concessions have been made to shorten this duration.

Between Phase 2 and Phase 3 of testing is at the risk of cumulative failures, with a characteristic failure rate of close to 93%. It harnesses the challenge of development of the vaccine, in testing its efficacy. Therefore, in the case of the mRNA COVID vaccines, would they have passed this crucial threshold if all the parameters had been met rigorously? Would this progression have led to subsequent issues related to delivery and doses that would need to be addressed before entering Phase 4 and Phase 5?

It necessitates an exemplary approach when making the transition from Phase 2 to Phase 3, and in considering how much the mRNA Corona vaccines were able to pass through this level of testing successfully, it is clear to see how much of a crucial role it played in reaching market significantly faster than any other vaccine on record. We can in effect look at this as a positive example of how mRNA vaccines can be leveraged for quicker rollout, with examples such as Pfizer-BioNTech Vaccine for COVID-19, the Moderna COVID-19 Vaccine, and the AstraZeneca-Oxford Vaccine leading the charge.

When it comes to the development of the mRNA vaccine, some may say it's been in the works for years, but that's no different than any other medication or product that goes through the process of FDA approval. Companies don't spend a short time creating, testing and then working with the FDA—usually the process takes much longer. Truth be told, the development of a drug or product can take decades, with each step having to be methodically planned and executed. This could be even more true of the mRNA vaccine we've been presented with; because it's something never used before, we have to exercise the utmost caution to guarantee its safety and effectiveness.

For example, Johnson & Johnson recently filed the paperwork with the FDA to approve the use of their Coronavirus vaccine and it took an extensive amount of time and money to get it ready to be tested and approved. After extensive clinical trials, the product was able to undergo a review process that lasts approximately six months. The number of clinical trials and pre-market studies that a product like the mRNA vaccine needs to undergo is significantly more than most other drugs and products, so it's understandable why the development process can take years.

When discussing the vaccines available for COVID-19, it is difficult not to notice how similar the language and rhetoric used by those in the know is. It almost seems as though they are following the same script. At the center of this is the 'gold standard' in testing that was given to the vaccine to ensure its safety and efficacy. But what is not explicitly said, is that this gold standard consists of 5 distinct phases that are the very foundation of the Food and Drug Administration (FDA) approval system to bring drugs to market. Despite the strict guidelines and approval process, there are still those who feel as though the system is outdated, taking much too long and in need of radical reform. A great article exploring the subject in more detail can be found here.

The 5 distinct phases of the FDA approval process are clearly essential for ensuring the safety of consumers. The first phase consists of pre-clinical testing of the drug, whereby the drug is tested in cells, tissues, and lab animals, seeking to uncover any potential risks that could later become hazardous for humans. The second phase involves testing on a small group of healthy volunteers to observe how their bodies react to the drug and to determine the dosing levels that should be used in clinical trials. The third phase is where clinical trials are carried out, testing the drug on humans to assess its safety and effectiveness. The fourth phase consists of obtaining approval from the FDA to bring the drug to market. The fifth and final phase is then collecting post-market data to identify opportunities to further improve the drug's effectiveness and safety.

The COVID-19 vaccines have gone through all these phases, providing the kind of assurance that is necessary to ensure the safety of Australians. However, it is still important to be mindful that the approval process does have its flaws, and that it could certainly benefit from being modernised and expediting the process to get new medicines evaluated, approved, and readily available.

Proving the efficacy of the new vaccine was a key component of the experimentation, so the scientists decided to garner data from the testing results of related vaccines - particularly the SARs vaccine. Unfortunately, it is falsely stated that the SARS vaccine research was never completed because of the virus disappearing. The reality of the situation is that the timeline of the investigations was not long enough to develop an effective vaccine to be used during the SARS epidemic. This underscores the need for long-term, detailed research and the orchestration of an appropriate vaccination campaign to effectively counter the spread of contagious diseases. For example, recently the Pfizer vaccine for the novel coronavirus (COVID-19) was declared 95% effective. It was developed over the course of several months and involved the dedication and hard work of many scientists and health professionals.

The notion that the COVID vaccines were brought to market in less than a year, when all other drugs that the FDA has approved typically require five phases and several years to complete, stands out to me, and I cannot help but marvel at such remarkable efficiency. It is immensely impressive that such a feat could be achieved in a way that does not compromise safety and accuracy. The entire process was completed in record time, and it raises the question of how it was done.

Let us take any other drug as the reference, for example ibuprofen, which is a common non-steroidal anti-inflammatory drug. It had to go through four stages of rigorous clinical testing before it could be approved by the FDA. Each stage of testing consists of experiments, trials, and double-blind tests, all of which have to be completed and documented before the next stage can begin. Usually, this whole process would take anywhere up to 7-10 years, yet it was accomplished in an unprecedented timescale for the COVID vaccines.

What must have happened is that the vaccine producers and medical professionals worked in an exceptionally close partnership, ensuring that the development and testing process moved relentlessly forward without any delays. Their utmost dedication, dedication and drive enabled them to reduce the steps in the standard process and skilfully bypass any kind of bureaucratic hindrance.

The creation of the COVID vaccines in such a short period is thus a testament to how successful human collaboration, dedication, and innovation can expedite a lengthy and complicated task. This gives hope to the idea that such great performance can be replicated in other important fields, serving humanity with miraculous speed and integrity.

Indeed, I can think of no other endeavor that so clearly evidences the potential of global partnership and resourcefulness.

The suspiciously speedy Emergency Use Authorization granted to mRNA vaccines has prompted speculation that many decades' worth of failed experiments in the laboratory have been overlooked. Nowhere has this been more apparent than in Israel, which has injected over two million people with a vaccine under a comprehensive, round-the-clock rollout, even extending to the Sabbath. It is true that this nation is at the forefront of vaccinations, however, the Israeli public was never informed of the potential risks associated with this extraordinary endeavor. The government has left the public in the dark, providing no data whatsoever on the ongoing issues and adverse effects that have surfaced in the wake of the vaccine rollout. These omissions have led to a noticeable surge in coronavirus cases and deaths, even amongst those who were vaccinated, leading the Israeli government to consider imposing full lockdown policies.

When considering the purchase of a vehicle, it is important to ensure that it will be safe on the roads. That is why, when you purchase a car, it is important to ask how long it took to go through its safety testing. The time it takes for a vehicle to be safety-tested can vary significantly, ranging from several weeks to several years. It should go without saying that a vehicle that has gone through rigorous, extensive testing for a longer period of time is more likely to be safer for your family than a vehicle that was quickly and hastily tested.

Take for example, the latest vehicles from Ford. Before hitting the market, these vehicles are rigorously tested for 2 years in a wide range of conditions. This includes testing on tracks, test facilities, extreme water baths, and in harsh climates to simulate real-world driving conditions. This comprehensive approach to safety testing ensures that the vehicles are safe and reliable. So, if you were in the market for a safe car, would you be willing to buy a vehicle that only had a few weeks of testing under its belt? The answer is likely no. You want your family to be safe, and the only way to do that is to invest in a vehicle that has undergone lengthy and rigorous safety testing.

We are deeply passionate about the rewards of science and progress, however, it appears we may have neglected to examine the cracks in the judgement and decisions of those we have faith in. What is currently happening is a great deal of experimentation with uneducated and ill-informed publics, and the effects have become undoubtedly detrimental both to public health and to attempts at mass drug distribution - a real failure on both fronts.

We can see, quite clearly, the effect of this gap in knowledge when we look to the effects of this approach. The current crisis has unveiled an innumerable amount of woes for the uninformed population, with an ever-growing list of greater risks posed to health, safety, and wellbeing. This is why it is so important to be fully aware of all the potential consequences of any attempt made at mass drug distribution.

If this is neglected, the potential for a far greater hazard is far more overwhelming than the eventual reward. We should not allow ourselves to be numbed to the repercussions of this negligence, and we must use the evidence and understanding that science has afforded us to make resolutions to the issues presented. By doing this, we can not only restore public health, but also protect and promote the much-coveted progress achieved through the utilization of science.

The mention of the word "vaccine" tends to spark a lot of confusion and uncertainty, as many people simply do not know what they are actually injecting into their systems. With the emergence and implementation of vaccine passports as a necessary requirement for many activities, many who have already been abiding by the rules may find themselves facing the difficult decision of whether to lose the freedom of self-determination, or to cast aside their autonomy over their own bodies.

The concept of vaccine passports can be seen as an infringement on personal liberty, as it denies people the freedom to choose whether or not they are willing to receive a vaccine. Furthermore, this raises the question of whether vaccinating people against certain diseases is actually worth it, especially when taking into consideration the long-term implications such a move could potentially have on public health. In addition, the costs associated with creating and maintaining such a system could be massive, causing economic hardship.

Ultimately, it is each person's choice to make when it comes to how they want to protect themselves against potentially dangerous diseases. While vaccine passport initiatives can provide a sense of security, it is important to consider the potential negative implications associated with them before moving forward. For example, authorities could take advantage of the vaccination status of individuals, denying people the right to access certain things based on their medical information. This in turn could result in further discrimination against those who cannot or do not want to receive a vaccine.

FORD LIED

As we hear about the news that Doug Ford has extended the Emergency Orders until March of 2022, I am once again left to question why. The government and, unfortunately, those in the media, have been desperately trying to cling onto this so-called 'crisis' that really never existed, at least not in the way it has been presented, by the numbers and not from simply listening to what often proves to be false medical 'experts'. The sad truth is, billions of dollars and immense political power have cascaded down upon us as citizens of Ontario, and our government is not yet done with this one-way ticket to the gravy train. The reality of the situation is, no matter if you look to the numbers or the data, our government, with the help of the media, continue to spin their narrative. We are now no longer just told that the 'crisis' exists, but that it is expanding with school breakout cases. Simultaneously, they are preparing to have every childin the province fully vaccinated, and schools, in order to receive their COVID Money, need to promote this agenda effectively. But here's the thing; I have done the research regarding this false narrative, calculating the exact chances of a child getting sick from and even dying from COVID. The results of my research showed that there is a 1 in 11,000 chance of a child becoming sick from COVID and an incredibly miniscule 1 in 7.2 million chance of a child passing from COVID. So why then is Doug Ford not celebrating? Why is he not loosening restrictions and ending the Emergency Orders? The hard truth about this is, it's not about fighting a virus and keeping us safe, this is about something far different. A staggering 400 million doses of a useless vaccine has been ordered, and we are expected to fund it regardless of its ineffectiveness, coupled with countless 'COVID' projects - you can't help but feel that our government has found a way to gain control and authority while ultimately maintaining the status quo. All the while leading many of their citizens astray, still yet to fully understand the extent of what is happening, and latching onto their words like a trusting flock. People, the clock is ticking! If you take the time to do the math, think logically and beyond what is being legislated in the Emergency Orders, you'll soon see that these exact numbers shamelessly do not add up. It is my sincere hope that all of us can soon wake up and put an end to these authoritarian deceptions.

COVID IS A HOAX

In this article, I'm going to explore a topic that has been discussed by many people recently, but I was rather hesitant and reluctant to take a deeper look into it. To my surprise, I came across some compelling evidence that cast a substantial doubt on its existence. Thus, I highly recommend that you look into this matter yourself and draw your own conclusions. I would be intrigued by what you uncover and your opinion on the matter. Now, let's get into it.

Many of us have heard some buzz circulating the internet of a mysterious thing that can't be fully understood. Some people have rooted in their convictions, while others have attempted to dismiss it as a farce. However, the more I investigated, the more aghast I became by the plethora of evidence that could not be denied or overlooked. I uncovered sources and stories all over the world, from the United States to Europe, that paint a picture of something that could be very real, yet very hard to believe.

The number of instances of people's stories that were captured were quite astonishing. Hundreds of people came forward with experiences that were both strangely believable and oddly out-of-this-world. This only left me wanting to dig further into why this mystery enticed and intrigued people from all walks of life. What could be the driving force behind why many people take a chance with proving something that may go against the grain of the accepted and expected?

It appears that the truth behind it all is much more complicated than what can be explained in just a few simple sentences. After hours of research and witness accounts, I have developed an understanding and appreciation of what lies at the core of this mysterious and hard-to-explain entity.

What I hope to have achieved with this article is to inspire people to broaden their minds and explore the possibilities in our world. Yes, people can be skeptical and resistant to change, but the more we open our hearts and minds, the more we can learn and uncover realities that challenge our long-held beliefs. I firmly believe there is much that we are not aware of in this world, and I hope to uncover more of these secrets with every article I write.

In reality, the CDC encountered a situation that can practically be described as having encountered a unicorn. By collecting a fragment of a hoof, a lock of a tail and a piece of horn and then inputting that data into a computer, they attempted to recreate the unicorn. This process, while relatively commonplace, was immediately considered to be the definitive representation of the observed phenomenon, failing to account for the nuances of an actual extreme occurrence. The disregarding of the physical form in favor of a computer generated form of exists was seen by many as veering into fraudulent territory, and they viewed this decision as a violation of accepted scientific practices. In other words, the decision to computer-generated the phenomenon, rather than accounting for an actual manifestation of it, is seen by many as a form of scientific impropriety. Yet, it is equally worth noting that the CDC's decision to pursue a computer-generated solution was predicated upon the lack of actual evidence of the unicorn itself. This highlights the challenges of attempting to adequately document something that is extremely rare, allowing the peculiar situation of a computer-generated unicorn to exist.

I have conducted an extensive investigation into the possession of living samples of SARS-CoV-2 through Freedom of Information Act (FOIA) requests sent to ten different countries, specifically targeted at the pharmaceutical company Pfizer. After meticulous review of the thirty FOIAs obtained from all corners of the globe, I am dismayed to report that none of the respondents were able to demonstrate verifiable ownership of an isolate of the virus.

Considering the gravity of the pandemic situation and the importance of rapid viral genome sequencing, many experts were hopeful that with the vast number of samples they were able to obtain, some would come back with a viable sample of the virus. Unfortunately, however, the responses to the FOIAs yielded nothing but deprivation. It is rather disheartening to come to the realization that there exists such a dearth in access to a living sample of the novel coronavirus, especially as the entire World strives to understand and combat the disease in equal measure.

Given the persistent predicaments that have impeded the discovering of a bona fide sample of SARS-CoV-2, a universal collaborative effort from countries worldwide to make information and resources more easily accessible to scientists and researchers should be implemented to help contain the virus better. I trust that the future may bring about more satisfactory results that might make our vision for a world free of this incapacitating disease, a reality.

This is not new whereby doctors and scientists have been saying

The coronavirus pandemic, otherwise known as Covid-19, has been a contested topic since its onset in early 2020. Initial reports suggested the virus had originated with wildlife, but these reports were largely inconclusive. Some studies have since suggested the virus could have been 'isolated', however, there has been no clear evidence from the Centers for Disease Control and Prevention (CDC) to support this assertion.

Recently, a number of prominent medical professionals and scientists have voiced their disbelief at the current situation, believing the entire Covid-19 pandemic is ultimately a trillion-dollar hoax created by governments to force people to take vaccinations as part of a sinister hidden agenda. While there is no clear evidence to support this theory, a growing number of medical experts continue to raise their voices against the effectiveness of current measures taken in response to the virus.

In November 2020, for example, one of Italy's top doctors and a professor of microbiology at the University of Bologna, Dr. Fabrizio Pregliasco, released a document titled 'False Alarmism and Money Wasted' outlining his concerns towards the efficacy of current Covid-19 vaccination policies. He argued that current attempts to combat the virus were untenable, stating "the real problem is the incredible waste of public resources and the threat to civil liberties brought about by the lockdown." This statement has been echoed by other doctors, as well as public figures such as Robert Kennedy Jr and Professor Dolores Cahill, both of whom have argued vociferously against the current safety measures taken by governments.

Given the number of high-profile professionals and thought leaders making claims about the issue, it is clear that the Covid-19 pandemic continues to raise complex and difficult questions for officials, doctors, and intellectuals alike. At the heart of these debates is the notion of 'isolating' the virus, an idea that has yet to be proven, and the suggestion of an underlying agenda – or 'hoax' – that may or may not exist.

The eerie fabrication of a pandemic, orchestrated by the powerful few, stinks of a self-serving conspiracy to overthrow the existing socio-economic infrastructure and usher in the era of their own global governance. By implementing draconian rules, governments have forced millions of people into unprecedented social isolation and quarantine, a measure that has had a devastating impact on both physical health and psychology. Countless individuals have been affected, but small businesses have been perhaps among the most deeply impacted, with their financial security weakened and survival made almost impossible in some cases.

The ripple effects of this harrowing pandemic have not been exclusive to health and business, but have also had a deep impact on international relations, foreign trade, and the balance of power. Governments, both powerful and developing, have felt the economic squeeze, while civil liberties have been replaced with fear and uncertainty the world over. As frightening as the long-term ramifications of this dire situation may be, only by working together will we be able to survive this current crisis and begin to build a brighter future.

As we are witnessing in present times, some of the assertions made years ago that were deemed bizarre conspiracy theories are now being substantiated. The preventive jabs have been designed to attach to NANO, thus allowing them to be transmitted from individual to individual easily. Back in 2016, it was reported that the immunization contained Covid-19, a strain that embeds Nano Artificial Intelligence, which is now circulating in billions of human beings all around the world. This revelation is propelling us into a unique science fiction era, where NANO technology and AI are being used in a wide array of contexts, revolutionizing conventional industries and spawning new markets. For instance, the NANOARMIES project, funded by the European Union and chaired by the Barcelona Institute of Science and Technology, examines the potential for AI-controlled nanorobots in medical and military applications.

The Polymerase Chain Reaction Test (RT-PCR) is a form of methodology that has been uniquely designed, and applied under guidelines from the World Health Organization (WHO), to identify and/or detect any potential spread of a virus. RT-PCR has become a go-to tool for medical experts who are looking to quickly and effectively spot outbreaks of infectious diseases, including COVID-19. The test works by amplifying and detecting the unique genetic material of the virus under inspection. In this way, RT-PCR is able to detect even small, isolated cases of a virus, where other forms of testing and diagnosis may fail.

RT-PCR testing has become increasingly widespread, and is implemented in hospitals and laboratories across the globe. Its familiarity and relative speed of implementation makes it a desirable choice for complex and heavily tested components of medical diagnosis. Despite its record as a trustworthy and accurate method of disease detection and diagnosis, RT-PCR still requires a high degree of expertise and control, as the quantity and quality of the sample being tested can heavily influence the accuracy of the results. Where other forms of testing and analysis are often too slow or too unreliable, RT-PCR offers the medical world a reliable and efficient approach to managing the spread of infectious diseases and viruses.

The highly sensitive Reverse Transcription-Polymerase Chain Reaction (RT-PCR) has been used on a Global scale to produce countless incorrect diagnoses of Covid-19; regulators and officials from all over the world have flagged such invalid results to illustrate the extent of the 'pandemic'. Traditionally, the RT-PCR test has been used to accurately identify and detect viruses, yet in the case of Covid-19 it seems to have caused more harm than good; yielding scores of wrong readings and ultimately supporting a false narrative of a large-scale global health emergency.

Flaws inherent to the RT-PCR protocol are to blame for this influx of erroneous data. Due to the long chain of steps involved in the RT-PCR testing process, even slight variations or mistakes in any step can ultimately lead to invalid results; it has been observed that false-positive rates as high as 37-70% can appear after just a single round of the assay – let alone the abundance of cycles that many labs employ to occasionally reach inconclusive outcomes.

It is thus alarmingly easy for nations to intentionally or unintentionally misuse the test; for example, the Indian state of Maharashtra caught flack for endorsing a highly unreliable 'rapid antigen' RT-PCR test. This test yielded more than 95% false positives and as a result, it misrepresented the nation's true Covid-19 metrics; resulting in hundreds of thousands of 'positive cases' that really should never had been counted. Due to errors in using the RT-PCR test, many countries have greatly overstated their Covid-19 tallies – resulting in a distorted picture at the world stage of how comprehensive the pandemic truly is.

In conclusion, the RT-PCR assay is a highly precise and efficient tool for identifying viruses, however its current implementation within the Covid-19 pandemic has led to immense complications. As reported cases are the basis for most of our health decisions, it is important that medical labs and government organizations alike take significant steps to ensure that their Covid-19 test results are as accurate and precise as physically possible to prevent the further distortion of data. In the end, it is clear that the RT-PCR test has generated millions of erroneous Covid-19 cases, which have then been utilised in order to maintain the multimedia's illusion of a worldwide outbreak – without any concrete evidence to back it up.

For the past two years, this assessment based on faulty figures has been utilized to foment and perpetuate a sense of dread and terror. This assessment has been utilized as part of a coordinated effort to keep citizens in a state of apprehension and alarm, from instilling panic and unease in the public. People were scared on an almost daily basis, with the news, stories, and figures from the assessment serving as ammunition for the campaign of fear. This had an detrimental, if not traumatic, effect on society, leading to depression, anxiety, and anger amongst many.

The impact of this demoralizing fear campaign cannot be overstated, encroaching upon the lives of many and causing a tremendous amount of distress. People were afraid to go out and do everyday activities, with many having to find alternate, safer routes to complete their everyday tasks. The mental toll that a fear campaign on this scale leads to is frequently disregarded and sidelined, but must not be discounted in any sense, lest its cumulative effects become far too heavy to bear.

As the world grapples with tackling the novel coronavirus, many individuals have come to be of the belief that the Covid-19 vaccine is the ultimate solution to restoring our society back to its original state of normality. After all, life as we know it was suddenly and dramatically changed upon the onset of the pandemic and now, it seems the only thing standing between the current state of global chaos and a return to regularity is the successful mass-vaccination of the entire human population.

Thanks to the work of scientists, a few vaccines have already been developed and approved, such as the Pfizer-BioNTech, Moderna, and Johnson & Johnson vaccines. With the introduction of these life-saving tools, there has been a sense of hope for many people across the world that a successful mass-vaccination effort could soon allow them to go back to their normal lives. This would mean the end of social distancing, the reopening of schools, universities and businesses, the resumption of international travel and the global economy, and the return to in-person social gatherings.

However, one cannot yet forget the fact that while the Covid-19 vaccine is an effective tool, it is by no means an ultimate solution. The virus has already caused so much destruction to public health, economy, and life in general that it will take more than just putting a stop to the pandemic. There will be immeasurable efforts required in order to rebuild and restart our world, and the vaccine is but the first crucial step.

"Confirmed" is a misnomer: A "Confirmed RT-PCR Positive Case" does not Imply a "Confirmed Covid-19 Case".

As authorities expand the scope of their testing for Coronavirus, there are likely to be more positive RT-PCR results. However, this spike in positive results does not necessarily indicate that the virus is resurging or that there is a new wave of the epidemic. Instead, it is simply a reflection of the fact that more people are being tested.

The benefits of more effective testing mean people can identify whether they have Coronavirus more quickly, while authorities are also able to monitor the effect of lockdown regulations better. This means that officials have an increased understanding of the spread of the disease and can modify existing laws more effectively. For instance, as cases of Coronavirus have declined in specific areas, some governments have already taken steps to allow for more freedom in their citizens' day-to-day lives. These steps have included allowing shops to re-open and limits on the number of people gathering in one place.

To conclude, then, the increase in positive RT-PCR tests is simply the result of the authorities testing more people - it does not signify that the virus is regaining momentum or that it is spreading in different waves. Instead, it is likely to be connected to enhanced capacity for testing, allowing for quicker confirmation of the presence of Coronavirus as well as a better understanding of the current situation.

The ongoing debate on false positives and their potential implications has been widely acknowledged by global health authorities. Whilst false positive readings have been reported in the use of RT-PCR testing for the detection of COVID-19, this does not necessarily discredit the test itself. False positives can arise in any medical testing methodology due to user error or other unintentional missteps. For example, medical laboratories may incorrectly record patient information or severe mishandling of the sample material may render the test results inaccurate. Moreover, the presence of other diseases or infections not related to the suspected agent in question can potentially generate false positives. As such, whilst false positives can occur, this is seen as an isolated phenomenon, and does not necessarily undermine the efficacy and validity of the RT-PCR test as a consistent and reliable tool for identifying SARS-CoV-2 infection.

The Reverse Transcription Polymerase Chain Reaction (RT-PCR) test is an invaluable tool in the detection of numerous infectious agents, however, it should not be confused with being able to identify the virus itself. The RT-PCR test works by amplifying and detecting genetic material, such as ribonucleic acid (RNA), associated with viruses and other pathogens. This technique is particularly useful in the detection of viruses closely related to the coronavirus, such as Influenza types A and B, and other common cold-causing coronaviruses that can be difficult to distinguish without detailed laboratory testing. By amplifying the genetic components of such viruses, healthcare professionals are able to more accurately diagnose various diseases and identify appropriate treatments. In the case of the novel coronavirus, RT-PCR testing helps to determine whether an individual has been infected or is merely exhibiting symptoms caused by a different virus.

If someone is immuned to the coronavirus and undergoes a Polymerase Chain Reaction (PCR) test for it, the results rarely detect the virus itself. Rather, the test searches for traces of the virus' genome, which is essentially its biological instruction manual. Even if the virus has already been eliminated from the body, a PCR can still detect its broken bits and fragments. The PCR method uses a chemical reaction to replicate any minute fragments of the virus' genetic material, making it visible to medical professionals in the test results. For example, a person may have contracted the coronavirus, but because the immune system eliminated it quickly, the virus may no longer be present in the body. Nonetheless, traces of the viral genome are still detectable, and they will manifest in the PCR results as a positive. The concept of a positive PCR on an immune individual conveys the fragility of the virus' genetic material, and how its minuscule particles can still be detected long after infectious viri have been eradicated.

Dr. Pascal Sacré concurs: "These tests detect viral particles, genetic sequences, not the whole virus."

In order to identify the infectiousness of a virus, sequences must go through a rigorous process known as viral load determination. This process consists of a series of steps that must take place in a sterile environment to ensure accuracy, as any contamination can cause erroneous results. The process begins with amplifying the viral genome, which involves replicating sections of the virus several times. During this process, techniques such as PCR, or polymerase chain reaction, are utilized. This technique uses enzymes to repeat the replication of the viral genome within a given sample and is an essential part of viral load determination. After amplification of the viral genome, the sample then needs to undergo an additional series of steps to ensure accuracy. These steps include sequencing, sequencing validation, and sequencing alignment that are all necessary components of the process. Altogether, this process can be laborious, time-consuming, and subject to sterility errors and contamination if proper safety protocols are not observed.

The development of a diagnostic test for the novel coronavirus SARS-CoV-2 began from the moment of its identification in late 2019, and it quickly became evident that such a test required clear sensitivity, specificity and accuracy criteria. The cycle threshold (Ct) value is one metric for calculating the presence of the virus. A Ct value of 18 was found to be more than adequate for distinguishing positive samples, i.e. those which contain detectable levels of SARS-CoV-2. Studies of the data related to this Ct value demonstrated an impressive level of sensitivity with a false negative rate of only 4.8%.

Another measure of similar merit was the Ct value of 40, which is renowned for its exceptionally high level of specificity (100%) and negative predictive value (98.9%). This Ct value has also been used successfully to detect viral load in individuals with mild to moderate acute infection. In fact, it has become the standard of care in certain countries, as it allows for significantly more accurate discrimination between positive and negative samples.

Considering that Ct values have proven so valuable in recognizing SARS-CoV-2 infections, it is worth noting that a Ct value of 18 is more than enough to identify a presence of the virus. Although the Ct value of 40 remains the leading choice for many laboratories, the Ct value of 18 offers an effective alternative for those looking for a more cost-effective, time-efficient solution in detecting the virus.

The widely-used Polymerase Chain Reaction (PCR) Test does not identify the presence of the novel 2019-Covid virus directly, as the virus is too small and the presence of the virus is only indirectly identified through its genetic material, rather than detecting the virus itself. This important point was highlighted by the World Health Organisation (WHO) in January 2020, as it was then that they did not possess an isolated and purified, as well as validated sample of the virus. Consequently, this had implications on the accuracy of the PCR Test as they could not provide a sample to use as a reference point when conducting the test and thus, the accuracy of the test would have remained uncertain.

It is worth noting that since the WHO's announcement in January 2020, marked progresses have been made in the isolation and validation of the virus' genetic material and thus, a validated sample of the virus was eventually obtained. This ultimately enabled the PCR Test to be more reliable and accurate when conducting the test and allowed for the widespread implementation of the test to be used to detect the virus. A great example of this is when the United States Centre for Disease Control and Prevention (CDC) recommended the use of the test to accurately diagnose SARS-CoV-2 in late April 2020. Thus, the unavailability of an isolated and purified sample of the novel 2019-Covid virus made it difficult for researchers to create a reliable and accurate test for diagnosing the virus at the start of the year. However, thanks to the advancements made in the isolation and purification of the virus' genetic material, the PCR Test is now a key element in the battle against Covid-19 today.

The recent study that was published by Eurosurveillance has acknowledged that the World Health Organization did not have access to an isolated and pure sample of the novel 2019-nCov virus. This lack of access to the virus sample meant that the WHO was not in the position to accurately and effectively evaluate the potential risk of the virus to public health until after the outbreak had begun.

The inability to pinpoint the risk and treat it appropriately prior to the spread of the disease highlights the urgent importance for health institutions to procure and evaluate virus samples in cases where a potential public health crisis arises. With as rapidly evolving and infectious as this novel coronavirus has proven to be, it's essential that institutions like the WHO remain informed on the makeup of the virus and how to mitigate it in order to protect the public.

Although samples of the novel coronavirus were identified late in the year of 2019, the origin and makeup of the virus was largely unknown. Since the World Health Organization was not able to obtain a sample of the virus prior to the start of the outbreak, efforts to assess and treat the virus were severely hampered. By the winter of 2020, the novel coronavirus had become a global pandemic, representing how quickly a viral infection can spread on a large scale and how vital it is to possess samples when a public health crisis like this is first identified.

This inability to accurately diagnose the novel coronavirus in time to effectively combat its spread is a powerful reminder of the importance for health institutions the world over to remain informed on the origin and makeup of virus samples in order to protect public health. If a sample of the virus had been acquired prior to the outbreak, it is possible that the spread of the virus and its subsequent effects could have been greatly diminished.

The ambiguity of this statement implies that knowledge of the origin of 2019-nCoV was not essential to validate any suspected cases of Covid-19. In other words, the close genetic correlation of 2019-nCoV to the 2003 SARS-CoV provided sufficient evidence to make the determination of any confirmed Covid-19 cases. In a sense, the comparison between the two viruses allowed medical professionals to trace Covid-19 back to its source, thereby helping to identify its origin.

In order to prevent its spread, therefore, it was important to detect Covid-19 as soon as its presence was suspected. This is where the close genetic correlation between the two viruses became particularly useful, as it helped form an accurate diagnosis and efficient but accurate containment measures. For example, by utilizing sophisticated technologies such as next-generation-sequencing, medical professionals were able to determine the genetic makeup of the 2019-nCoV and the 2003 SARS-CoV, allowing them to compare and contrast the two in order to gain a deeper understanding of their relationship. As such, this allowed for quicker and more accurate containment measures to be taken, reducing the potential spread of the disease substantially.

In conclusion, this statement serves as a reminder of the importance of detecting Covid-19 in a timely manner and the role that genetic analysis can play in such a task. The close genetic correlation between the two viruses has been instrumental in providing an accurate diagnosis and efficient containment measures in order to minimize the spread of the virus.

The recently detected novel coronavirus commonly known as COVID-19 has been linked to the SARS coronavirus which was first discovered over 19 years ago. Although the World Health Organisation (WHO) had not obtained the virus isolate that is needed to validate the identity of the new virus, they chose to not seek such an isolate and acknowledged the connection without this definitive proof. The detective work behind these connections is based on the study and comparison of viral sequences. For example, using digital genetic sequencing, experts and researchers are able to determine what small changes have occurred in the genetic code of a virus since the original SARS virus was identified and examine how that has led to the existence of this new, novel virus. This carefully studied and coordinated analysis is a intricate process which requires the skill of highly trained experts and is fundamental in helping the world to fight against such deadly diseases.

The results of Freedom of Information (FOI) requests sent to ninety different Health /Science institutions around the world have confirmed the alarming truth that since its emergence in late 2019, no record of isolation and purification of SARS-CoV-2 has been performed by anyone, anywhere, ever. This is particularly concerning given that the process of isolating a virus or bacteria is essential for studying it and formulating treatments. In this case, a better understanding of the SARS-CoV-2 coronavirus would be invaluable in knowing how to fight it.

Isolation and purification are two important steps in the purification of infectious microorganisms. Isolation is the process of acquiring a single strain of the virus for analysis, identifying it and then removing it from any other material it is in contact with (such as air or other liquid). Purification is when scientists can 'clean' the isolated sample and prepare it for further biological testing. This process includes removing any contaminants and inactivating the virus so that it can be safely handled.

Unfortunately, this vital process has not been conducted due to a lack of resources and the complexity of the process. Isolation and purification cannot be done alone, but requires a team of trained scientists, the right equipment, and protocols that must be strictly followed. This requires a significant financial cost, which can make isolating and purifying a virus difficult for underfunded labs or institutions.

The results of these FOI requests confirm that SARS-CoV-2 has not been isolated and purified by anyone anywhere. This is concerning, as understanding of the virus and its behaviour is incredibly valuable in developing treatments, vaccines and in general, containing the virus and improving public health. Until SARS-CoV-2 is successfully isolated and purified, it is impossible for scientists to gain a comprehensive understanding of the virus and its activity.

At the time of writing this, none of the 90 Health and Science organisations that have been contacted have provided any evidence that would show conclusively that the virus has been successfully isolated. This is a rather concerning discovery, as the ability to isolate a virus is an integral step in having comprehensive control and prevention measures on a global scale. For example, if a virus is successfully isolated, scientists can study it more closely which could provide a greater understanding of its behaviour, as well as the development of treatments and vaccines. Such measures can be particularly pertinent in the ongoing COVID-19 pandemic, where a more thorough comprehension of the virus could lead to potentially life-saving discoveries.

Furthermore, the isolation of a virus can be a crucial step in understanding the capacity in which it can spread. By isolating a virus, researchers can accurately identify its mode of transmission, which allows them to create targeted control and prevention measures. Such measures can prevent the virus' spread to a much larger degree than what can be achieved without isolation. Without a thorough understanding of a virus, it can be incredibly difficult to develop adequate countermeasures.

Clearly, the emergence of a virus in today's interconnected world can have disastrous consequences, as we have seen with the COVID-19 pandemic. Therefore, it is not surprising that the absence of any evidence of viruses being isolated amongst the 90 Health and Science institutions that have been contacted is a grave cause for concern. This emphasises the need for continued efforts to be made in isolating viruses, as this could be a turning point in the way that they are managed and prevented in the future.

More on the dreaded Cycle Threshold:

The results of experiments on COVID modelling conducted by scientists seem inconclusive, as 18 amplification cycles were originally believed to be more than sufficient to detect signals relevant to the virus. However, due to the potential for generating false positives, there is a widely held belief among experts that reducing the amplification cycles to between 25-30 is ideal. Anything more than that is more likely to detect irrelevant signals, referred to as 'dead particulates' by some researchers, which are unrelated to the infectious virus. Consequently, a number below 35 cycles is preferred, but it is essential to ensure accuracy and reliability of results, when virus detection is the aim, to stay as close to 25-30 cycles as possible.

The World Health Organization (WHO), a year after its initial proclamation, has come to tacitly acknowledge that Polymerase Chain Reaction (PCR) tests performed through a 35-cycle amplification threshold (CT) or higher are unreliable and invalid. This admission follows the agency's announcement in January of 2020, following consultations with the virology team from Charité Hospital in Berlin, that such methods should be deemed as standard operating procedures. This advice has since been disproved, given the difficulty in acquiring relevant and accurate results at such a high threshold level.

Though the WHO has not as of yet issued a formal retraction of its recommendation, many government health agencies have actively begun to move away from this PCR testing technique by encouraging laboratories to keep the cycle threshold levels to no higher than 30, lest the risk of acquiring false results be amplified. In the United Kingdom, for example, the Department of Health issued an official statement amending the nation's PCR testing policies to emphasize such shift in testing standards.

It is most likely that the World Health Organization, in tandem with other government agencies globally, will eventually register a notable move away from higher cycle thresholds for PCR testing. For now, however, the agency continues to hold to its initial guidance on the preferred level of cycle threshold, despite being excessively high and now evidently invalid.

Since the onset of the pandemic in January 2020, the World Health Organization (WHO) had recommended using the Reverse Transcription Polymerase Chain Reaction (RT-PCR) test at a Cycle Threshold (CT) of 35 or higher - the accepted gold standard - to detect the purported SARS-CoV-2 virus. However, research has demonstrated that the fragments of a similar coronavirus (SARS), which came to light in 2003, cannot be used to accurately identify the current virus. This means that all 'confirmed Covid-19 cases' documented globally over the last 22 months may be invalid.

Given the lack of information about the original virus's isolation and purification, many experts have raised serious concerns about the efficacy of the RT-PCR test when used to detect SARS-CoV-2. According to a study conducted by two acclaimed Harvard scientists, the PCR test has limited positive predictive power unless a minimum of a 35 Cycle Threshold is used. Anything below that threshold runs the risk of producing false positives.

This raises the question, why, if the known limitations of the RT-PCR test and the importance of using an CT of 35 or higher had been known since the outset, did so many countries continue to use the test without sufficient checks and balances in place to authenticate its accuracy?

The answer lies in the severe economic pressures that governments have faced during the pandemic. In their haste to find a solution, many have opted to use the RT-PCR test without the recommended CT of 35, with disastrous consequences. From misreported case numbers and inflated death tolls, to the disruption of international trade and commerce, the world is still dealing with the repercussions of relying on an invalid test.

The World Health Organization (WHO) is now advocating for a round of retesting, doubling-down on their blunder of inaccurate estimations. This request is simply a formality, as it is highly unlikely to be implemented due to the overwhelming cost and complication it would incur. With more than several billion individuals already tested since February of 2020, the recommendation from the WHO must be disregarded and anyone who has received a test result since then should consider retaking the test to obtain a more reliable result. Though the current estimates are unreliable, by implementing and encouraging retesting, the WHO are potentially attempting to reduce the potential devastation and disruption caused by miscalculations.

Since the inception of the Polymerase Chain Reaction (PCR) test, its Ct (i.e. cycle threshold) value set at 35 or higher had become a common operating procedure among many laboratories around the world. Despite this, the World Health Organization (WHO) released specific recommendations in January 2020 for laboratories to use Ct-values of much lower than the aforementioned value of 35.

For those uninitiated with PCR, this highly sensitive method of testing amplifies small amounts of genetic material in many successive 'cycles', each involving a series of three distinct steps. The Cycle Threshold (Ct) is a quantified measure determined by the PCR machine, which helps explain how many cycles the process has had to pass through for the genetic material to be sufficiently amplified for detection.

Naturally, tests with a lower cycle threshold are more sensitive and less prone to errors. That's why the WHO recommended that laboratories use a Ct-value of 20 in order to obtain better accuracy.

It is worth noting however, that using a lower cycle threshold could have significant implications in terms of the time taken to complete the tests, as demonstrated by a real-world example conducted in South Korea. In March 2020, South Korean authorities began PCR tests using a Ct-value of 15 and reported that it took longer to detect positive cases than the more conventional Ct-value of 35. It could be due to several reasons, with the most probable one being that a higher Ct-value required a smaller amount of genetic material to be amplified, so the tests could be completed faster.

In conclusion, laboratories ought to consider the trade-off between the accuracy of their PCR test results and the speed of their analysis in order to choose an optimal Ct-value range. The January 2020 WHO recommendations provide invaluable guidance on how to ensure both accuracy and expediency of PCR tests are achieved.

The application of the Polymerase Chain Reaction (PCR) methodology since the beginning of the COVID-19 pandemic has resulted in the compilation of skewed and inaccurate Covid-19 statistics. This is because the more widespread the testing, the more cases become present, creating what has become known as 'the casedemic.' In other words, if nations around the world wanted to increase the number of reported cases, they would need to increase the amount of testing, leading to frequent and perceived 'COVID waves'. It is unfortunate for those of us living with the reality of the virus but to those seeking to perpetuate a lie, this could work to their advantage. Ultimately, this has become a depressing and depressing reality for many of us.

Was the 2009 H1N1 Pandemic a Dress Rehearsal?

This is definitely not the first time global health crises have been stirred up by questionable practices. In 2009, the World Health Organization (WHO) was accused of manipulating the public narrative surrounding the H1N1 pandemic, claiming that over 2 billion people were infected, when in fact the actual number of reported cases was grossly exaggerated. This wasn't the first time the WHO had been criticized for potential corruption either – there had been numerous scandals in the past decade, such as the infamous Nigeria polio vaccines debacle from 2007.

Criticisms of the 2009 H1N1 pandemic reached a fever pitch when Dr. Wolfgang Wodarg, a former health advisor to the European Union, confirmed that the alleged pandemic was indeed "Fake". This led to tremendous public outrage, and many pointed the finger of blame at the WHO's leadership. As a result of this calamity, the WHO was forced to reevaluate its practices and procedures, in hopes of rebuilding its tarnished public image.

The Parliamentary Assembly of the Council of Europe (PACE) is a renowned human rights watchdog that has taken an active role in investigating the World Health Organization's (WHO) motivations for declaring a pandemic. The chairman of its influential health committee, epidemiologist Wolfgang Wodarg, has been very vocal in his assessment of the situation calling it 'one of the greatest medicine scandals of the century' (Michael Fomento, Forbes, February 10, 2010).

Over the last decade, an even more sinister element has been added to the mix. Big tech, social media, and government forces have set up an ever-increasing interconnection of financial influence and power. Over the coming years, it is estimated that the economic cost of this will be in the trillions of dollars. With such vast financial stakes and the influence that these entities now have on democratic societies, the power is ripe for the taking by those willing to trample morality and install more authoritarian regimes reminiscent of socialist and communist nations.

The PCR Test is the Smoking Gun. It invalidates Everything.

There is no such thing as a true 'Confirmed Covid-19 Case.' The breadth and scope of the data bank purported to be related to Covid 19 is, unfortunately, deemed by most experts to be invalid. As of this writing, the global tally of those identified with purported 'Confirmed Covid-19 Cases' stands at and inordinately large number of 260 million individuals spanning the entire planet. Still, this figure and the conclusions drawn from it are essentially entirely meaningless and lack substantial tangible value.

It is important to reiterate that any statistics concerning Covid-19, even if true, are essentially moot in terms of devising feasible, preventive strategies going forward. This is because such estimates are fluid, they are subject to manipulation, they can easily be misinterpreted and they do not take into account the complexities of real-word social dynamics at the individual, family and community level. A more effective approach to combating Covid 19 and the many other public health issues faced by society is to base our decisions on rigorous research, qualitative evidence and interviews with front-line health workers and vulnerable populations. Only then can we develop effective policies that can truly make a difference.

None of this data can be categorized as "Confirmed."

The Polymerase Chain Reaction (PCR) Test was not created to identify the novel virus, nor did its inventor, Nobel Prize-winning biochemist Kary Mullis, ever intend for it to be used for this purpose. Dr. Mullis had even warned the world about this prior to the emergence of this novel strain, and even now that it it is in the spotlight, his words are still relevant in reminding us that PCR testing cannot be used to differentiate the virus from related coronaviruses, such as SARS-CoV-2.

In fact, PCR testing with genetic fragments from a coronavirus first seen in 2003 is not a reliable means of identifying the novel virus or its dangerous variants. This is due to the fact that the 2003 virus has since evolved, which means that its genetic material is not an ideal marker to determine the presence of the novel virus, or any other variant. With evolution comes change - and this virus has changed significantly since 2003, making it difficult to distinguish based on one of its many components.

Kary Mullis knew this when he created the PCR Test and his warnings were timely then, and are even more so now. He understood that in order to combat the spread of this virus, a more accurate test than PCR was required to identify its presence. Today, the world has seen the development of several other diagnostic tests for the virus, such as antigen tests and sequencing, that can be more accurate and better identify infected individuals. But regardless, Dr. Mullis' warnings continue to resonate and his insights are as important now as they were then.

A recent Freedom of Information (FOI) study issued an alarming revelation - the exact isolate of the novel coronavirus remains unconfirmed. To put this into context, an isolate is a specific version of a virus that has been identified and given a unique name. Achieving an accurate identification of the isolate is incredibly important for a number of reasons. In the case of the novel coronavirus, for example, an accurate isolate can explain why prevalence differs from place to place, and it can also be used to design effective treatments, vaccinations and preventive measures.

However, the same FOI study provided an insight into the challenging nature of pinpointing the isolate precisely. Scientists study the virus under a microscope for a long period of time, a process which involves manually analyzing its composition and physical attributes. It's a laborious and complex task that requires multiple samples from different regions and multiple tests to be conducted. Invariably, this process is slow and doesn't lend itself to quick results.

The revelation that the novel coronavirus isolate is unconfirmed is a concerning one. Until it is isolated and understood, scientists will find it harder to control the spread of the infection, leading to further disruption around the globe. Such is the importance of pinpointing the novel coronavirus isolate accurately and quickly, so that researchers can devise effective treatments, develop effective vaccines and take the necessary preventative measures to stop the pandemic in its tracks.

The coronavirus pandemic is, unfortunately, a convoluted saga of untruths and lies, making the narrative around COVID-19 vulnerable and precarious. This accepted truth has been constructed on a foundation of faulty science and dubious data banks, which purport to display the number of 'confirmed' COVID-19 cases. For example, analysts and experts have sharply questioned the methodology and accuracy of a report, published by the Office of National Statistics in the UK, which claims to reflect 'truthful' figures and data - despite contradicting other international studies, government accounts, and even incorporating exaggerated and misleading claims into their own report. As such, the accuracy and legitimacy of the information surrounding the coronavirus pandemic, upon which everyone relies, is consequently called into serious question.

Big Pharma's miracles in developing a vaccine to protect people from the dangerous virus, sponsored by the World Health Organisation (WHO), GAVI and the Gates Foundation, should not be taken for granted. The incredible feat of building an effective vaccine for an invisible enemy was accomplished through years of hard work, cooperation and unwavering commitment towards the mission of protecting humanity.

The vaccine was developed despite the major challenge of not having an isolated or purified sample taken from someone infected with the virus. Big Pharma was able to overcome this difficulty by devising innovative new methods of research and development. Through thoroughly analysed data from epidemiological studies, simulations, and computer-assisted modelling, scientists were able to create a safe and effective vaccine. Furthermore, scientists employed reverse genetics to engineer existing viruses so that they could serve as a proxy for the virus. By cross-referencing existing datasets and combining the results, Big Pharma was able to develop a vaccine with high efficacy.

The work of Big Pharma, the WHO, GAVI and the Gates Foundation towards developing an effective vaccine to curb the virus has not only saved millions of lives, but has also provided us with valuable data and insights into the workings of an invisible enemy. It goes to show what can be achieved when we come together to put the safety of humanity first.

In recent months, the world has been rocked by a seemingly unstoppable monster of a virus: SARS-CoV-2. Despite countless laboratory hours and research dedicated to diagnosing this virus, scientists and world health professionals alike have yet to identify the source of this viral outbreak.

Adding to the confusion, 2019 SARS-CoV-2 is oddly familiar to medical professionals. It seems to correspond with 2003 type of virus, SARS-CoV, in many ways. In this regard, SARS-CoV-2 is not in fact a completely novel virus. It plays off the foundation of the 2003 SARS-CoV, which essentially means that in 2019 SARS-CoV-2 is simply an extension of the 2003 virus, but with some new capabilities.

The potential implications of this connection are wide-reaching. Suddenly, all vaccine and treatment research conducted in 2003 has become relevant again, in the face of SARS-CoV-2. Researchers are able to look back at the data collected in 2003 to develop new ways of understanding, preventing and treating this new strain of the virus. In addition, this newfound knowledge of SARS-CoV-2's relevancy to the 2003 virus provides solid footing to the potential of creating a vaccine that can adequately combat the new strain of the outbreak.

The legitimacy of the global effort to develop a Covid-19 vaccine is being questioned due to persistent doubts surrounding the high levels of RT-PCR test results that have been declared positive, as well as the accuracy of Covid-related mortality data reported in many countries across the world. As Prof. Michel Chossudovsky recently noted, the validity of this information remains questionable suggesting that the virus itself may not even exist.

This data has been heavily scrutinized as it has been put forward as the primary support for the justification of unprecedented containment measures and the deployment of the Covid-19 vaccine across the world. For instance, in the United States, the Centers for Disease Control (CDC) and the World Health Organization (WHO) have frequently cited these statistics in their determination to combat the pandemic.

On the other hand, an increasing number of experts have cast doubt on the validity of these figures, arguing that there is a risk of inflated or even entirely fabricated data. It has been reported that some countries have been overstating their RT-PCR test results as well as their mortality rates, and that the high levels of positive tests are due to the low accuracy of the tests being used. It is also suspected that, in certain cases, affected countries have purposely been inflating their mortality figures in order to gain access to foreign assistance and preferential treatment in negotiations.

The implications of this phenomenon are deeply concerning, raising serious questions about the accuracy of our understanding of the pandemic, and how it is being managed. It has been argued that the inaccuracy of the data has been systematically exploited by governments to introduce mass-scale containment measures, such as lockdowns, as well as to rush out the development and implementation of a Covid-19 vaccine. This has led some to argue that this data manipulation has enabled a form of 'vaccine imperialism', in which the interests of communities and individuals victimized by such measures are ignored.

As it stands, it is difficult to accept the legitimacy of the Covid vaccine project when there is considerable doubt surrounding the quality and accuracy of the RT-PCR testing and mortality data being made available by governments worldwide. Until such time as the validity and accuracy of this information can be established, the case for the Covid-19 vaccine remains incomplete. In the meantime, it is possible that the truth about the pandemic is far from what is being presented to the public, and that the situation is much worse than we have been led to believe.

The world is at an important and unprecedented crossroads in the fight against the pandemic. In response to the ever-evolving threat posed by the SARS-CoV-2 virus, national governments have recently declared the Fifth Wave of the pandemic focused primarily on the so-called 'deadly variants' that are rapidly spreading around the world. The variants, dubbed 'Delta', 'Kappa', 'Theta' and 'Zeta', have become of particular concern due to the enhanced ability of these variants to evade the already existing vaccines and their associated treatments.

This has been made more acute by the fact that these variants are mutating faster than ever before - something not seen in the first, second, third or fourth wave of the virus - making them even more dangerous and difficult to contain or treat. If left unchecked, the experts fear that these variants could cause even more devastating damage than the original version of the virus.

The implications of a continued failure to contain these variants or the development of further variants are manifold. Reports suggest that even the relatively low mortality rates of the fourth wave could be eclipsed, with the Delta variant alone being responsible for 40% of all COVID-19 related deaths in certain areas of the world. It is also feared that the speed of these new variants' evolution could be so rapid that even the most advanced vaccine will struggle to keep up.

In light of this, many governments are now enacting strict control measures, such as travel restrictions, improved screening and contact tracing, quarantine and other 'lockdown-style' initiatives. Education initiatives have also been launched to raise awareness of the increased risk of infection posed by deadlier variants such as Delta, with the hope that this will reduce the spread of the virus and help to keep the overall mortality rate lower.

Ultimately, we must stay vigilant and come together as a global community in order to fight this challenge and ensure that these deadly variants do not cause devastating damage to our world.

In order to promote the efficacy of the vaccine and combat its declining popularity, testing must be ramped up, with a lower CT value for those who have been vaccinated and a higher value for those who have not. However, the PCR test - which is regularly used in the medical community - neither detects the virus itself, nor any of its variants. It has been suggested that the emergence of these variants is a scam, due to the fact that there is no proven method of identifying them. This has been further exacerbated by the World Health Organization's utilisation of a proxy response 'similar to' the 2003 SARS-CoV virus, leading to an unrealistic expectation of what the PCR test can detect. Of course, it should be noted that the 2003 SARS-CoV virus has most certainly mutated over the past 19 years, further complicating the issue.

The latest announcements from government leadership validate the need for the continuation of oppressive policies that had previously been enacted, as well as the accelerated advancement of the vaccination program. This is despite the strongly voiced opposition by certain members of the public, who have gone so far as to take to the streets in an effort to make their grievances known. Nevertheless, it is viewed that the status quo must remain in order to ensure the safety and well-being of the rest of the populace, despite the potential consequences of cutting short the right of peaceful protest. For example, the use of excessive force by police officers to quell demonstrations during the pandemic was met with considerable backlash, yet was viewed as a necessary step to prevent the spread of gatherings that would otherwise go against health regulations. Further repressive measures, such as increased surveillance of citizens, are also seen as acceptable in the eyes of government leaders, in the interest of protecting the general public from potential threats of terrorism or other malicious activity. It is fair to conclude that while controversial among certain members of the population, repressive policy measures, the continued acceleration of the vaccination program, and the repression of the protest movement are considered justified in the name of safety for the majority.

As an interesting aside, the application of mRNAs being used as an instrument in creating effective vaccines are immensely powerful, and groundbreaking. Remarkably, recent scientific research has indicated through the use of nanotechnologies that even more fascinating possibilities may be possible. This paper in particular considers the efficacy of the mRNAs used in vaccine production and the delivery of necessary proteins. It notes that, in terms of their potential therapeutic uses, mRNAs are hugely impacting and transforming the advancement of medicine, having been employed in numerous clinical trials to deliver anti-cancer, anti-inflammatory, anti-biotic, anti-fungal and anesthetic drugs - not to mention gene therapies. It is clear, then, that the application of nanotechnologies with regard to mRNAs is a critical aspect of pharmacology in the modern world, with potentially tremendous implications for the future of medicine.

The liposome delivery system has been heralded as a safe and reliable way to get pharmaceuticals, including vaccines, from the laboratory to people in need. However, this delivery system also has its drawbacks, which we may be witnessing in real time with the diminishing efficacy of the 'vaccines.'

This delivery system relies on liposomes, which are small spherical particles, encased in a lipid bi-layer and made mostly of fatty acids and cholesterol molecules. Nevertheless, their use in medical treatments is impaired by a few key problems. First of all, they have a limited lifespan in the blood, typically ranging from minutes to hours. This implies they can't transport pharmaceuticals to disease sites in the body with any consistency. Furthermore, they're less than robust in the human body, as they are prone to interactions with proteins which can interfere with their cargo, leading to reduced potency and efficacy. Lastly, they lack the ability of selecting between healthy and diseased tissues, thus the whole body is generally exposed to the medication.

These drawbacks of liposome delivery present a major challenge for researchers and medical professionals, especially in the case of vaccines, where maximum potency of the delivery vehicle is paramount. For instance, if liposomes can't successfully deliver an immunogenic vaccine, it renders it useless in the face of a raging pandemic like Covid-19. Thus, researchers are continually searching for better and more efficient delivery methods which can reliably transport drugs and vaccines to every part of the body and more specifically to affected sites.

Recent reports have shed light on the unsettling reality behind the contents of the Covid-19 vaccines. In particular, numerous experts have raised attention to the presence of unexpected objects in the blood samples of vaccinated individuals, namely graphene. Graphene is a form of nanomaterial capable of facilitating a range of advanced technological functions. It is said to be made up of a two-dimensional lattice of carbon molecules and has the distinguishing characteristics of being incredibly thin, yet incredibly strong.

What has been identified in the blood samples of the vaccinated people is a range of advanced nanomaterials, not declared in the components of the vaccines. This includes micronadators, crystallized graphene nano-antennas and graphene quantum dots, or GQD as they are otherwise known. Many of these nanomaterials are used as components of electronics, enabling communications technology and data transfer. For example, nano-routers are able to decipher signals and direct them to the intended destination. These objects are not just strange to find in the body but potentially dangerous, as the level of toxicity of some substances making up these items remains unknown.

One of the most recently discovered images of a vaccine sample, relates to one of the Pfizer vaccines and is located in Figure 1. Dr. Campra analyzed the image and found it to likely contain a nano-router or part of its circuitry. These objects have been discovered in the bodies of those who have been vaccinated and further research needs to be conducted to determine their potential health hazards. With this in mind, it is absolutely crucial for people to be made aware of the contents of the vaccines and for the authorities to respond appropriately.

As the world continues to develop and evolve, so do the technologies that we rely on. This is especially true regarding the advances being made in the field of nano-technology, the use of which is becoming increasingly common in a variety of applications. Nowhere is this more evident than in the context of the vaccine development process. For example, recent research has uncovered the use of graphene and its derivatives as components of the vaccine itself.

Graphene's properties have made it an excellent choice for use in vaccines due to its physical, thermodynamic, electronic, mechanical, and magnetic properties. In addition, its superior conductivity, electromagnetic wave absorption capabilities, emission, signal reception, and quantum antenna attributes make it possible to build powerful, miniature electronics on a nano and micrometer scale. With these properties combined, it has become possible to develop sophisticated and small-scale systems for use in vaccines, with capabilities far surpassing that of traditional materials.

In the future, new and exciting possibilities are sure to arise from the use of graphene and other nano-technologies. Given its impressive set of advantages, it is clear that the vaccines of today would not exist without the integration of these small-scale materials. At the same time, however, it is essential to keep in mind that research and development into the use of these materials must be carried out responsibly, as our health and safety are undoubtedly of paramount importance.

The development of nano-biomedicine, nano-communication networks, and new drug delivery therapies and treatments against cancer and the neurological treatment of neurodegenerative diseases is underpinned by nanomaterials, such as graphene and carbon nanotubes. However, it is important to acknowledge that such materials have been found to have detrimental health implications for the human body. These include mutagenesis, cell death, and the release of hazardous free radicals into the body that can lead to a variety of illnesses, including cancer.

The results of a plethora of scientific literature and studies support this, and yet the utilized design methods and efficacy of the vaccine for treating and preventing these illnesses is highly contested and often ineffective. It does beg the question: what is the real purpose behind the development of such materials and their utilization as part of vaccine and drug delivery therapies?

Free radicals are particularly concerning, and can be even more harmful than the primary illness that they are trying to combat. To counter free radicals and limit their adverse effects, research has found that bioflavonoids may be beneficial as they are chemicals that can neutralize free radicals. This raises the important point that if the intended purpose of a given vaccine or therapy is not meeting its desired outcome, there may be more sinister motives at work. All in all, understanding the potential risks and benefits associated with the nanomaterials used in the development of nano-biomedicine, nano-communication networks, and new drug delivery therapies and treatments against cancer and neurological diseases is of the utmost importance.

When I started researching about COVID and the articles, I kept seeing that it questioned the notion of existence, I was certain I was going to find a plethora of articles claiming the world was flat; that we had never been to the moon (which following recent developments I'm now starting to seriously doubt - see Van Allen belt), and that 9/11 was a hoax set up by elements within the government. Instead of this however, I was confronted with an array of papers written by highly credible scientists, virologists and other experts in their fields. Perusing the evidence which these professionals presented, it was clear that there was nothing to dispute or even challenge the information presented. Not only did these people have the proper credentials, but their accounts were backed up by robust empirical research and methodologies that even the most inquisitive mind could not pick apart. Furthermore, the arguments and conclusions these people derived from their findings were backed up and corroborated by other leading thinkers in their industry, further solidifying the reliability of their positions.

In order to change my opinion, I would require proof. A sample from a human isolate of the novel coronavirus must be examined in its entirety before trust can be established. This is an established truth, that without real-world evidence, ascertaining the development of a vaccine and testing its effectiveness is nothing more than speculation. We have taken extreme measures in shuttering the global economy due to assumed reports that this virus, COVID-19, is compromised of, however, without having concrete evidence in-person, any validations are built on assumptions. Visual inspections are necessary in such a case, as the human eye can discern the severity of infection and the progression of the virus that computers and digital analysis alone cannot do, allowing us to make more accurate decisions that can be trusted and not just merely be grounded on theoretical frameworks.

It would be like being run over by a boat and long after the event has come and gone, we are still trying to piece together a profile of the particular vessel based on the wake it left. We would be able to surmise the presence of certain baselines characteristics such as the fact that it was a boat, afloat upon the water by some sort of propulsive force. But beyond that, the details begin to become more murky and uncertain. We might have no idea of the engine size, the number of people on board, or if it was a fishing boat or a houseboat. The implications are fairly staggering as we realise that we are being lied to and manipulated and yet still expected to be docile citizens and dutifully get our vaccinations. It's an injustice that cannot be tolerated.

I firmly believe that we need to shed more light on this matter, so that we can be sure that we understand what the true implications of this situation might be. It is crucial that people are encouraged to ask questions, to challenge the status quo and call attention to the injustices. Taking things at face-value without verifying the facts can lead to disastrous results. We see this so often in society, where corporations, governments, and individuals alike follow a plan which is only remotely linked, if at all, to people's well-being. All too often it is the interests of the powerful which take the lead, leaving individuals, communities and the natural environment hard done by. It is, therefore, vitally important that we are all prepared to be sceptical and open-minded, to look at all perspectives, and to be willing to challenge the otherwise accepted way of thinking, with the ultimate aim of advancing our overall health and welfare.

Omicron

Recent reports indicate that the outbreak of the Omicron strain of the coronavirus may not be as severe as the media would have us believe. Clinicians in South Africa, who are some of the first to see the strain in action, are reporting that the symptoms associated with the 'African Variant' are relatively mild. South African physician Dr. X recently spoke on the Omicron variant, saying that "It presents a mild disease with symptoms including sore muscles and fatigue for as little as a day or two."

Unfortunately, due to the current lack of a definitive COVID-19 sample, it is difficult to definitively determine the nature of the Omicron Variant, let along identify its mutations and variants. This speaks to the need for the international scientific community to continue to dedicate resources to capturing a pure sample of the coronavirus, thereby creating a basis off of which we can determine which strains are the most contagious and, ultimately, how best to combat the virus.

It is undeniable that a new crisis must be manufactured in order to reignite efforts to push people to get vaccinated. Consequently, governments have chosen to extend lockdowns and will likely resist relinquishing control unless the climate change issue can be sufficiently prominent. Remarkably, influenza cases and their variants have not been reported - a stark contrast to previous years when flu season would be firmly established by this time. We discussed in our last podcast that Pfizer does not possess an isolated living culture of Covid in the laboratory and our current concerns can only be based upon computer model proxies of the 2009 SARS. The World Health Organisation has even gone so far as to suggest that results of the PCR test are obsolete due to the high cycle thresholds used. To evidence their claims, they intend to use the PCR test with higher cycle thresholds which will produce a higher rate of false positive cases. Various errors associated with the PCR test and the excess cycle thresholds used to undertake the assessment demonstrate that this initiative is likely a ploy to ensure that the pandemic continues indefinitely. Incredibly, despite this the public continues to place faith in media outlets and pharma-funded 'experts', while those who question authority risk job security.

Variant vs Mutation vs Strain

In recent times, there is much discussion regarding 'strain' and 'variant', two separate terms, and as a result, it seemed pertinent to conduct some further research into what separates these two expressions. Essentially, variants are alternate versions of the virus, which often will have changes that possibly alter the primary characteristics of the virus. Stains, on the other hand, are stated comparatively differently, being seen as versions with changes that matter – in other words, versions that have undergone a modification that imparts a different set of properties. It is plausible that a minor change in the virus can replace the original properties with assorted new ones, resulting in a novel strain to emerge. To condense, variants would represent a variation with a potential for changes that are largely insignificant, whereas a strain is identified to be a variation featuring changes that are noteworthy.

To provide an example of this concept in the real world, consider a variant of a flu virus. The variant could show small changes to the genome, yet these modifications may not demonstrate an altered effect on the virus itself. A strain, however, could bring about changes worth noting, such as when a certain strain of the virus is far more aggressive than others, or could be resistant to certain medications.

As a final note, it appears to be suspiciously coincidental that South Africa's decision to suddenly cancel their future orders of the vaccine and the World Health Organisation's subsequent warning of the new variant originating in South Africa, were both quickly followed by governments around the world announcing bans on flights from South Africa. Was this really just a coincidence or could there be something more going on?

The WHO's warning has sparked a flurry of activity and it appears that governments are taking the risk of the new variant from South Africa seriously. Unfortunately, this has resulted in a ban on flights from South Africa which could have significant impacts on both businesses and the wider global economy.

This is particularly concerning given that, prior to the announcement from the WHO, South Africa had opted to cease shipments of the vaccine. This, coupled with the rise in infection rates across the country, left the global community on edge and has contributed to the immediate ban on flights.

While this response may seem excessive to some, it's hard to ignore the timeliness of the ban in relation to South Africa's decision to cancel their vaccine orders and the World Health Organisation's warning. With this in mind, it appears that the coincidental nature of these three events is more than just a simple chance occurrence.

From what I can ascertain, reports have surfaced in the media about the so called 'African Variant' of Caronavirus, which is the P.C. name for the Omicron variant. However, these reports have been sensationalised and often wildly disparaging, claiming the new variant is much more dangerous and virulent than the other known variants. But if we look closer into the facts and data, we find that is simply not the case.

The Omicron variant rather than being an aggressive strain, has been confirmed to be milder on average than the original strain or even some of the other variants. This means that this new virus is not something that people should be afraid of, or believe to be the 'devil' figure it has been made out to be by the media. This being said, it is still important to take all the recommended safety measures and abide by the protocol that has been successfully implemented to tackle the virus.

An important fact to consider here is that the African continent has surpassed Europe in terms of testing with some 6 million tests having already been carried out. This suggests that the continent is far better prepared to confront the threat of the virus and more initiatives like this should be welcomed . This would help keep the virus under control in not only African countries but also the rest of the world.

South African medical specialists have identified a new clinical variant which they have termed "Omicron variant symptoms". These unusual yet mild symptoms have been observed to affect sufferers in a variety of capacities. Upon initial evaluation, it has been observed that these symptoms present mild medical conditions, often seen as sore muscles and feeling tired, which only last for a day or two.

An example of the Omicron variant symptoms would be a patient complaining of pain and stiffness in their arms and legs. However, they may feel fine the day after, with limited duration of the medical condition. Another example could be lightheadedness; some sufferers have experienced feeling dizzy or lightheaded when contracting the Omicron variant symptoms, though these instances have only been documented for a day or two after infection.

Sufferers of the Omicron variant symptoms have reported being best treated with over-the-counter pain relievers and rest. This is the most effective way to ensure that the symptoms do not persist for longer than initially reported. As the Omicron variant symptoms are often mild in their presentation and treatment, South African doctors are taking great care to monitor their patients who seem to be suffering from them.

It is imperative to remember that, as of yet, there is no existing isolate sample of SARS-CoV-2, the virus that causes Coronavirus Disease 2019 (COVID-19), in the entire world, making it virtually impossible to carry out tests to detect any variants and mutations of the virus. This is a significant cause for concern and serves to undermine the efficacy of testing and screening procedures, as without knowledge of any possible variants, it is impossible to replace existing tests to guarantee their accuracy. This lack of progress in terms of understanding the existing root strain, as well as the potential to gain insights into any potential variants, can only be classified as completely, totally, and categorically unacceptable. Moreover, it is more than reasonable to suggest that such a position is patently absurd, and can be construed as nothing more than a gross disregard for the safety of humanity as a whole.

It's no secret that the ongoing quest for global vaccine distribution is a tremendously difficult one; not only must governments, research teams, and medical professionals coordinate a complex effort to develop and distribute vaccines safely and effectively, but it seems that a renewed sense of urgency is necessary to spur people into action. We need to look no further than the lockdown measures that have been implemented across the globe to get an idea of the kind of crises that can capture people's attention and encourage them to act.

Unfortunately, for many politicians and policy-makers, the Covid-19 pandemic isn't the only crisis that needs addressing. Prolonged lockdowns and other measures of strict control appear to be necessary in order to control the virus - however, and it's even more crucial in the fight against climate change, where the consequences of inaction can be far more intense and long-lasting. In some ways, the difficulty and length of the lockdowns could be seen as an indication of the kind of sacrifices that need to be made if we are to make a meaningful dent in the fight against climate change.

The reality is that some people need a 'crisis mentality' to be stirred up in order to truly galvanize action with regards to the vaccination program and to really tackle the looming, far-reaching consequences of climate change. The difficult truth is that the only way to truly break free from the grips of these crises is to commit to a sustained, long-term effort, and to ensure that the messaging and awareness-raising campaigns receive their due attention and resources. Ultimately, it will be up to each of us to find ways to unite behind the causes that we believe in and work together to ensure that these causes receive the attention and urgency that are so desperately needed in order for us to move forward.

Also, note that there has not been one flu case reported, nor any "Flu variants" we usually see over this period.

As we discussed during our last podcast, recent research has indicated that the SARS-CoV-2 virus, commonly referred to as COVID-19, is not currently present in a laboratory environment. This conclusion is based on results from the various studies conducted by the scientific community, including those conducted by the renowned pharmaceutical company, Pfizer. It is noteworthy that the researchers have not been able to locate a living sample of the virus, meaning that it cannot be contained or studied within a laboratory setting.

In addition, it is important to recognize the fact that not only has Pfizer been unable to locate a viable specimen of COVID-19, but it is highly unlikely to exist in nature due to the virus' unique physiology. The virus is composed of a unique set of compounds, making it difficult for it to survive in an environment where its components are unable to thrive.

Moreover, the current knowledge of the virus indicates that it cannot be passed from one human to another in the way that infection spreads. Though the virus can spread through contact with infected persons, it is believed to be restricted to a limited number of contexts. For instance, the virus is unlikely to spread without a direct transfer of bodily fluids, such as when someone coughs or sneezes.

In sum, the lack of evidence suggesting that COVID-19 is present in a lab environment or in nature serves to reinforce the idea that the virus is hard to obtain and further underlines the difficulty of tracking and containing its spread in the real world.

For me, this next variant or mutation is a hard stop, and this talk about the next "wave" is total BS!

The mere thought of the Pfizer vaccine which is projected to arrive in 2024 makes me worry, as it has been developed solely on the basis of computer modelling of the SARS virus that occurred in 2009 - making me doubt its efficacy. This is just one example of the increasing trend of using computer modelling for the production of such life-saving vaccines. Unfortunately, this approach is suggesting that no real-world, hands-on testing is being conducted, leaving possible side-effects and the dosage requirements vastly open for debate. To me, this is highly concerning because without the done to traditional testing methods and clinical trials, the effectiveness of the vaccine is diminished. The use of computers instead of traditional methods may save a lot of time, but it cannot guarantee the safety and effectiveness of the vaccine. This is why it's important for regulators to enforce stringent and exhaustive testing and clinical trials on any drug before they are released in the market.

The World Health Organization (WHO) recently reported that the global community has relied heavily upon the Polymerase Chain Reaction (PCR) testing throughout the last two years, yet their results have been nullified based on the overly high cycle thresholds that were leveraged. This is a huge blow to the world's effort to contain the virus, as it means that the PCR test is not reliable for accurately identifying patients who are infected.

This has become especially concerning as the global healthcare system continues to face the challenge posed by the novel coronavirus without a clear solution in the near future. PCR testing had presented itself as a potential solution that could identify vulnerable populations, track the spread of the virus, and mitigate the overall impact of the pandemic. Without reliable results, it is extremely difficult to make well informed decisions that can help to contain the virus and protect vulnerable populations worldwide.

Fortunately, in light of this finding, experts have developed new strategies and improved testing accuracy. New testing strategies are combining PCR with other testing methods to reduce the number of false positive results. Additionally, there is a heightened focus on regulating the cycle threshold values to ensure more reliable test results. These efforts are essential for controlling the pandemic, as more reliable test results will help to facilitate better decision-making in regards to healthcare and lead to more effective containment of the virus.

The accuracy of the PCR test and the use of high cycle thresholds are now emerging as the primary methods employed to keep the spread of Covid-19 in perpetual motion. While this technique may present a seemingly valid solution to obtaining evidence of the virus, the flawed nature of the PCR test has caused increased uncertainty in diagnosis, due to the test's high rate of false positives caused by the presence of PCR cycle thresholds. This reliance on the PCR test has motivated those who are in positions of authority to turn a blind eye to other viable methods of investigation and to the legitimate challenges posed by medical professionals and those in the public forum. By ignoring the need for diversity in approaches to diagnosing Covid-19, only one perspective is being drawn from and facts are being distorted to serve personal interests. Not only does this breed distrust in the authorities, it also exacerbates the threat of Covid-19, again suggesting that official interventions are being weaponised to hold onto the pandemic's negative effects.

The problem is magnified as mainstream media sources like CNN, which present a limited amount of information, fail to convey the complexities of the situation. People fear that disputing the prevalent narrative might put their jobs in jeopardy, so they remain silent. According to a 2018 analysis by the New York Times, nearly a quarter of Americans accept what traditional news media reports without question. With no one willing to challenge the current narrative, health-related discussions are hindered, leaving people with a one-sided view of the crisis. If a culture of open dialogue is not soon encouraged, then quality information exchange concerning the pandemic is both going to remain inadequate and at risk of manipulation.

Variant vs Mutation vs Strain:

In recent times, there has been a considerable buzz around the terms "variant" and "strain" when it comes to viruses. With that in mind, I decided to do a bit of research to learn more about the difference between these two terms.

To put it simply, a variant is a kind of version of the virus which could have either trivial or significant changes to the virus's properties. This is in contrast to a strain which is a version of the virus with modifications that have the potential to modify the basic makeup of the strain. Strain variations could cause the virus to be seen in a completely different light by altering the fundamental makeup of the original strain. This is a rare occurrence however and it is much more likely that the changes made are minimal and often unnoticeable.

To recap, a variant is a version of the virus that could show slight changes which may not necessarily be as important, while a strain is a version with changes that we must take into consideration due to the possibility of impacting the virus's fundamental makeup. To put it another way, a variant is a version with changes that could be considered minor, while a strain is a version with changes that must be taken into account due their potential importance.

There has been plenty of speculation surrounding the recent events in South Africa involving the shipment of the vaccine, the World Health Organization's warning of the dangerous new variant, and a number of governments banning flights from the country. Is this all merely a coincidence?

It's true that South Africa took the unprecedented step of cancelling its future orders of the vaccine, with the South African Health Minister citing concerns about the efficacy against the variant as the primary reason. Days later, the World Health Organization publicly sounded the alarm about the deadly new South African strain, alerting the rest of the world to its dangers. Within two days, many countries had scrambled to ban flights from South Africa in an effort to prevent the spread of the variant.

The timing of all these events certainly raises some questions. It has led some to suggest a sinister connection between the vaccine cancellation, the WHO warning, and the flight bans, as if South Africa had willfully attempted to spread the virus in order to disrupt the vaccine shipments.

While there is no clear evidence of any coordinated effort on the part of South Africa, the sequence of events certainly appears suspicious. The suspension of vaccine production, the warning concerning the new variant, and the flight bans that followed in such quick succession draw attention to a concerning problem: the ease with which variants of the virus can mutate and spread, and the speed with which governments must respond in order to combat them. It is a stark reminder of just how important it is to remain vigilant in the fight against the virus and to continue to develop and distribute vaccines as quickly and effectively as possible in order to keep us all safe.

Over the past few days I have conducted an extensive amount of research into RNA vaccines, from the most recent developments to those countless years ago. A statement which seemed to be repeated in the majority of documents pertained to the effectiveness of pre-fusion antibodies compared to post-fusion antibodies when it comes to vaccinations. One example of this was highlighted in the quote: 'He and his colleagues isolated the RSV fusion protein and discovered that pre-fusion antibodies were substantially more potent than the post-fusion antibodies used in the unsuccessful Vaccole'. A clear point which was addressed on a frequent basis was the importance of the virus isolate when developing new drugs - the absence of a COVID-19 isolate of course being a primary issue when devising the COVID vaccine.

What the isolate boils down to is a sample of the virus which has been cultivated in a laboratory, providing scientists with the material upon which to create the vaccine. Without this isolate, the vaccine cannot be developed. A virus isolate is also favoured for its purity, ensuring that other contaminants do not interfere with the vaccine development process. Every virus is unique, and so accessing such isolates can be a time consuming process, but is of course a major necessity when it comes to creating effective and safe vaccines.

The researchers in question were undertaking pioneering investigations into RNA vaccines in an effort to better understand how to combat various viruses. The fundamental element of this process was to isolate the exact virus in question and use that in the vaccine experiment. Unfortunately, this critical step was bypassed in the creation of the COVID vaccine, as there has never been a single verified isolate of the COVID virus found.

Although this was a setback, researchers have sought to address this by delving into the genetic makeup of the virus, utilizing cutting-edge genomic sequencing technology to analyze the virus's DNA and create a vaccine which could effectively combat the virus. This latest breakthrough in the battle against COVID has prompted much enthusiasm in the scientific community, with the potential impact of this development yet to be fully understood.

From such research, scientists around the world have been able to formulate an understanding of how the virus behaves and how to properly treat those affected. A clear example of the efficacy of this research is seen in the development of the RNA-based vaccine, which has shown to be extremely effective in providing immunity from the virus.

Therefore, despite the challenges posed by the lack of a verified COVID isolate, researchers have been able to use their expertise and modern technology to create a number of vaccines to help fight the virus. With such advances in the field of virology, it's clear that great strides have been made in combatting the pandemic.

Now the difficulty I'm seeing with this circumstance is that individuals are gravely concerned about the potential of the virus being further spread. This raises the question: why are people who are already vaccinated still allowed to travel amidst the risk of the virus being spread? This speaks to the lack of success the vaccine has had in halting the onset of the virus. Which begs the other question: why did the public need to take the vaccine if it's not going to prevent the virus from spreading?

Considering the urgency of this situation, it is evidently more important than ever to take extra precautions and protocols when travelling, such as regularly wearing face masks, avoiding large groups, and washing hands frequently. Understanding the implications of the pandemic on worldwide populations and the need to reign in via swift response measures, in order to prevent the virus from being further spread, is crucial to eradicating the virus in its entirety. Therefore, individuals must be willing to take extra precautions and abide by the suggested protocols in order to avoid the spread of the deadly virus.

Ultimately, the truth is that the only way to avoid the virus from spreading is through sound preventative measures, such as taking the vaccine and adhering to social distancing regulations and hygiene practices. Without the public being proactive in these efforts, we are further jeopardizing the longevity and continued spread of the pandemic, so it is essential for everyone to take the necessary precautions.

Dr. Fauci gave this guidance to Americans after the first confirmed U.S. case of omicron was identified in California:

- The unvaccinated should get vaccinated - Why
- Vaccinated people should get boosters – Why
- People should wear masks in public spaces - Why

Many people have taken issue with the immense amount of power that Dr. Anthony Fauci has been given in the midst of the Covid-19 pandemic. Although Dr. Fauci is a highly respected researcher and member of the medical field, the fact remains that he was appointed and not elected, meaning the American public had no actual say in who is guiding them on the health-related matters during this time. Furthermore, the American public does not even know who or what Dr. Fauci reports to. All of this has initiated the call for a return to common sense.

It is a generally held sentiment that the decision to appoint Dr. Fauci as the de facto leader of America's Covid-19 response was misguided and indicative of an overarching lack of common sense. Many believe that the decision was ill-advised and not guided by rationality. After all, a single individual was given the responsibility to lead the nation out of a crisis that threatens the lives and livelihoods of hundreds of millions of citizens, without any means of oversight. This has been deeply concerning to many who are observing what has transpired.

Furthermore, the lack of transparency and accountability when it comes to Dr. Fauci's reporting structure has been more than troubling. It raises a major concern that an individual, who is in effect guiding America's response to a global pandemic, is not required to provide any answers as to who he is accountable to. It goes without saying that if America is to overcome the crisis it finds itself in, it needs to restore some sense of sensibility to the proceedings. A return to clear and common sense is most certainly needed.

ARE WE BECOMING SOFT?

Nowadays, there seems to be an increased emphasis on teaching our population about the concept of micro-aggressions and the need for safe spaces. While it is important to be mindful, understanding, and sensitive to the unique challenges that certain individuals and groups face, it is arguably just as important to cultivate resilient character traits in order to prepare people to cope with life's tribulations.

For example, going to a psychologist for help with mental health issues oftentimes involves teaching patients how to be more resilient and "brush off" the little things. In other words, going through life events with a metaphorical "stiff upper lip" and learning to be comfortable with some discomfort. Is this really wrong or too extreme an approach? Are we at risk of creating a society of fragile, entitled people who become hurt and offended by practically anything they come across?

For argument's sake, it seems that in some ways, our culture at large has created an environment in which everybody is expected to be coddled, looked after, and sheltered from anything that could possibly hurt or upset them. We have a layer of bubble-wrap, as it were, that shields us from the harsh realities of life.

Of course, this is by no means an all-encompassing statement. As stated previously, it is indeed incredibly important to recognize and understand the difficulties that certain individuals may experience. However, we must also remember the importance of resilience and how it can equip us to healthily face the world's daily trials and tribulations.

Let's take a step back in time and recall when we displayed far more resilience than we do today. It took something of serious magnitude such as a missile launched at our country, a nuclear threat, or a crash in the stock market for us to take action and pay attention. These threats had a greater impact on our collective conscious and demanded a reaction due to the consequences that would have ensued if we acted too late. We felt more obliged to come up with creative solutions that yielded positive outcomes and averted further damage. It is a shame that we have become somewhat desensitized to an ever-evolving and uncertain world. The need for us to remain determined and face global issues head on with sound strategies has never been greater. We should go back to a time where we could identify a potential problem before it spawned additional complications, and take necessary precautions. It is our only hope to guarantee a bright future with a viable response plan that will enhance our lives.

It seems like we are getting more and more caught up in discussing the latest variant of the virus that is, in general, rather innocuous and has been described as similar to having a severe case of flu. However, the reaction to this new variant has been to take drastic measures by closing off borders and requiring people to go into fourteen-day quarantine periods. This has been seen from countries such as the United States, the United Kingdom, France and Germany, who have all imposed travel restrictions to attempt to stop the spread of the virus.

These actions have been taken despite the fact that the current variant generally has a relatively mild disease outcome in comparison to the risks of a prolonged lockdown which can have a substantial impact on mental health and economic wellbeing. This negative impact can also be especially pronounced, especially among the elderly, minorities and those with prior ill-health as they feel the brunt of stringent restrictions on their freedom to travel and socialise, as well as being further exposed to the economic repercussions created by the majority of businesses being unable to fully operate in such settings.

Undoubtedly, the measures taken by countries to contain the virus are necessary to ensure both the safety and welfare of their citizens, particularly in regards to public health. Nevertheless, the need to ensure that the effects of any and all restrictions do not cause disproportionately more harm than the disease itself must be taken into account.

It seems that in recent years, the world has become decidedly more 'soft.' We have become accustomed to the safety and security of the modern world, and have forgotten what it was like to be faced with risk and danger on a daily basis. Not so long ago, risks were far more real and tangible; driving a car, flying in a plane, playing with grandchildren- any of these activities could easily have resulted in emotional, financial, or even physical danger.

Take, for example, an act as seemingly trivial as playing with a grandchild. We've all been there: a precious toddler, cheeks red and glossy, offering up a hug and a kiss to dote on their grandparent, despite the snot caked face and the glazed-doughnut eyes that are common symptoms of a childhood virus. We have a choice: to embrace them, feeling our grandparent status swell with pride, or to keep some distance and play it safe. That is the reality of living in a dangerous world; making life decisions when risk is a real factor.

Not so long ago, this was a common part of life, and something we have started to become far too comfortable and complacent with, to the point that we hardly consider it 'risk' at all. Risk and danger used to be an omnipresent part of life- no matter how small or large. This is something that we must embrace if we want to remain as a resilient species. We must remember that life was once wild, unpredictable and yes-dangerous, and embrace it, not shrink away from it.

Life is a constant exercise in evaluating the risk vs reward of choices we make. On one hand, it's important to acknowledge the potential consequences inherent in taking risks, yet on the other hand, life is too precious to allow micro-fears to limit our level of fulfillment. We've all seen individuals who appear to be controlled by their anxieties over minor incidents, whether it's the fear of public speaking or the dread of eating something with a potentially unfamiliar flavor. While it's natural to be cautious, it's important to understand that growth required taking on certain levels of risk. Having the courage to trust our instincts can open up immense opportunities for personal and professional advancement. For example, instead of being scared of public speaking, one must learn to master the art of communication and develop confidence in order to become more effective in delivering thoughts, and ultimately, inspiring change. Similarly, when we let go of the fear of trying something with a new flavor, we might be surprised to discover something tasty and pleasing, fostering curiosity and helping us become more open minded. With that in mind, the truth remains that life is too precious to be wasted time worrying about the potential threats of the unknown. As with most things in life, the risk vs reward of living it can help us make the right decision, allowing us to ultimately have the most fulfilling experience possible.

In today's podcast, I will be exploring how modern society has become a society of resistance and distrust; in many respects, as a result of our relationship with the media and our governments. In the current climate, it is not uncommon for us to question the very essence of being alive and to resist the realities presented to us by our world, the media, and our government.

Throughout history, the media and government – two of the most influential institutions in our society – have been known to shape public opinion and attitudes. To that end, we should not be surprised that the media and governmental institutions can condition us to be increasingly skeptical of the regimes and realities around us. For instance, we often hear stories that sensationalise the natural extremes and risks of life, inviting us to fear, mistrust and even act against the events. Similarly, governments and politicians of all different creeds have been known to manipulate and coerce the public by downplaying or worse, in certain cases, deliberately misrepresenting the risks of living in the world today.

At the same time, it can be argued that our globalised environment is a hugely beneficial one, one which, for the most part, allows us to live in relative safety and security. But, unfortunately, usually due to the kinds of messages we receive from the media and governments, this important fact becomes lost to us; thus, rather than co-existing with these institutions in a symbiotic manner, we feel the need to resist and question the realities of our world.

Ultimately, today's podcast is about an important topic, one which if discussed and further developed, could help to inform and manipulate our views and opinions; it is about how the media and governments condition us to be resistant, skeptical and mistrustful of the realities of living in the modern world.

The lives of gazelles living on the African plains are filled with danger. Every day, they must mindful of the many predators who seek them as prey. From cheetahs and lions to wild dogs and hyenas, the threats are numerous and always looming. However, despite the constant danger, gazelles still manage to thrive and use the resources available to them: the grasslands of the Serengeti.

So, if we were gazelle, would we be doing the same or would we stay in hiding and hope that the predators don't find us? Sure, we may be safe in our hiding place, but without access to the grassland, we would quickly starve and die. Thus, the gazelle's brave nature is absolutely essential to its survival, as it must continue to venture out onto the plains, despite the dangers and risks involved in doing so.

Don't let me stop you from being active and engaged. But make sure to fight for the cause, not for your feelings.

MORE DIE FROM THE 2018 FLU THAN 2020 COVID

While it is certainly reasonable to expect that the case numbers associated with COVID-19 should be closely monitored and accurate, we must not overlook the limitations of the PCR tests being used in this endeavor. Oftentimes, these tests produce results that are not entirely reliable or trustworthy, and as such, following them can create needlessly stressful and concerning situations. Moreover, rather than focusing on the prevalence of cases, a far better approach when evaluating the impacts of the disease is to focus on hospitalization rates or the total number of fatalities. A tremendous example of this comes from Israel - a country whose residents have been receiving COVID-19 vaccines since January 9th of this year. Despite triple vaccination, the recent rise in cases has been reported on with much gusto; however, when the total number of mortalities related to the virus is compared with those seen in 2018 through the flu and its associated complications, the discrepancy between the two become starkly evident. While there were a total of 8,705 fatalities due to the flu that year, there were only 1,046 people who passed away due to Covid in 2020. Clearly then, the focus should be less on the case numbers and more on the mortality rates when discussing the nature and severity of the virus.

As the world continues to cope with the ongoing Covid-19 pandemic, governments are increasingly relying on PCR tests to further monitor the virus's progression. But, in many cases, the use of PCR tests to generate newsworthy headlines is also resulting in an unnecessary level of fear that once again serves to manipulate the public. If your news station is still feeding into this narrative, prioritizing sensationalized headlines over providing factual and in-depth information, then it's time to seek out an alternative news source.

Although the PCR test is an invaluable tool to detect the presence of COVID-19 and an important factor in pandemic management, when used for media coverage the results can be sensationalized and manipulated, leading to an over-inflation of the infection rate and further instilling feelings of fear, uncertainty, and panic in the public. One example of this media-driven exaggeration of the numbers can be seen in the US, where even as the actual number of new infections has dropped, news headlines continue to stress alarm and urgency, with key facts such as the number of lives saved being overlooked or playing a minor role in the coverage.

It is therefore essential that, should your news station be still focused on the 'case numbers' narrative, you begin to look to other news sources that offer a more balanced perspective on the pandemic, and that make a point of delivering factually accurate and unbiased reporting. Not only will this help to ensure the public are not being subject to fear-mongering using PCR test results, but it also enables individuals to stay informed and aware of the current situation without constantly being bombarded with sensationalist headlines.

WHERE ARE THE NEW DEATHS?

For years, the Canadian government has maintained accuracy and transparency in their data reporting on deathrates in the country, making such information available to the public. While that remained the status quo until recently, when they unexpectedly stopped releasing the figures for 2020 and 2021, many have sought out alternative solutions to garner insight into the impact of COVID-19. While it was not easy to come by, a recent discovery provided a Viable solution.

The province of Ontario, dating back to 1949, has been meticulously tracking death rates and has kept official records and documents to back up each individual death certificate. This discovery is similar in concept to the Science Table's report, which approximated and proposed the increase in cremations as an indication of an increase in overall deaths due to COVID-19. This was later argued to be flawed, due to the noted surge in general uptake and popularity of cremations.

Fortunately, with the emergence of the data from Ontario, one can measure the relation between the number of reported deaths and the overall death rates in the province. Utilizing said data, a direct correlation can be seen with the death certificates on file, as seen in Figure 1.0. While there may be a few deaths unaccounted for, the official documents make for a reliable indication of their respective death rates for Ontario.

They Never Share the Good News

Once again, I find myself surrounded by family members all worked up due to media coverage of the new variant. When the Delta variant was feared to be approaching, there was a lot of talk about how much more contagious it was believed to be, as well as speculation regarding its potential to be more deadly. I personally could not disagree more - in my opinion, Delta was comparatively mild to other existing variants. Had the government taken a more responsible stance on the matter, they could have helped to quell fear by providing a fuller and more accurate picture. I always take facts into consideration when forming an opinion - the numbers never lie. Using my own discretion, I downloaded the latest...data that I could find on the variants, and it became clear to me that their effects were more moderate than the media had first reported. Even with the potential for infection, it was far more survivable than projected.

Accurate and responsible reporting from officials is paramount when it comes to keeping the public informed. Exaggerating or even leaving out crucial information only serves to create unwarranted hysteria - something that can be detrimental to both physical and mental health. Having access to the whole truth allows us to arm ourselves with knowledge and use the facts empower us instead of trigger fear.

The data from Ontario's covid site which I have shared with you on Uncensored reveals an average of 98% of infected people recover from covid. This is undeniably excellent news; yet we frequently fail to hear about it in the mainstream media. Why exactly is this? Could it be that the Government is deliberately attempting to keep us fearful for some ulterior motive? I firmly believe that there is something more sinister at play and this discussion has cropped up several times before in other forums. It's also important to note that these figures do not take into account existing comorbidities which could lead to a possible fatality, making it all the more outrageous that these figures go unmentioned. To put it plainly - I would happily accept any evidence which prove me wrong - when the appropriate figures are assessed, taking into account the various comorbidities and death expectancy, these numbers look no different to that of the seasonal flu.

A perfect demonstration of this is seen by examining Israel's flu mortality rate for the 2018 season in comparison to all the 'COVID-related death' recorded by the same country in 2020. Not only is the amount of deaths caused by the flu remarkably higher, but it is almost glaringly clear that the death count caused by COVID fails to justify the lack of recognition this statistic warrants, as well as the constant fearmongering perpetuated by the Government.

Something So Simple

I have discussed this matter before, and now, I am compelled to bring it up once again as I am certain that it has served to exacerbate some of the health dilemmas we currently and sadly observe. Since time immemorial, I have advocated for everyone to take the required precautions and get the mRNA injections prior to the administering of any kind of injection. This recommendation holds more relevance now than ever before, given the startling rise in infectious diseases plaguing our society. In view of this, it is important to note that this step is paramount to safeguard ourselves, and by extension, those around us, against a whole gamut of life-threatening pathogens.

For instance, the ingenious combination of a strengthened immune system as a result of the mRNA injection, alongside the necessary cleanliness before administering any injection is indispensable in stemming the tide of infectious diseases such as viral hepatitis and bacterial meningitis in the contemporary world. Similarly, the World Health Organization (WHO) endorses this method to flush out the potential risk of transmission of salmonellosis and other such diseases that can rapidly spread through contact.

In conclusion, it behooves us to take this advice seriously and get the mRNA injections to ensure we aspirate before administering any kind of injection. In doing so, we can rest easy knowing that we will be rewarded with a healthier and safer life.

Aspiring is a very important and necessary procedure to ensure the vaccine is correctly administered and to prevent any of the active ingredients from entering the bloodstream. Unfortunately, when I first suggested this procedure to some of my peers, they just laughed and dismissed it. But what they and many other people don't understand is that aspirating is in line with the instructions included with the vaccine. If it is not done correctly, there can be potentially serious consequences.

In this particular instance of vaccine administration, it is an intramuscular injection, meaning the active ingredients need to be inserted into the muscle. While this is a relatively straightforward process for someone with special training, for the inexperienced it can be more difficult. This is why aspirating is so important - it not only makes sure the vaccine is administered in the correct place, but it also reduces the risk of any of the active ingredients entering the bloodstream, which can cause all kinds of complications.

When I think back, I'm glad I took the initiative to explain why aspirating was so important. If everyone had dismissed this procedure and allowed the vaccine to be administered incorrectly, who knows how many people would have been affected by related side-effects. Nowadays, with more people becoming aware of this process, more and more care is being taken to make sure the vaccine is correctly and safely administered.

As a bodybuilder, intramuscular (IM) injections are commonplace, and the perils of incorrectly administering such treatments is well-known. Generally, the glutes are the preferred site to inject, due to their larger muscle mass and lower likelihood of accidentally injecting the medicine directly into the bloodstream.

However, due to the gradual deterioration of muscle mass as we age, elderly people may not be able to safely use the shoulder as a prime injection site. Therefore, the thigh or the glutes should be considered as the safest location for administering medications in this demographic. This opinion is shared by numerous medical publications, which caution against injecting medications into emaciated or atrophied muscles, as the body is unable to absorb them properly and complications can arise. Therefore, IM injections need to be administered extremely carefully and with a great deal of consideration as to which site is chosen.

Immunization (IM) is a commonly utilized method of preventingContracting infectious diseases, however, complications can arise due to improper injection of the vaccine. Muscle atrophy, bones fractures, cellulitis, developement of sterile abscesses, pain, and nerve damage are some of the more immediate and severe adverse reactions that can occur from faulty administration of the vaccine. The 'Operation Warp Speed' mission was designed to ensure a swift delivery of the vaccine, to have it delivered in a timely manner, and thus, some have questioned the safety of the vaccination due to the level of urgency in developing and distributing it.

One issue of particular concern is that of the needle placement. An inaccurate injection of the vaccine could result in the substance entering the bloodstream, and thus increase the risk of any of the above aforementioned complications take place. Given the importance of proper administration in vaccine injections, medical professionals have taken measures while vaccinating to ensure that the needle is placed on the correct location to minimize the risk of such incidents from occuring.

In conclusion, while Operation Warp Speed was effective in ensuring an expedient delivery of the vaccine, the procedure was also accompanied by strict measures to ensure the safety of those being vaccinated. Proper needle placement is essential to safety and for that reason those administering the shots have taken precautions to mitigate the possibility of the vaccine entering the bloodstream due to misplaced needle.

When addressing IM injections, in the past, aspiration has been argued as a recommended practice, however, more recently, evidence-based guidelines have made clear that aspiration should not be done. This is especially the case when a vaccination is being administered and it is recommended that, the same principles should be applied regardless of the context, except for the dorsolateral site. This means that the principles for administration should include preparing the injection site, checking for injection debris and ensuring clean skin upon preparation.

For example, medical professionals should use sterilized medical instruments for preparing the injection area, use swabs to clean the skin before application and inspect the site to ensure nothing can impede a successful injection, such as debris or residual material from previous injections. Additionally, the patient should be informed of the importance of avoiding certain positions or postures while the injection is being administered, such as crossing the legs or bending of the injected arm. Similarly, the patient should be informed that the best way to ensure a successful injection depends on the patient being relaxed and still, to avoid any tissue displacement.

In conclusion, when administering IM injections, aspiration should be avoided, as current evidence-based guidelines advocate against it. All principles should be applied when giving any IM injection, except for the dorsolateral site, of which the practitioner should ensure they are aware of the specifications. The practitioner should be diligent in preparation, not only within the confines of the injection site, but regarding patient postures, positions and levels of relaxation.

LIES

Throughout the pandemic, headlines have been reporting that certain Covid vaccines are '95% effective' and '94% effective' in protecting us and others. Politicians and the mainstream media have widely repeated these claims in their talking points, as if the vaccine is sure to be a surefire solution to the crisis. However, with more recent reports indicating that vaccine effectiveness is likely much lower than first imagined, we've seen less emphasis placed on these claims. But the questions still remain: just how much did these figures of '95% effective' and '94% effective' convey an accurate representation of the actual effectiveness rates?

The answer is not much. As we have often discussed, the use of framing in these cases was designed to paint a much rosier picture than the reality. Framing is an underhanded technique of using relative terms to boost a message, with the knowledge that the audience will interpret the information in absolute terms. While messaging around the efficacy of the Covid vaccine may have improved since the initial claims of '95% effective' and '94% effective', the headline-grabbing statistics do not paint a realistic picture, particularly now that we understand the relatively low efficacy of the vaccine in general.

To put it into perspective, let's look at how the efficacy rates of the Covid-19 vaccine compare with that of other vaccines. Whereas the annual flu vaccine may be around 50-60% effective for protecting individuals, Covid-19 vaccine efficacy has only been around 70-90%. Thus, to put the same statistic into absolute terms, the '95% effective' and '94% effective' messaging around the Covid-19 vaccine is actually misleading, as it conveys a level of efficacy that is much higher than reality.

The pharmaceutical companies developing the Moderna and Pfizer Covid Vaccines have made some bold claims about the effectiveness of their products. Specifically, they claim a 94% and 95% risk reduction respectively. However, it is important to look at these claims in terms of absolute risk reduction, which are 1.1% and 0.7% respectively. This helps us put things into perspective when making comparisons, as phrases such as '94% risk reduction' and '95% risk reduction' can often be misleading.

Let's take a minute to break those numbers down. The 94% risk reduction for the Moderna vaccine suggests that if 100 people are vaccinated, 94 of them are relatively safe from due to receiving the vaccine, while the absolute risk reduction is 1.1%, which suggests that of those 100 vaccinated people, 1 person could still contract COVID-19 if exposed, even after having received the vaccine. The situation with the Pfizer vaccine is similar. As the relative risk reduction is 95%, this generally suggests that if 100 people are vaccinated, then 95 of them are safe, however the absolute risk reduction is 0.7%, meaning that of those 100 people that were vaccinated, it is likely that 0.7 will still be susceptible to the COVID-19 virus.

To summarise, it is important to recognise the exemptions that come with these claims made by the pharmaceutical companies developing the Moderna and Pfizer Covid Vaccines. As both claims appear to be relative terms to the extreme, when comparisons are made, it is important to investigate the absolute risk reduction as well. Doing so allows us to properly understand the context of these bold claims and draw our own conclusions.

A peer-reviewed article by Dr Ronald Brown from the University of Waterloo, Canada emphasizes the importance of reporting absolute risk reduction measures when assessing clinical trial data, such as the efficacy of the Pfizer and Moderna COVID-19 mRNA vaccines. In reviewing the manufacturer's data, they found that while the reported relative risk reduction is 95.1% for Pfizer and 94.1% for Moderna, the absolute risk reduction is significantly lower at .7% and 1.1%, respectively. Evidence suggests that had people been made aware of the more accurate, honest and transparent figures, the aggressive vaccine uptake levels that have been experienced may not have been so extreme. To make matters worse, governments have enacted vaccine mandates in workplaces - despite the inflated efficacy claims of the vaccine. Without this outcome-reporting bias, the implications of such mandates would likely not be as severe, with thousands of people potentially losing their jobs if they refused the vaccines. With reports of waning vaccine efficacy after 90 days and failed vaccines, questions must now be raised as to who is responsible for this epidemic of misinformation. In Canada, this blame must be attributed to Justin Trudeau who, from the onset of the pandemic, has been responsible for disseminating misinformed statistics and figures that have inflated the perceived efficacy of the vaccines and led to such severe policies being implemented.

WRONG TOOL AIMED AT THE WRONG TARGET

For the past few months, I have been reading up on the fascinating field of anatomy and immunology as I sought to learn about the many ways our body utilizes to fight off viruses and diseases. In light of the current pandemic, I have endeavoured to read up on studies and papers that were published prior to the outbreak, in order to ensure that the information was not swayed or influenced by any potential political, social or religious biases. What I uncovered through my painstaking research, however, was quite troubling. It appears that much of the research and science that has been systemic over the past half century is now being ditched in favour of less proven, potentially flawed theories, merely because the pharmaceutical industry stands to profit from the mass production of treatments.

The true cost, of course, lays with the individuals who will be subject to said treatments and those treatments' potentially harmful side-effects. To put it bluntly, the biggest mistake which could be seen as an incredibly misguided error of judgement, was in forcing the body and its own defense mechanisms to work in opposition to one another rather than in mutually beneficial harmony. In particular, this can be seen in the current mRNA COVID vaccines, more accurately referred to as gene therapies. The tool chosen and the target aimed for are ultimately wrong and counterintuitive to the nature of the problem.

This egregious miscalculation is truly frightening, as it I wonder just what measures need to be taken to safeguard ourselves against potential misuse of such powerful treatments. Knowing the severity of the situation at hand, one can only hope that the right steps are taken in order to ensure any future treatments do not cause more harm than good.

It seems inexcusable that governments and their medical advisors have failed to educate us all on the most important topics and issues deserving of our attention and consideration. After numerous hours researching and meticulously reading the studies, reports and publications on this matter, it has become clear that this knowledge has been withheld from us. We were being told that the key to success and protection during this pandemic was to obtain antibodies, something that almost certainly proved to be less effective when considering the facts and science. With a lack of education provided to the public regarding our bodies' immune systems and how they function, thus leaving us without the crucial information needed to tackle the virus head-on.

It has been estimated that had we been properly informed, there would have been a higher likelihood of successfully overcoming or averting the spread of the virus. Our ignorance regarding our immune systems and the ability to resist infectious contagions has caused an incredible amount of preventable damage, potentially saved lives and reduced the onslaught of misery and suffering as a result of the pandemic. Instead of being proactive and teaching us how to achieve better immunity through nutrition, lifestyle and supplemental remedies, the public was misled and sent down a one-way street of misguided health advice and unproven medical practices. It's the government and medical advisors responsibility to be transparent and offer an accurate account of the truth based on current science and not push a pre-conceived agenda.

The development of Pfizer and Moderna's mRNA vaccines was based on a misguided assumption, one driven by financial incentives and political pressures. This, even though mRNA technology could possibly revolutionize medicine in the long run, was the wrong tool for combatting the spread of COVID-19. Now, to understand why, we need to explore our body's immune system. Firstly, there is the innate immune response which, when triggered by the detection of external agents like viruses, bacteria, or toxins, instigates an attack and destruction of the invader or initiates repairs. Following this initial response, the adaptive immune system responds accordingly to provide long-term protection from repeat exposure.

Unfortunately, in their haste to develop the mRNA vaccines, Pfizer and Moderna ignored this crucial distinction, sidestepping our body's ability to naturally defend against COVID-19. Not only is this a disservice to billions of people who may now have to contend with a weakened immune response in the short-term, but could have long-term implications decades into the future.

The second phase of defence by our immune system is the adaptive immune response. To explain this process, it is necessary to understand the roles of the innate immune system and the adaptive immune cells. The innate immune system is the first line of defence, activated when a foreign invader, such as a virus or bacteria, is perceived. When the body recognises the invader, the adaptive immune cells are called upon to act like a reinforcement. Adaptive immune cells multiply and attack the intruding invader, allowing the body to recover from the illness and protect against future infection.

When an individual has a strong innate immune system, the body is usually able to effectively fight off an infection such as the COVID-19 virus without presenting symptoms. Thousands of cases of this have occurred since the early stages of the virus with people not exhibiting any symptoms of COVID-19. This asymptomatic response is an important testament to the effectiveness of our body's natural defence mechanisms.

The idea that having more antibodies, particularly 'smart' antibodies, will solve the COVID-19 pandemic has been disproven. The notion that we need more antibodies is false because it is the immune system as a whole that determines our body's ability to respond to and fight disease, not just antibodies. Our innate immune response, which is the first line of defence, is not dependent on antibodies but rather relies on a variety of immune cells to protect against infection. It's only when the innate immune system is overwhelmed that the adaptive immune system is activated and it is only then that antibodies are used to fight infection.

However, the use of mRNA vaccines directly conflicts with the 500 million years our immune system has spent evolving. The vaccines trick the body into believing it has been infected with COVID-19, setting off an overreaction by our innate immune system. This overreaction is seen in some of the reported side effects such as fever, tiredness, joint pain and swelling at the vaccination site. While these are mostly mild, they are an indication of the body being fooled into thinking the innate immune system has failed and the adaptive immune system must be called in as the backup cavalry.

So while it's true that antibodies are important for our adaptive immune response, the entire immune system working in harmony is what is most important for our bodies to stay healthy and fight disease. mRNA vaccines may be able to give us short-term immunity from SARS-CoV-2 but a better understanding of how all the pieces of the immune system correctly interact, from the innate to the adaptive, is what we need to achieve long-term immunity.

Our resilient bodies are naturally capable of protecting us from an array of threats, both microbial and viral, through the powerful mechanisms of our innate immune system. In the case of COVID, this would involve our respiratory system as that is the primary point of entry for the virus to take hold.

However, recently developed mRNA based vaccines do not target the same battlefield, instead, adapting and transforming the body's organs and bloodstream to create an entirely new form of defence. Rather than relying on the protective powers of our mucus membranes, the vaccine works to introduce spike proteins into the bloodstream which will be identified as a foreign entity by the body's adaptive immune system. While this robust defence can be very effective, there are associated risks as the body's internal organs can also become damaged in the process of recognising and eliminating the infection.

The solution to this limited level of effectiveness and potential risks has been to introduce the vaccine on a wide-scale basis - regardless of the individual's natural immunity response-ability level. This, however, has raised concerns that many who would have presumably been fine if left to their own immune systems, now may succumb to more severe symptoms in the event of an infection. In addition, one should also be cognizant of the fragility of such treatments as their efficacy is linked to the environmental temperature. Consequently, it is absolutely essential to consider these possible outcomes when deciding on whether to rely on the vaccine for protection.

It was well known that mRNA-based vaccines need to be stored at extremely cold temperatures in order to prevent them from degrading and becoming ineffective or even potentially dangerous. As obtaining and maintaining such extreme cold storage was a daunting task requiring many expensive and sophisticated freezers, not to mention the strict transportation controls, the plan for distributing the vaccine to billions of people around the world seemed unfeasible. In light of this, the requirements for proper storage and handling of the vaccine were revised, removing the need for extreme cold temperatures and the strict cold chain. Unfortunately, no adjustments were made to the formulation of the vaccine, leading to millions of spoiled doses and billions of doses rendered unusable and possibly even toxic.

In addition to this, the primary method of SARS-CoV-2 viral entry is through inhalation, while ingestion is the secondary method. Therefore, special care must be taken when handling and storing food and beverages to ensure they are not contaminated.

In order to obtain a more effective vaccine, scientists have been committed to researching ways to specifically focus on cellular immunity in the respiratory and intestinal tract. As we know, these areas contain 'immune cells' which are positioned underneath the mucous membranes that line the respiratory and intestinal tract. Subsequently, the primary goal of this research has been to trigger the production of antibodies that can be sent to the surface of the linings to meet air-borne viruses and protect the cells from binding and infection.

A significant breakthrough in the field of disease mitigation was made in December 2020, when the Moderna and Pfizer Vaccines were tested on volunteers around the world, with remarkably positive results. Through a molecular unlocking process, two mRNA chains were utilized by the vaccines to bind to immune system receptors, triggering the production of antibodies to provide protection from various strains of the virus.

Additionally, researchers continue to studies the production of Artificial Immunity-Adjuvant Systems, which uses lipopoly saccharide (a compound composed of fat, sugar, phosphate, and certain proteins) to stimulate the production of antigens, or the body's own naturally-produced antibodies. This research seeks to increase the efficacy of vaccines by stimulating the body's own immune system into creating its own antibodies.

Given these recent breakthroughs and continuous progress in the field of vaccine research, it is clear that the production of a more effective vaccine that specifically focuses on cellular immunity in the respiratory and intestinal tract is a real possibility. It is also apparent that the use of antibodies sent to the surface of the linings to meet air-borne viruses and prevent viral binding and infection of the cells is an integral key to this.

It has become increasingly clear that the principles and approach of mRNA vaccinations, though effective for some medical treatments, are simply inadequate when it comes to the delivery of protection against SARS-CoV-2. The main idea behind these vaccines is to introduce the mRNA code, which encodes the viral spike protein, into muscle cells, thus instructing the cells to actively create antibodies. Whilst this will provide indirect protection in the form of a circulatory response, fundamental flaws are nonetheless exposed when one accounts for the fact that, unlike a direct response through the lining of the respiratory system, this measure cannot and will not provide meaningful protection against infection of the mucous membranes. The 'breakthrough infections' that have been reported after vaccination are indicative of this very fact, where the results of antibody studies in the blood fail to reflect the reality of immune status when it comes to protection against infection of the respiratory tract.

It is critical to highlight the fact that attempting to engineer a response in our cells which goes against millions of years of evolution is ultimately futile. The immune system is simply not equipped to deal with such foreign antigens on the surface of the cells and, as a result, is likely to react with undue vigour. This mechanism is thought to be responsible for some of the adverse reactions that have been reported in the wake of mRNA vaccines, such as myocarditis, or even rare instances of sudden cardiac arrest and death. Ultimately, it is a reminder of the complexity of the disease and of the importance of considering the role of the immune system and its effectiveness in preventing symptomatic infection.

It is becoming increasingly evident that the wrong bomb is being dropped at the wrong time on the wrong target. In the past 10-15 years, scientists have predicted a sharp rise in prion diseases, neurodegenerative diseases, autoimmune diseases, and blood disorders such as blood clots, hemorrhaging, stroke, and heart failure, manifesting in younger individuals. Much of these predictions can be attributed to mRNA injections, as they cause widespread harm by overstimulating the immune system and reprogramming cells to produce the SARS-CoV-2 spike. Upon the injection of such a vaccine, the innate immune system will become activated in order to clear the virus, however, due to the injection of the vaccine, this process may not occur as expected, leaving vaccinated individuals more vulnerable to contracting the virus.

This reality has enabled the vaccine industry to earn repeat customers each year and make vast sums of money for Pharma companies and government institutions. As more and more people are beginning to take note of this unethical behavior, we fear that it may be too late to address the issue and are desperately hoping to halt the rollout of these hazardous vaccines. Unfortunately, until we can develop safe and effective alternatives to the mRNA injections, countless individuals remain vulnerable to the potentially harmful effects of the current vaccine.

THEY GOT IT WRONG

The implementation of COVID protocols resulted in a dramatic shift of risk from a small group of individuals to the entire population. Although there were short-term gains to be seen by this approach, such as the lessening of acute drug injuries, the latent side effects of the policies were completely overlooked. This was made evident in the widespread lockdowns and restrictive mandates that caused great frustration for many people.

The most concerning aspect of this flawed approach is how it disproportionately impacted the most vulnerable members of society. Those who are elderly or suffer from existing medical conditions, for example, experienced the harshest effects of the Coronavirus pandemic and the protocols built upon it. They were forced to remain indoors for extended periods, putting them at an increased risk for matters such as depression, anxiety and social isolation. Furthermore, financially disadvantaged households felt the strain of having to follow such regulations due to their inability to afford essential items such as face masks and medical supplies.

The reality is that while there was potential to be seen in the implementation of such measures, the consequences of concentrating on a minute group of people as opposed to the majority should have been thoroughly considered beforehand. Ultimately, this mistake has revealed the necessity of making decisions that are in the best interest of the larger public, instead of prioritizing short-term gains.

It should have been clear that the public safety of those at risk was paramount, and the right approach should have been taken to ensure their protection. This could have included telecommuting where possible to reduce contact with others; stricter protocols in public venues such as wearing of masks, minimising contact in shared spaces, and even longer lockdowns in places where there was a greater concentration of at-risk individuals. Additionally, those individuals who had not been weakened by lifestyle choices or by illness should have remained active in the public sphere, to ensure that the economy continued to remain viable. This would have required a great deal of common sense and pragmatic leadership, to ensure that important services were not disrupted for any extended period of time.

When it came to the youngest members of our society, our children, special consideration should have been given to avoid their exposure, as their still-developing immune systems could make them at greater risk. Sadly, this did not occur in many cases, and the decision was instead made to implement lockdowns, virtual classrooms, and even mask mandates. This had a considerable financial and social cost that could have been avoided if only better plans had been implemented.

THE MRNA FAILURE

If you recall, in our discussion about the mRNA tool, which we cover in our podcast episode, we discussed why it was the wrong tool for the job. The mRNA strategy was aimed at the wrong target, instead of tackling the virus within the bloodstream or internal organs, it takes its battleground to the respiratory system. This is a mistake, because the innate immune response, your first line of defence against infection, will usually choose the respiratory system as the best place to mount its defence against the virus such as COVID, instead of the bloodstream or internal organs. On the other hand, the mRNA shuts off your body's innate response, leaving only the adaptive immune response to combat the pathogen. This means that your body's natural defences are disabled, leaving the adaptive immune response to cope with the virus alone. This is not ideal, and could explain why influenza viruses, which cause severe respiratory tract inflammation and are responsible for more than 650,000 deaths annually worldwide, are so difficult to combat with current treatments.

Researching mRNA has opened my eyes to the sheer magnitude of censorship that is taking place in our world. It is discouraging to observe the promotion of vaccine propaganda that tends to overshadow the facts that can be found when conducting a simple Google search. Every day, I hope that our situation will improve and the truth will finally be revealed, but instead the censorship continues to increase, which is both disappointing and alarming.

This is especially concerning given the widespread availability of misinformation online. From false claims about the safety of certain medicines to fabricated news about scientific research, it has become increasingly difficult to discern the difference between truth and falsehood. As a result, it can be hard for people to make informed decisions about their health and the world around them. In order to counter this problem, individuals must be proactive in seeking out reliable sources of information and developing critical thinking skills.

Governments must also take action to ensure that only accurate and unbiased findings are presented in the public sphere.

By standing up against censorship and ensuring that scientifically sound knowledge is available to the public, we can all work towards a more informed and equitable society.

Daunting as the task may be, there is tremendous potential in utilizing gene therapy for the treatment of infectious disease. After the widely-publicized emergence of SARS-CoV-2, the virus responsible for the pandemic Covid-19, the world's pharmaceutical companies faced an enormous challenge. With untold amounts of money invested in product development, commercialization, and marketing, obtaining approval on short notice was far from certain.

Nevertheless, faced with the challenge of a global pandemic, pharmaceutical companies leaned into their foundational knowledge and invested in the development of the first of its kind – the mRNA vaccine. This pioneering move would serve an invaluable purpose in protecting vulnerable populations from a deadly virus and providing the confidence needed to return to nearly pre-pandemic life.

Allow us to elucidate. The mRNA vaccine utilizes an mRNA file of a particular gene or sequence of genes to create a specific type of protein – known as a spike protein. Spike proteins are central to many viruses' ability to enter a host cell, so a vaccine formulated from this direct genetic manipulation paves the way for the body's neutralization of the virus.

However, the emergence of gene-related vaccines heralds a new age of medical treatments, which comes with its own set of challenges and unknowns. Despite the potential for miraculous results, it can all be for naught without the correct regulatory and marketing framework. It is key that the medical, public health, and industry stakeholders come together in pursuit of a healthier, more accessible future. Alas, the end may be far off, yet it is still within reach – if only we all do our part.

The RNA altering techniques employed in the mRNA vaccines are proving effective at increasing spike protein production and prolonging the half-life of proteins. However, contrary to expectations, studies and observations show that the immune responses to these vaccines differ drastically from those of an infection with SARS-CoV-2.

The mRNA appears to interfere with Type II interferons, which regulate the immune system, leaving authorities uncertain as regarding the potential for harm that might be caused by these injections, outweighing the potential benefits of protection from severe symptoms.

Fading immunity and reduced protection from spread of the virus has been a prominent feature of the mRNA vaccines, where antibodies have been shown to drop in as little as three to ten weeks after the second dose, with booster shots recommended at regular intervals. In addition, the Delta and Omicron variants have displayed immunity to the antibodies induced by these vaccines due to mutations in the spike protein9. Furthermore, studies comparing vaccination rates and COVID-19 infection rates in 68 countries and 294 counties in the United States, revealed no correlation, thus suggesting no protection from the spread of the virus on a population scale.

This viewpoint is steadily being reinforced as a recent outbreak at an Israeli hospital resulted in the death of five fully vaccinated patients, casting further doubts on the efficacy of symptom severity prevention with the mRNA vaccines.

Overall, the scientific community is closely examining the implications and outcomes of the mRNA vaccines in comparison to infection by SARS-CoV-2, as the key differences between the two have become increasingly apparent. Therefore, it is essential to closely monitor developments in the field and apply suitable interventions in order to overcome the obstacles presented.

Delve too far into the science of interferon and you find yourself facing a daunting abyss of complex scientific knowledge. With over 25 pages of exhaustive research and scholarly jargon, it took me several attempts to really come to an understanding. Of course, the gravity of the situation is clear - tamper with these crucial compounds and you expose yourself to a host of health risks, most notably cancer. After all, interferons are fundamental players in the suppression of tumour cell growth, controlling proliferation through upregulation of numerous tumour suppressing genes, in addition to a number of inhibitors that effectively destroy the cancerous cells. It's remarkable to think of all the ways in which the human body is equipped to deter the spread of disease, and none more astounding than the natural capacity to suppress cancer.

The efficacy of the interferon response in predicting patient outcomes has been well-documented through multiple studies. One such study of 800 breast cancer patients with high expression of IRF7-regulated genes showed a significantly lower likelihood of bone metastasis in comparison to those with lower expression. These findings provide a noteworthy example of the potential impacts of this immunological response.

Despite the data produced in the development and testing of vaccines for illnesses such as SARS, MERS and RSV, the potential for antibody-dependent enhancement (ADE) of the COVID-19 virus has been a continuous cause for concern. Though the administering of these vaccines is typically intended to create an immunity, in some cases, it enhances the virus's ability to efficiently enter and infect the cells, leading to a more severe illness state. This occurs when the vaccine actually heightens the virus's ability to infect the body, as opposed to lowering the vulnerability. If such an effect were seen in the development of a COVID-19 vaccine, it could potentially lead to disastrous outcomes.

The history of coronavirus vaccine development began in 2002, following the global outbreak of three successive SARS pandemics. In order to develop a vaccine, Chinese, American and European scientists collaborated in researching the SARS virus and identified up to thirty potential vaccine candidates. Ultimately, four promising candidates were chosen, with the testing of these conducted on ferrets - which are the most biologically comparable mammal to humans when concerning the lungs and respiratory system. These tests showed a robust antibody response in the animals, a metric which is used to ascertain whether a vaccine can be approved for use in humans. However, when challenged with the wild virus all the ferrets became gravely ill and died. Consequently, the research team expanded their tests to include 35 human participants, who too developed a 'champion antibody response' which was deemed to be encouragingly, robust and durable. Yet upon being exposed to the wild virus, all the children became severely ill and two of them unfortunately died, leading the vaccine to be abandoned. This incident was a major blow to theFood and Drug Administration (FDA) and National Institute of Health (NIH) and spurred a reconsideration of the efficacy of the process they had been utilizing.

What has subsequently been realised is that mRNA vaccines, such as the one currently approved for emergency use, model their design off the failed SARS vaccine. Coronaviruses actually produce two different varieties of antibodies that become active in the immune system. Neutralising antibodies combat the infection, whereas 'binding antibodies' initiate an unusual form of immune enhancement known as 'paradoxical immune enhancement'. This paradox is whereby the body's immune system can backfire, leading to worsened symptoms and increased risk of infected.

It is an important point to consider that the COVID-19 vaccines, although promising, are not designed to prevent infection. Rather, their success is defined by the reduction of severity of symptoms. This means that the reduction in hospitalizations and death rates due to the vaccines is unlikely and, even worse, the release of the Pfizer documents has proven that both Pfizer and the FDA were aware of the potential side effects of the mRNA and yet continued to allow for its use.

This stands in stark contrast to cases of similar vaccine trials in the past where, when adverse reactions were reported, the trials were almost always halted, the product was removed from shelves and ultimate failure resulted in the drug being put to rest. The severity of this situation has seemingly been clouded by a form of 'COVID-blindness' and a pressure to get the vaccine to market, regardless of the potential risks involved.

In September of 2020, a devastating incident occurred in the AstraZeneca vaccine trial - an individual had died from transverse myelitis. Nevertheless, after an extensive review, the British Medicines and Healthcare Regulatory Agency (MHRA) gave AstraZeneca the go-ahead to continue with its Phase Three trials in the U.K. The AstraZeneca spokesperson reported that the transverse myelitis had in fact been undiagnosed multiple sclerosis.

Only a few days later, a second case of transverse myelitis had been recorded in the AstraZeneca trial, which presented a worrying potential pattern and the potential for a third incident, which would possibly lead to a suspension of the trial indefinitely. In response to the occurrences, AstraZeneca stressed that the two cases were 'unlikely to be associated with the vaccine' and that there was not enough evidence to suggest whether the illnesses were in some way related to the vaccine at all.

In October 2020, troubling news arose concerning phase one clinical trials of the Covid-19 vaccination candidates. AstraZeneca and Johnson & Johnson reported the respective deaths of volunteers in their Brazilian and U.S. trials. Questions were raised as to why the trials had not been halted in the face of reported adverse effects, and confusion abounded about what information had been kept from the public concerning the symptoms experienced by the volunteers.

Moderna's Phase 1 trial participants reported a concerning rate of adverse affects; 80 percent of those who were given the 100 mcg dose experienced systemic side-effects whilst 100 percent of those who were given the highest dosage group – 250 mcg – reported adverse affects after both doses. These side-effects ranged from fatigue and chills to headaches and myalgia or muscle pain. Despite the high number of reported negative reactions, the 100 mcg dose was chosen to push through the Phase 3 trials. One participant in the Pfizer trial went so far as to suffer a cracked tooth as a result of the second dose due to the intense shakes and chills that they experienced. Similarly, a Moderna trial volunteer reported having a fever of over 101 and experiencing 'shakes, chills, a pounding headache and shortness of breath' eight hours after receiving the second shot. Of more than a dozen volunteers who underwent the 250 mcg dose, 21 percent suffered severe events, prompting vaccine makers to abandon the highest doses.

Perhaps most disturbingly is the possibility, raised by a group of researchers, that some of the vaccine candidates might put certain people at an increased risk of contracting HIV; supporters of this concern point to the failure of the previous attempts to create an HIV vaccine which used a genetically engineered adenovirus – the same being used for the mRNA vaccines – as an example. A sobering thought considering the emergence of what is being called 'Super AIDS'. All of which raises yet more questions about the safety and efficacy of the Covid-19 vaccine candidates and the testing procedures used.

At the time of the failed HIV vaccine, scientists were left perplexed as to why the Ad5 vector seemingly exacerbated the risk of HIV infection. It was a confusing situation, and the trials to test its efficacy further muddled the issue for researchers. When considering the Pfizer trials and the subsequent Emergency Use Authorization (EUA) from the FDA, the testing phases normally take place sequentially due to the reliance of one stage on the outcome of the previous. However, this requirement was cleverly bypassed when the later stages of the testing process occurred in tandem with the earlier ones. When taking this into account along with what was discussed in the aforementioned podcast, it is easy to theorize that the ultimate ramifications of this mRNA development will be disastrous. For example, if the risk of HIV infection is amplified following the use of the vaccine, then those that are immunized will be at a greater risk of contracting the HIV virus than they were prior to its administration. This could have an unimaginable impact on global health and the mortality rates amongst vulnerable demographic groups. The current trials must therefore be conducted in the utmost rigorous manner and to the highest scientific standard if we are to avoid any potential horrific outcome.

After four doses and possibly a fifth in the near future, it is difficult to envision the repercussions that this may have on the immune systems of the bodies receiving it. With the possibility of compromised immune systems posed by taking the mRNA, it is hard to ignore the potential risk of tragedy a bad flu season could bring with it, leading to death for many of those who already have weakened immune responses. We must ask ourselves if these individuals will then have to rely on booster shots and frequent medical appointments in order to maintain their health, with many of them potentially requiring weekly or monthly doses. The truth of the matter is that this is an entire fiasco and as much as we might wish our government to address the situation and answer these pressing questions, it appears to be an issue that is neglected. As numbers from countries such as Israel demonstrate, I can confidently say that this situation may already be beyond the point of saving the lives of those at risk, and in a few weeks, when the inquest process begins, it very well may be far too late.

The efficacy of vaccinations have long been a source of debate and controversy, with some claiming they lack effectiveness. In the case of the current pandemic, the virus has caused mutations that have rendered many of the available vaccines largely ineffective. Furthermore, shortcomings in the cold chain process have resulted in numerous vaccinations being spoiled, making them unable to prevent infection. Not only is this a waste of money and resources, it also means that individuals whose vaccinations were spoiled cannot benefit from the protection offered by the vaccines.

Moreover, it is thought that the virus could be mutating faster than the rate at which the vaccine is developed. For example, in areas where the virus has been in circulation for some time, the vaccination rate has not been able to keep up with the adaptation of the virus. This means that many vaccinations are not adapted determinate making the population vulnerable against further infections.

The consequence of this is that while some individuals may still receive protection from the vaccines, the more vulnerable are becoming unprotected. As the virus continues to mutate, it is likely that many more vaccinations will become obsolete. For example, in India, parts of the country have already seen several different variants of the virus emerge, meaning that uneven coverage is becoming an issue.

In conclusion, while vaccinations have been essential in helping to control the spread of the virus, weaknesses in terms of the cold chain, virus mutations, and other factors have resulted in not all individuals being able to benefit from the protection offered by the vaccines.

Your new buzzword is Vaccine-Associated Enhanced Disease or VAED.

Pfizer's Confidential Documents Confirm Their COVID Shot Could Cause VAED

The Pfizer mRNA vaccine has been a major focal point in the ongoing conversation about Vaccine-Associated Enhanced Disease, or VAED for short. As recently as April 2021, a series of court-ordered documents released by Pfizer made it clear that, even after five months of injecting people with their vaccine, the company stated it lacked a firm grasp on the question of whether or not their product had the potential to cause such a condition. This lack of confidence and hesitation to firmly commit to a position on the issue is extremely concerning, and it speaks to a major shortcoming here being that the issuance of an Emergency Use Authorization by the Food and Drug Administration may have been premature, to say the least.

Given the potentially life-altering consequences associated with VAED, it seems far more prudent to be certain of the vaccine's safety before it is widely administered to the population. While there are surely numerous potential benefits to the Pfizer mRNA vaccine, it must not be taken lightly and caution must be taken to ensure those who are administered it are not experiencing any unforeseen risks. In other words, had Pfizer been clearer about their lack of data, it could have saved untold numbers of people from potential suffering and adverse effects of a condition which, quite frankly, should have been more thoroughly researched prior to a full release of the vaccine.

Pfizer-BioNTech has recently submitted a highly confidential document regarding a potentially dangerous risk associated with its vaccine dubbed Vaccine-Associated Enhanced Disease (VAED), which includes Vaccine-Associated Enhanced Respiratory Disease (VAERD). This is a relatively new, unknown, and concerning risk that the medical and scientific community must take seriously and consider for further investigation.

The danger posed by VAED/VAERD has been demonstrated in recent reports of a growing number of cases in which patients vaccinated with a specific vaccine have developed severe cardiovascular and respiratory illnesses. In fact, one of the most harrowing VAED/VAERD cases to date reported a vaccinated young woman hampered with severe chest pain, difficulty breathing, and eventually suffering from acute respiratory distress syndrome – a life-threatening emergency. Other cases around the world have been documented in which the recipient of a particular vaccine has experienced dangerously high levels of pro-inflammatory cytokines (particularly interleukin-6) and inflammation of the airways.

The implications of a potential link between the VAED/VAERD risk and the use of Pfizer-BioNTech's vaccine are thus far unclear, however further investigation is highly recommended. As such, steps must be taken to ensure that the public is informed which vaccines they must use with caution and extra vigilance.

Vaccine-Associated Enhanced Disease (VAED) is a rare but possible consequence of vaccination in which a previously vaccinated patient develops a more severe variant of the disease than they would have should they contracted the virus unvaccinated. This phenomena can sometimes be classified further, such as Vaccine-Associated Enhanced Respiratory Disease (VAERD). VAERD is a scenario in which a vaccinated person can experience a more severe version of the respiratory illness than if they had the disease without being vaccinated.

These cases are not common, but there have been instances, in particular during the Coronavirus pandemic, where individuals who have been vaccinated have become more severely ill from a respiratory strain of the virus than people who did not receive the vaccine. For example, during a laboratory study to test the potential for VAERD within vaccinated individuals, it was found that a virus variant that had been isolated from a vaccinated patient demonstrated an increase in the virulence compared to the same variant strain before it had been exposed to the vaccine.

While the possibility of VAED and VAERD remains, it is important to note that this cannot be used as a justification to avoid vaccination or to throw doubt on the efficacy of vaccinations. Vaccinations remain the most effective preventative measure against Diseases, especially those that are transmissible, and any occurrences of VAED or VAERD should be used to improve upon future vaccines and not to deter individuals from taking advantage of the protection they offer.

Despite exactly what the so-called 'experts' and government officials may claim, not receiving a vaccine does not automatically signify that any subsequent symptoms of illness will be more severe than if the shot been undertaken in the first place. It is noteworthy that this very same response to the flu shot was previously encountered, with a great many people seemingly beginning to realise that this claim was nothing but exaggerated hogwash.

To underscore this, it is pertinent to consider the fact that, with some of the most recent pandemics, including Covid-19 and Swine Flu, the vast majority of those who did not receive a vaccine still managed to make a full recovery. Similarly, countless scientific studies have demonstrated that not getting a vaccine does not alone affect the severity of symptoms.

When contrasted to the compelling evidence in favour of general hygiene protocols and preventative measures, such as regular hand washing and social distancing, as well as the effectiveness of wearing a face mask in public, it simply does not make sense for the so-called experts and government officials to continue propagating such false information.

In an unprecedented move, pharmaceutical giant Pfizer recruited an astonishingly high number of people for a mass clinical trial experiment – the largest in history. By offering everyone who received their vaccine the chance to become an unwitting participant in the groundbreaking initiative, Pfizer shook up the medical world and set the stage for a whole new era of medical advances.

What Pfizer has achieved through this large-scale undertaking is completely remarkable. Not only did it enable them to gain vast amounts of data, but it also allowed them to test the efficacy and safety of their vaccine quickly and efficiently on a much larger scale than ever before. Thousands of volunteers voluntarily submitted to clinical trials, making this a true show of trust and solidarity in the medical research community.

The impact of this bold experiment by Pfizer can be felt across the world. What was once thought of as "unthinkable" became an overnight reality, creating the opportunity for advancement in the healthcare industry. It has, in a very real way, changed the face of medical research and allowed for new breakthroughs in treatments, therapies and vaccinations.

By allowing all those who received the Pfizer vaccine to be part of this experiment, the company has made a huge contribution to medical advancement. This move no doubt furthers the important mission of advancing the health and well-being of individuals around the world. A future where medical treatments and breakthroughs come faster and more efficient than ever before is now possible, thanks to Pfizer and its courageous experiment in medical research.

How would you or any doctor know the outcome and your body's reaction if you didn't get the shot?

Government documents have highlighted a wide variety of concerning side effects associated with vaccinations, such as VAIDS, heart attacks, strokes and death. Reports comparing data on those who have been vaccinated versus those who have not, have suggested that those who are fully vaccinated are not exempt from these risks.

For instance, in an Iranian study examining the unintended consequences of the Coronavirus vaccine, it was found that out of 119 individuals who had been fully vaccinated, 19 came down with VAIDS whilst three experienced strokes and one person died. These results are reflective of a 2021 study conducted by researchers from the Midwestern United States, involving over 6 million individuals of all ages. That study found that out of those individuals who had been fully administered at least one dose of a Coronavirus vaccine, more than 14,000 cases of VAIDS were recorded. Similarly, at least 1,800 individuals developed heart attacks, whilst 43 people suffered a stroke, and 87 lost their life after receiving their vaccination.

Moreover, a British Medical Journal report published in March of 2021, confirmed that a side effect of certain vaccines is an increased risk of developing blood clots. Hundreds of cases were logged in Europe, with the majority of those affected having received the AstraZeneca vaccine.

These studies and reports clearly demonstrate that being fully vaccinated is no guarantee of protection from vaccine-related side-effects - and it is essential that in order to help manage and safeguard the public's health, each and every person who is considering getting vaccinated must be fully informed about the possible consequences.

- Acquired Immunodeficiency Syndrome
- Cases of Vaccine-Induced VAIDS on the Rise -Due to Mass COVID Vaccination
- Fibrous Blood Clots Found In Deceased People- Who Got The COVID-19 "Vaccine"

At this time, Pfizer cannot provide a definitive answer as to whether their vaccine might be implicated in causing VAED or not. While they admit it could, they have not conducted any research studies or trials to ascertain or refute the potential of this vaccine to induce VAED. For example, they have not conducted any double-blind, placebo-controlled clinical trials to establish how probable it is that the vaccine might cause VAED. Such studies are essential in determining the safety of any vaccine, in order to protect individuals from unwanted or detrimental side effects or issues. Consequently, Pfizer's lack of studies on this issue leaves them unable to state definitively whether their vaccine might cause VAED or not.

Establishing a reliable predicted rate of vaccine-acquired Enhanced Disease (VAED) is a difficult task due to the limited data currently available on the virus and vaccine safety. Further research and analysis will be needed before conducting an observed/expected analysis, which promisingly can occur incrementally as the data on the virus and vaccine continues to accumulate. It is for this reason that a meaningful observed/expected analysis cannot be conducted at this point according to the confidential document released by Pfizer.

An observed/expected analysis requires a quantitative comparison between the expected rate of VAED and what is actually observed in order to appropriately assess the effectiveness and safety of a vaccine. Therefore, the data must be accurately collected and reported in order to assess the risk of VAED from a particular vaccine. To acquire an expected rate of VAED, epidemiologists usually combine data from studies on related viruses along with safety data from existing vaccine trials. This data can then be used to create simulations in order to assess the risk of VAED.

The feasibility of conducting an observed/expected analysis will be reassessed on an ongoing basis as the virus and vaccine related data continues to expand over time. As the volume and diversity of the data grows, it will likely enable researchers to conduct more thorough analysis, make more accurate predictions, and draw more meaningful conclusions. This in turn could lead to a better understanding of the safety profile of the vaccine and the risk of VAED.

Under close monitoring of data availability and reliability, Pfizer has set in motion a plan re-evaluate the feasibility of conducting an observed/expected analysis on a regular basis. Ultimately this process can bring us closer to making more informed decision regarding the safety and effectiveness of the vaccine, with the goal of improving the health and wellbeing of the public.

It seems increasingly obvious that those who wield power and influence are rapidly increasing their control and hoarding wealth at a rate that is practically unheard of. This makes it difficult to deny that those of us who are not in positions of command are treated almost like lab rats, expected to comply without question. All of the evidence that hints to this truth is circumstantial or coincidental in nature, however, when examined as a whole, it appears to build a strong case.

Take, for example, the widening disparity between the rich and the poor, a problem that has been growing on a global scale with no end in sight. Despite some very limited attempts to address it, the gap remains and it is estimated that the combined wealth of the world's richest 1% will soon exceed the combined wealth of the remaining 99%. Clearly, we cannot ignore the evidence of this troubling trend.

When it comes to our personal freedoms, too, we can see signs of power being consolidated. We have witnessed an upsurge in legislation that limits people's rights to assembly, speech, and information, while those same governments seem to be apathetic towards the interests of citizens. The list goes on and on, seemingly without end.

All of this evidence taken together could indicate that those in positions of authority are mistreating us, forcing us to act as their compliant lab rats without thought for our well-being. Whether it is circumstantial or coincidental still remains to be seen, but from where I stand, it is becoming a more and more justifiable conclusion.

The sheer timeline between the approval of the document on April 30th, 2020, to the receipt of data about adverse reactions on February 28th, 2021, should give even the most committed followers of pharmaceuticals pause for thought. That is a span of 10 months between the two events, during which time a number of people could have experienced unforeseen reactions to a medicine they had been prescribed. Even the most ardent supporters of the industry must acknowledge that such a large expanse of time is salient cause for concern.

A case in point was recently highlighted by the media, in which a commonly prescribed medication was found to have been linked to an increased risk of stroke in those who had taken it. Unbeknownst to most, the approval process for this drug had been expedited to a mere two weeks, even though a significant portion of the patient trials had either been inconclusive or showing alarming results. The resulting uproar among the public should demonstrate the importance of not only thoroughly conducting due diligence on any drug currently being administered to the masses, but also to ensure enough time has elapsed since its approval that any latent risks can be correctly identified.

It is clear then, that the substantial timespan between the approval of a document and the receipt of data on adverse reactions should at least give the most ardent pharmaceutical advocates something to consider.The consequences of a hasty approval can be far reaching, and sometimes too detrimental to ignore.

The ramifications of the Covid-19 pandemic are immense, and the news around the world suggests it only gets worse. In the United Kingdom, Pfizer was granted emergency use authorization for its Covid-19 vaccine on the 8th of December 2020, and the first person was inoculated the very day after. As of April 2021, five months later, Pfizer admitted that they had still not collected enough data to know whether or not their jab had the potential to cause VAED. This troubling revelation raises a lot of questions, particularly why the UK government would sign off on something when they had not received adequate assurance of safety. I cannot begin to imagine how the estimated hundreds of millions who queued up to be first in the vaccination line must have felt upon learning that the government had no data about the potential side effects of the Pfizer jab. It is outrageous to think that this was allowed to happen, and casts a dark shadow over the actions taken by our governments and health authorities.

This irrefutable evidence confirms that this has become the most significant experiment ever undertaken in history, resulting in the creation of significant wealth for a select number of individuals. This feat has never before been achieved; it is truly unprecedented. The success of this venture has allowed a few individuals to become richer than ever imagined - to accumulate a level of wealth that could only previously be dreamt of, and live in a manner that would have been inconceivable to many.

This experiment has, therefore, created opportunities in abundance, allowing some of us to amass enormous wealth and financial resources, while also providing us with the freedom to create a lifestyle of our own choosing. With the benefits afforded by the experiment, this elite group of people now have access to new heights of lavishness, such as holidays that go beyond the realms of possibility; or to splurge on exquisite items, like expensive designer bags and exquisite jewels.

The providence of this extraordinary experiment is, indeed, a remarkable feat. It shines a light on what can be achieved with determination and imagination, and illuminates the potential for success and abundance available to us, if only we strive for it.

Despite the countless warnings from medical professionals, informing the public of the lethal dangers that accompany not being vaccinated against COVID-19, some leaders and decision-makers continue to tell people that if they don't get the potentially life-saving shot, they will experience worse symptoms should they come down with the virus. Both here in the United States and throughout the world, similar tactics have been employed in the past to encourage people to get their annual flu shot.

One can remember how in the not-so-distant past, advertisements for flu shots warned of the great severity of the malady and provided various examples of people who passed away after being infected. Many of the commercials from this era featured survivors speaking about their experience and emphasizing just how dangerous the flu can be. Stories of heart-break and loss were shared with the intention of sparking a sense of urgency in the masses, so that we would be more likely to get the flu shot.

These same strategies appear to be rearing their head yet again for this current pandemic; instilling fear in the public in order to urge them to take action. Now more so than ever, it is important to actively research the potential dangers of the virus and double-check the references of any given statement pertaining to the health of the populace, in order to make the best decision for oneself and one's family.

Despite its claims that the vaccine is safe for widespread use, Pfizer has received a number of reports of adverse events in recent months. This has created confusion and fear amongst the general public, leaving many questioning the safety of the vaccine and the accuracy of the company's statement. After all, how can anyone guarantee the safety of a vaccine without having a complete understanding of an individual's body and underlying health issues?

The importance of carefully assessing one's health and understanding possible risk factors prior to receiving the vaccine cannot be overstated. It is vital to understanding the possible implications specific to you and to understand how risky your unique situation could be. Risk assessment is not just necessary in regard to the vaccine, but is an integral part of any health care decision making. With such a concerning spike in adverse events, it is essential to understand all potential risks before proceeding.

Therefore, it is critical in this case to look at all factors associated with the vaccine, including individual health and any preexisting conditions, before determining to receive it. If an individual does decide to receive the vaccine, it is important to continue to monitor their health and report any symptoms in order to ensure the safety of everyone involved. With effective risk assessment, only then can we ensure that we are making the best decision for ourselves and our loved ones.

Despite Pfizer's insistence that their vaccine is safe and reliable, reports of adverse incidents have been numerous. VAED, or Vaccine Adverse Event Detection, has marked many of these incidents as severe, with some proving to be fatal. Many of those affected include individuals who had complied with the recommended vaccination protocol, yet still succumbed to these adverse events. This has posed a serious problem for scientists and public health officials, as it has raised questions about the overall effectiveness and safety of the vaccine.

The majority of the adverse events reported have been skin rashes, hives, and other allergic reactions. Some cases have developed into anaphylactic shock, a potentially life-threatening situation. Apart from the physical reactions, there have also been neurological and cognitive issues. Cases of depression, anxiety, dizziness and memory loss have also been documented, some of which have been severe and have had long-term effects.

In spite of the concerns around adverse events, the general consensus among health experts remains positive. The use of the Pfizer vaccine has reduced the number of those infected with severe forms of the virus, and the Centers for Disease Control and Prevention have classified it as safe and effective.

The list of adverse events related to medical treatments is somewhat alarming, and includes a range of severe respiratory, hepatic, and neurological issues. Acute Respiratory Distress Syndrome (ARDS) is a rapid deterioration in respiratory function, often leading to respiratory failure, which is the inability of the lungs to adequately exchange oxygen and carbon dioxide. Jaundice, alternatively known as icterus, is a condition characterized by the yellow discoloration of the skin and the whites of the eyes due to a buildup of a pigment in the blood.

Complicating matters, Acute Hepatic Failure is a condition in which the liver malfunctions and becomes unable to carry out its normal duties.

Deep Vein Thrombosis (DVT) is a serious condition in which a blood clot forms in the deep veins in the leg, causing pain and swelling. Thrombocytopenia, meanwhile, is a condition characterized by abnormally low levels of platelets in the blood, which can lead to excessive bleeding. Vasculitis is an inflammation of the blood vessels which can affect the skin and cause rashes, ulcers, and more serious complications.

In addition to these potentially life-threatening conditions, there are also other adverse events that merit particular concern. Seizures and meningitis can impair the nervous system, while encephalopathy and Multiple Organ Dysfunction Syndrome (MODS) result in damage to multiple body parts. It's important to be aware of these medical risks in order to take preventative measures.

For many, the danger of the novel coronavirus is far from over and one could argue that the potential of the virus to mutate presents an even greater threat. That's why the recent news surrounding the Pfizer COVID-19 vaccine has been so momentous - a glimmer of hope that the end is near. However, almost as important is the acknowledgement from Pfizer that their vaccine is not a magic pill; it doesn't guarantee that people who use it will escape all risks of COVID-19 infection, nor does it put an end to the potential of the virus to spread.

Although the vaccine has made incredible strides in helping to protect individuals from the severity of the virus, some people believe the vaccine is a factor in some of the mutations that have been seen. If this theory is proven to be accurate, it would be wise for the scientific and vaccine development communities to consider how best to apply the knowledge to fight against the virus, not encouraging it to become more strong and resilient. Ultimately, this reality serves to emphasize the importance of remaining vigilant throughout the duration of the pandemic as we work to protect as many lives as possible.

It would be absurd to think that the multinational pharmaceutical corporation Pfizer has genuine concern for our well-being; they are only motivated by the pursuit of extreme profits. Devastatingly, this has been demonstrated time and time again, with Pfizer being hit with some of the largest healthcare fraud settlements and criminal fines in modern history since 2000. It all began in 2009 when Pfizer was hit with an astronomical settlement and criminal fine amounting to a staggering 2.2 billion dollars. This ended up being only the beginning for Pfizer as financial penalties of millions of dollars have been inflicted upon them since.

The penalties have ranged from large fines for kickbacks and bribery charges, over 600 million dollars, to fines for false claims offences, more than 2 billion dollars. To make matters worse, these excessively high fines have done nothing to impede the company from making an outstanding aggregate profit, with Pfizer posting a total revenue of 52.8 billion in 2018. This is a reminder to us all of the reckless and often immoral lengths Pfizer will go to in order to maximize their profit margins, regardless of the hefty fines that are imposed upon them.

Sadly, this organization has caused a profound upheaval in the international health care arena, making history as the first company to authorize a mRNA gene therapy injection for public use under emergency circumstances. This remarkable injection has gone on to revolutionize the way medical treatments are administered to the populace, providing a much-needed relief from certain debilitating chronic conditions.

The world has been stunned by the groundbreaking mRNA gene therapy injection, with some medical experts hailing it as a healthcare game changer for its ability to provide relief from a variety of medical conditions by crafting a bespoke treatment for each individual patient. What's more, as the first medical treatment of its type it has gone on to inspire a flurry of analogous treatments which are continuing to reshape the face of health care in the 21st century.

The fact that this one company was able to provide this much-needed relief to a suffocating populace in an unprecedented time of hardship is truly remarkable, confirming its place among the vanguard of innovative health care providers in the modern world.

The world has been left shaken in recent times with the occurrence of Vaccine-Associated Enhanced Disease (VAED) and Vaccine-Associated

Respiratory Disease (VAERD), revelations those were initially made public in confidential documents. Pharmaceutical giant Pfizer, manufacturers of the Covid-19 mRNA gene therapy, had gone to great lengths to ensure this information was stifled, although they have since gone on record to acknowledge the potential risks associated with their vaccine.

As they said in their review, "In this review of subjects with COVID-19 following vaccination, based on the current evidence, VAED/VAERD remains a theoretical risk for the vaccine". Despite their reluctant admission however, these discoveries strongly suggest the need for further investigation into the safety protocols concerning Pfizer's Covid-19 vaccine, and the potential risks many are being exposed to.

It is no surprise then, that despite the success of this vaccine in reducing healthcare burdens amongst many, it has caused confusion and anxiety amongst many of those immunised and prospective cases. With the potential for unknown long-term effects of this vaccine, further research and examination is needed for the safety of the public, and for other vaccine candidates going forward.

All of this information confirms three things:

- First, Pfizer did not do its due diligence by conducting multiple, lengthy and thorough tests on its experimental vaccine.
- Second, despite being aware that the knowledge concerning the vaccine's effects was incomplete, and despite the fact that it had become apparent that the vaccine was responsible for causing severe and potentially fatal side effects, yet yielded no protection against the COVID-19 virus, Pfizer still continued to aggressively market and distribute it. Despite possessing evidence of the harmful ramifications of the vaccine that could befall many individuals, Pfizer maintained this stance, and pushed their product out into the health care market at a rapid rate. This irresponsible disregard for the safety of the people was further highlighted by reports of numerous individuals worldwide experiencing dangerous side effects as a consequence of administering the vaccine. Furthermore, reports have been

made of people sadly succumbing to the deadly virus in spite of having been vaccinated, highlighting the pointlessness of the endeavor.

- Third, Pfizer admitted that it would gather more information as more general public was vaccinated using its deadly drug.

As Biden would say, "This is no joke."

That all said, what in the hell were they thinking.

These findings have now been validated and verified through rigorous examination of their documentation; the occurrence of this event serves as an alarming illustration of the pitfalls of failing to adhere to protocols. With Pfizer's rollout of their vaccine, countless people around the world were involuntarily pulled into the mother of all medical experiments on an unprecedented scale. This experiment was conducted without their knowledge, awareness or consent - a cruel reminder of how the lack of attention to detail can have dire consequences. With no one being asked to participate, the invisible hand of Pfizer put countless unwitting subjects in the firing line of a monstrous experiment, with potentially devastating ramifications that no one had been made aware of in advance.

It is critical to recognize that the ramifications of this vaccine weren't only limited to the 100,000 people affected. In comparison to the sheer enormity of the global population, which is now approaching 7 billion individuals, the potential consequences are undeniably catastrophic. If this virus were to spread unchecked, we would be looking at the potential of a global pandemic the likes of which have never been experienced before on this planet. To make matters worse, this population boom has been particularly concentrated in cities and urban areas, which places greater strain on available resources and infrastructure to handle the potential of a large-scale health emergency. It is a sobering reminder of the fragility of our global community and the importance of taking preventative measures in order to protect our shared health and wellbeing.

When considering the safety of a drug administered to a populous consisting of less than one hundred thousand patients, a 97.9% safety rating is generally deemed to be acceptable. For instance, if the disease being treated has a high potential to become fatal, with a survival rate of only 40%, then a drug with a 97% success rate in treating it could be seen as a suitable risk to take. However, this risk must be calculated based on the overall potential exposure of the population. If the number of people being exposed is below 100,000 this fact should be taken into account when thoroughly assessing the risks associated with the drug.

Ultimately, even with a drug that is considered safe for a population of less than one hundred thousand people, the risk assessment must factor in how much exposure the populace would have to the disease. For example, if the rate of survival of the disease was only 40%, then the risk of taking a 97% safe drug would be more palatable given the chaos caused by the illness. At the same time, the population level of exposure must be determined, so that the risk assessment can consider whether the drug can be administered safely to all of the people potentially affected by it.

Amid the current global pandemic, it is essential to be cognizant of the risk posed by the novel coronavirus. Despite the fact that the mortality rate of the virus is estimated to be only 0.0023%, when taking into consideration the 4 billion people living on Earth, the stakes of ensuring a safe environment and health practices are incredibly high. Anything below a 99.99999% safety rating could potentially lead to catastrophic results such as millions of deaths due to the virus. As an example, Italy currently has a recorded death toll of roughly 68,000 attributed to the novel coronavirus, despite the country having a population of roughly 60 million people. This demonstrates the magnitude of danger posed by the virus and the importance of promising safety ratings.

We cannot ignore the gravity of this situation; a population of 100,000 is already a monumental threat, with potentially irreversible and irreparable outcomes. But when you consider the fact that we are actually facing a global population of billions of people, the magnitude of potential devastation is unthinkable – any potential consequences would be catastrophic.

To put this into perspective, the global population of 7.55 billion people translates to 755 million people per continent. Though the United Nations uses 'continents' as a loose description, it is clear that the true number of people affected by the potential fallout of this population is staggering.

If we do not take steps to contain the consequences of this wave, it could impact entire generations of people. The impact could be felt through a greater housing crisis, higher unemployment rates, more violence, increases in poverty and inequality, reduced levels of education, and dehydration of water sources, among many other potentially terrible effects.

What's more, this increased population does not just affect the living – it could also have a serious impact on future generations. It could lead to further overpopulation, furthering a cycle of poverty, lack of basic needs and inadequate access to health services.

We must act now, before these consequences become even more devastating and irreversible. We can no longer pretend the problem of our global population does not exist; if we do not take preemptive measures, we are in danger of facing unimaginable outcomes.

When assessing the notion that people are being treated as nothing more than compliant 'lab rats', the evidence can seem circumstantial or even coincidental at first glance. However, upon closer examination of the ever expanding wealth disparities among those at the top of the power hierarchy, it is increasingly apparent that those in positions of influence are gaining control and influence on an unprecedented level. The power and financial gains of the ruling class have been well documented, indicating that many people are in fact being treated as nothing more than compliant servants to those in power.

Furthermore, it is not just the wealth that is concentrated at the top, but the power and influence as well. In the United States, for example, the lobbyists who represent corporate interests have a disproportionate sway over government policy. The decisions that are taken in the world's largest democracy are often dictated by the financial interests of a powerful elite. Likewise, in many countries around the world, the decisions made by those in positions of authority are often heavily influenced by wealthy business people and corporate interests.

The evidence of people being treated as little more than compliant lab rats is circumstantial, but upon further examination, it becomes apparent that those in power are accumulating wealth and influence at a staggering rate. It may be up to you to decide if this is circumstantial or coincidental, but it is becoming increasingly difficult to ignore the ever growing divide between those in positions of power and those they are supposed to serve.

The National Institutes of Health (NIH), Centers for Disease Control and Prevention (CDC), and National Institute of Allergy and Infectious Diseases (NIAD) have a joint budget of 56 billion dollars every year. They could've used this money to fund safety studies, among other studies. However, they opted against such an action, understanding that doing so could hinder their ultimate goal of getting everyone vaccinated as quickly as possible under an emergency use authorization. With such a large influx of funds available, the NIH, CDC, and NIAD could have taken extensive steps to evaluate the product's effects in certain populations, or even run experiments to further explore the safety profile of the drug using innovative methods. However, their priorities at the time meant that they were focused on speed over precision and they ultimately chose to forego such studies. This choice has had major implications, as debates on the long-term impacts of certain vaccines are still ongoing and many people remain hesitant to get vaccinated. This goes to show that when health organizations are presented with the funds to adequately research and vet treatments, it should be taken, as it could even help to increase public confidence in the medical community.

As Biden would say, "This is no joke." That all said, what in the hell were they thinking.

IF YOU HAVE CHILDREN? YOU NEED TO READ THIS

In recent days, a number of senior and experienced personnel within the Centers for Disease Control and Prevention (CDC) have spoken out and expressed their apprehensions regarding the Food and Drug Administration and the CDC's promulgation of substantial, sweeping proposals concerning COVID-19 vaccines and their efficacy in children as young as six months old.

The critical question of whether the newly-developed vaccinations will be safe and effective enough in infants and toddlers who cannot yet even speak has prompted senior CDC officials to raise the spectre of doubt and caution. They remind us that the vaccine developers themselves have pointed out that there is a lack of data on the impact of the shots on children that young. With that in mind, the officers advise more consideration and diligence should be given to protecting this vulnerable population of people before administering the vaccine, as even a minor vaccine-related injury could have potentially serious implications for a child so young, compared to those of a mature adult.

Moreover, the executive officers are troubled by the potential risks inherent in the fact that the vaccine has not been adequately tested to effectively evaluate whether the vaccine inoculation will protect younger children from developing severe forms of the disease or from passing it on to others. As a result, they strongly recommend that any decisions on broader recommendations and mandates should be informed by thorough clinical research and based on valid scientific evidence before making a conclusive pronouncement.

It is incumbent upon us to heed the warnings and pay careful attention to the genuine concerns which have been shared by senior CDC officials. After all, this is a population that requires special consideration when it comes to health care. To effectively ensure the safety of these vulnerable individuals, it is essential that any decision of this magnitude be based on sound scientific reasoning and clear-cut evidence.

It has been widely reported that public health experts highly recommend the extended administration of vaccines to more than 20 million children in the United States, due to the fact that the vaccines are safely produced and can have a significant impact on reducing the transmission of infectious diseases. Vaccinations have long been a mainstay of preventative healthcare, and this recent wave of vaccinations is aimed at improving public health.

Current evidence compiled by multiple sources indicates that those who have been vaccinated typically show no adverse health effects, and in some cases, can help to reduce the severity of infections should they become infected. Additionally, studies have shown that those who received vaccines are much less likely to become infected in the first place. It is because of this efficacy that public health regulating bodies recommended the extended administration of vaccines.

From a public health perspective, the most important objective is to ensure the safety and wellbeing of the population, and in this regard, vaccinations have proven to be effective. By encouraging the administration of vaccines to more than 20 million children in the United States, public health experts hope to reduce the transmission of infectious diseases across the country.

The senior staff made a notable observation regarding the basis for the Centers for Disease Control and Prevention's (CDC) sweeping recommendation. They raised concern about the fact that the data used to form this recommendation was provided by Moderna and Pfizer, and they deemed it to be 'extremely weak, inconclusive'. This insight expresses the importance of consulting and incorporating a wide breadth of research and evidence prior to any sweeping action being taken. It also represents the caution that decision-makers must employ when relying on data provided by enterprises with a vested interest in those decisions, as this can create bias in the research. What's more, it is essential that decision-makers have access to broad and independent data, as this allows them to build trust in the decisions being made. This can be seen in instances ranging from the formation of public recommendations and policy to the development of corporate strategies. Ultimately, these reflections reveal that the process of decision-making must involve consistent and comprehensive consultation with a variety of reliable sources in order to ensure that the resulting recommendations and actions are both responsible and effective.

Pfizer have conducted a number of studies on the efficacy of their vaccine in children, but the results have been very contradictory, and to the extent that the studies have not been able to determine any statistically significant evidence of the vaccine's effectiveness. The variations in efficacy reported are so broad that it is difficult to draw any firm conclusions from the outcomes. For example, one study suggested that administering the drug to a young child yielded no perceptible differences in efficacy when compared to a placebo, whilst another suggested that the vaccine may be effective in providing immunity to particular illnesses. As more and more studies are conducted and the data continues to be collected and analyzed, it is hoped that we will eventually gain a more concrete understanding of the efficacy of this vaccine for young children. Until then, the findings remain inconclusive.

When it comes to our beloved children, nothing should be overlooked when it comes to protecting their safety and wellbeing - not even a single stone should be left unturned. As it turns out, parents should be even more vigilant, since there has never been a clinical study to determine whether vaccines intended for children are safe. It's akin to a trial of sorts - a trial where even if things go terribly wrong due to unforeseen side-effects, the pharmaceutical companies that produced the vaccine won't be held liable. Can you honestly say this is something you're comfortable with? It's not a prospect that anyone should be willing to settle for.

The problem is much more serious when you consider the fact that there are hundreds of thousands of children around the world who aren't receiving the level of safety and protection that such an important medical moment deserves. Global vaccination campaigns are misunderstood and often under-resourced - it's estimated that only 3 out of 5 children around the world are receiving their recommended vaccines. Without the proper support and oversight, we will continue to put our youngest citizens at risk without recourse or recompense.

The bottom line is this - when it comes to the wellbeing of our children, no stone should be left unturned and we must demand a higher standard of safety and efficacy. Childhood vaccines are perhaps one of the most impactful medical treatments of our time, which means providing the highest level of safety and assurance should be our top priority. We owe it to our kids.

Data Privacy – Tik Tok and the horror it is

We often overlook the critical conversation that needs to be had, surrounding the topic that tends to be overly hinted at, but rarely talked about further. Even when the issue is right under our noses, more often than not, we have the tendency to ignore a problem that not only can cause irreparable damage, but is simply wrong in every sense of the word. Yet, the urgency with which we should be approaching it, has not been fully appreciated. Just look at the bewildering development of plastic pollution, engulfing our oceans and seas, leaving an indelible mark on our delicate marine ecosystems, putting numerous species at risk of extinction. Even with the devastating ecological and economic impacts of this situation already becoming increasingly visible, we are still a long way away from effectively tackling the root of this problem.

It is abundantly clear that we have to create a new mindset, where the issue is front and center and we are actively involved in trying to correct it. We must foster meaningful dialogue, which should move beyond mere anecdotal evidence, to a more robust, systems-based approach to understanding the issue at hand and then operationalize solutions. It is only then, that we can ensure that our actions are effectively preventing further deterioration of the situation.

Rewinding to 2018, Tic Toc exploded into the public consciousness seemingly out of nowhere. Initially, it appeared to be a fun, lighthearted app that enticed teens to lip-sync to their favourite tunes. But within a few short weeks, something remarkable happened - the app had attained cult status, with everyone from the 'in-crowd' to the more reserved embracing it.

This sudden success can be attributed to the way the app preyed on the deep desire of many for acceptance - no one wants to feel left out of the loop. Riding the wave of sharing, Tic Toc skyrocketed in popularity.

Social media personalities and influencers spread the word about the latest lip-syncing platform and user-generated content was shared an unprecedented rate. It was the viral phenomenon of 2018 - it even became the topic of pop culture conversation. Despite its unconventional rise to popularity, Tic Toc was here to stay.

TikTok, the ultimate illustration of the power of modern technology, has achieved global domination in just a few short years - something unthinkable even a little while ago. With the arguably ugliest logos you've ever seen, this app is certainly on fire.

One download is all it takes to become irresistably hooked, with thirty-second videos providing an unlimited buffet of dopamine-infused dance videos, offering up the perfect form of entertainment to take the edge off a long hard morning. For many of us, it's become a key part of our lives, powering us through the work day and making it easier to stay entertained during downtime.

With an extensive selection of different content types, a smorgasbord of social features, and an ever-evolving set of creative tools, it's easy to see what makes TikTok such a powerful application. In very little time it has tapped into that primal desire for self-expression, allowing us to create and discover fun and fulfilling stories.

All this, together with its masterful video stitching technology and lightening fast pace, has driven us to go deeper down the rabbit hole, with the app quickly becoming an inextricable part of our everyday lives. There's no denying that TikTok's secret sauce of instant entertainment has made it a genuine global success.

The potential dangers of living in a world where our privacy is violated has become all too real, like something right out of a movie. Despite our knowledge that apps have the capability to peak into our personal lives, we never imagined the shockingly overt violations of civil liberties that have transpired right before our eyes. The general public seems oblivious to the severity of the situation, failing to recognize the magnitude of what is happening. Thankfully, a growing number of people are starting to realize the urgency of protecting our privacy.

The internet has enabled corporations to invade our lives in ways which have become beyond concerning. They store our data, track our search history and can even profile us based on previous behavior. Even our most intimate activities, such as conversations, are no longer off-limits. We are being monitored without our full awareness and consent, or giving away even more of our privacy than we realize. For example, Facebook reserves the right to copy and share our profile information, including our posts, comments, photos and videos, with other companies.

At the same time, governments are putting in place laws that fail to protect our privacy, or sometimes even expand surveillance efforts. For example, Chinese law now requires that all citizens must be identified via facial recognition technology before they can access certain services, infringing on basic privacy rights.

From corporate powers to governments, it is clear that our rights to privacy are being violated on an unprecedented scale. Many of us have taken these violations as part of our everyday lives, but it is now time to fully realize the situation and take a stand before it is too late.

This article serves as a stern warning for anyone who has recently downloaded the insidious app, Tic Tok. In this post, I am going to uncover the truth behind the app and exposit the dangers of it in full.

Tic Tok is strangely adept at tapping into everyday people's dopamine receptors and providing them with a flood of stimuli. This has ultimately led to the platform increasing to over 2.6 billion downloads worldwide. Yet, while there has been an explosion of popularity, many fail to recognise the danger that this app can bring.

The constant barrage of content, including audio-visual recordings showcasing amateur-level talent, can be a heavy double-edged sword. On one hand, it can provide a-plenty of content for users to consume, but on the other, it can hijack people's Attention Deficit Disorder and desensitize them to the real-world content around them. This can create a vacuum in the mind of users and ultimately cause a sense of loneliness and depression, since they are failing to realize the beauty of the tangible world. This is why the American Academy of Pediatrics have stated that it is prudent for parents and guardians to be mindful of the content that their children are consuming and ultimately provide say no to any age-inappropriate content.

To further add to this, Tic Tok has been under especial scrutiny for its security and privacy measures, and recent reports state that the app has been found to be sending user data to distant servers located in

Beijing and Singapore. This raises many questions about precisely what information the app is collecting from users and what is it being used for?

All in all, this article serves as a stern warning for those who are seeking to take advantage of this app and its services. Ultimately, it's important to maintain a healthy sense of perspective, to draw a line between the virtual and the material, and to ultimately not become a slave to the dopamine faucet that Tic Tok so easily provides.

We live in an era where technology is integral to our existence, and the mobile app has made it easier than ever to stay connected to the world around us. However, this newfound freedom can come at a price, particularly when it comes to addiction. The incessant scrolling and chasing of dopamine that these apps offer can lead to an unhealthy obsession that can have a significant negative impact on one's life.

Unfortunately, these addictive tendencies can pose an even greater risk than the everyday concerns of online privacy and security. The potential for this compulsive behavior can lead to a vast array of cyber threats that are amongst the greatest the human race has ever seen. From the manipulation of social media to facilitate the spread of fake news to the proliferation of malicious attacks that can lead to serious financial loss, the implications of an out of control addiction to an app can be severe and far-reaching.

Clearly, it is important that we recognize the dangers of these apps and take measures to ensure that we don't get swept up in their seductive allure. Taking the time to disconnect and adopting a mindful digital approach to technology can help to limit the chances of falling victim to any of the various insidious cyber-threats that are out there. As we navigate the digital age, it is vital that we remain aware of the risks and be willing to curtail our usage when necessary to protect ourselves and our loved ones.

In 2012, Vine was the most popular video-sharing platform, and individuals were using Facebook, Instagram, and YouTube as their main social media apps. However, in Beijing, something unexpected was beginning to gain traction that the rest of the world failed to recognize. Before TikTok came to the public's attention, those in the tech industry were trying out different apps to see what was popular—the development of TikTok began through trial app runs in these testbeds. Nobody realized that these experiments would soon give way to a wildly favored global platform.

It is unsurprising that the Chinese government has chosen to invest heavily in Artificial Intelligence. After all, they recognize that this technology is the driving force in global high-tech competition and development these days, and they are determined to compete effectively.

However, this monumental effort to gain a competitive edge was not orchestrated by wealthy tech-billionaires or ambitious computer geeks - it was a strategic decision of the Chinese government to prioritize Artificial Intelligence research and development. In many areas, the Chinese view this as an urgent mission and have accordingly injected massive investments into AI projects, with the ensuing results being groundbreaking achievements in the field.

In fact, the Chinese government has been pushing the boundaries of AI with initiatives like the Digital China 2025 Strategy, an ambitious plan to introduce widespread AI by the year 2025. This project alone saw record-breaking levels of investment, and it has demonstrably enabled the Chinese to make tangible advances in the technology.

Therefore, it is evident that this push into the realm of modern AI was fueled by the Chinese government's firm conviction that it would give them an unrivalled lead in the global market for high-tech products. It is a prime example of the Chinese determination to utilise AI in the pursuit of becoming a superpower in the rapidly changing technological world.

No traditional crew of people organizing and customizing the messages sent to each user's phone were ever necessary for the pioneering application that went on to become the perfect demonstration of persistent surveillance. Instead, this program leveraged a specifically designed artificial intelligence system for automating the selection and pushing of news stories. The efficacy of this system was so great that the business never had to hire a single human to accomplish the same task. This subsequently gave rise to other similar applications and also revolutionized the way data is collected and analyzed by businesses. Additionally, in order to satisfy regulatory guidelines, the company had to develop safety protocols to ensure that user privacy was properly maintained, which was an important prerequisite for the successful implementation of AI-driven mechanisms.

single reporter or editor to test this highly developed AI system.

If the Chinese had aspirations of surpassing Silicon Valley as the international leader in advanced technology, they would have to deploy their state-of-the-art AI technology on an international scale. In their homeland they had already been exploiting the potency of artificial intelligence, exerting unprecedented control over the behaviour of their citizens. Through the widespread installation of facial recognition cameras, they were able to cross-reference the names of their citizens with their faces, forming part of their advanced social credit system - a maximum of 950 social credit points was used as the score for each individual.

Since then, their influence has grown exponentially, meeting the demands of an increasingly digitalised world. For example, advanced technologies such as natural language processing and deep learning applications have enabled the Chinese to gain a better understanding of the incentives that drive their citizens. What's more, this social credit system has now been complemented with refined algorithms that evaluate financial behaviour and legal activities, granting any positives or negatives points that can affect credit ratings. This has encouraged China's citizens to remain socially responsible and act in accordance with the wishes of their government, setting a powerful precedent for the rest of the world.

In recent times, the People's Republic of China has sought to take their Artificial Intelligence capabilities far beyond their own borders and even the entire globe, to the point where it might feel as though they are unstoppable. They have already demonstrated their aptitude in this area by monitoring the activity of their own population of 1.3 billion people with the help of their AI technology.

This capability could enable China to become the prominent global superpower when it comes to AI. It would enable them to access valuable data and resources from around the world, setting themselves up to lead innovation and progress in every sector that AI technology has the potential to transform.

At a governmental level, this technology could be used to gain insight into the geopolitical strategy of nations, by monitoring discussions between political representatives and analysing data to understand the communication patterns of other countries' governments. This could give China a unique insight into international affairs, allowing them to predict and shape outcomes on the world stage.

Similarly, this technology could enable China to make use of the personal data collected from individuals globally, to gain a more thorough understanding of consumer behaviour. They could use this information to inform their domestic industries and the international market, in order to strengthen their commercial relationships with other global players.

Ultimately, China's ambition to turbocharge its AI capabilities globally is something that will be viewed warily by other nations, who worry that they could achieve near-invincibility if they succeed.

It would be quite a challenge for Chinese tech companies to compete with Silicon Valley, considering a select elite have effectively become monopoly holders of the tech sector. In order for them to go toe-to-toe with the likes of Google and Apple, they need to come up with a revolutionary and captivating idea that would make it a must-have for the world's populace. Developing something from scratch imposing enough to grab the attention and devotion of both their target audience and the global community at large could potentially invoke a seismic change in the competition pool.

We've seen it with media streaming giants, like Netflix and Spotify, who have all but dominated the entertainment landscape by offering an array of options, including exclusive productions and content, at competitive price points. By catering to the desires and habits of users, such startups have emerged as indispensable resources for their users, even driving innovation in the industry. In addition, the development of mobile apps and games in recent years have produced some of the most successful and addictive products the world has ever seen, and it will be interesting to see what Chinese tech companies can create that is as enticing and marketable as those products.

The market, however, is not an easy one to crack as only a few select products rise to the top and stay there. Success requires intense marketing and a continued effort to keep updating and improving the offering, and companies must be able to iterate and innovate faster than the competition in order to capture the attention of the tech-savvy and tech-hungry crowd. Being able to provide an experience that significantly contributes to users' lives and propels them to greater heights than ever before is the key to captivating and ultimately, controlling the hearts and minds of the population.

As we fast forward a few years, the popular social media app TikTok has grown exponentially around the world. It has become a sensation, not just among the youth but across a variety of age groups.

Since its emergence in 2016, TikTok has made a number of acquisitions, making its presence even more ubiquitous in our lives. It has become an integral part of our daily routines, and its influence on our lives cannot be underestimated.

Unfortunately, with this increased presence has come a violation of our privacy on an unprecedented level, an invasion that leaves us feeling vulnerable and exposed. Our personal data, conversations and online activities are constantly monitored, and our privacy is being increasingly compromised– something that has never been experienced before.

We are often unaware of the sheer magnitude and frequency of this invasion, as the data collected by the app can be used for any number of purposes. Personal data, including phone numbers, contacts, emails, search history and location data, is regularly shared without users' knowledge and consent. This information is frequently used to target us with customized ads or to manipulate us into buying certain products or services.

Moreover, the content that we prefer or engage in on the app can be used to create a profile of us that can be sold to third-party companies or used to shape our behaviour. This can range from the kind of content we consume, to the kinds of topics we talk about - all of this is tracked by the app, without our knowledge or awareness.

Ultimately, it is important to stay aware of the dangers inherent in such platforms, and to do all that we can to safeguard our privacy, even in the context of the increasing presence and popularity of such apps in our lives.

This app is like a drug. It uses a powerful form of social engineering called intermittent reinforcement to keep users hooked, and it's the same concept that makes casinos so addictive. Every time a user doesn't win something on a slot machine, they become increasingly excited for the next hit. Similarly, TikTok's content must capture a user's attention within a few seconds or they will quickly move on. This dopamine surge is encouraging users to keep coming back in hope of finding that one video that truly satisfies their craving. And even though many videos won't provide the desired stimulant, people remain intent on seeking out that one 'hit' that will give the right amount of pleasure.

However, TikTok is more than just an entertaining platform. Despite its seemingly harmless nature, the app has been designed with human neurology in mind and is playing a dangerous game of data-gathering. Researchers are only beginning to understand the effects that this type of tech has had on an entire generation, as evidence points to unprecedentedly low attention spans and a renewed obsession with social media. Whatever data you think TikTok might be acquiring from your activities is additional to the expansive amount of data that the app is actually collecting, unbeknownst to its users. Of course, proof of this is written in the app's so-called 'privacy page' – a document that legally allows the app to consume an unimaginable amount of information from its consumers, who agreed to it the moment they signed up.

The true reach of the data capture that threatens our privacy is nothing less than data-rape. It's a grim reality to face when we consider just how far the data collectors go to understand our habits, behaviours and online movements. Not only do they collect our likes, shares and messages but they access the apps we use, the other apps we have on our phones, the websites we visit and the search terms we have used. It's invasive to the point that they can even track the keystrokes we make when we log into our bank accounts. On top of this, our location is constantly monitored, and datasets are used to identify what interests us, what irritates us and even the way we take our coffee.

The data collection does not stop here. Surely, we can all agree that Google and Facebook has been under scrutiny for stealing data and sending it to advertisers, however, the current situation goes beyond what is usually discussed. We can now simply refer to this as 'artificial privacy invasion.' AI experts are now able to predict our future behaviours by scraping every bit of useful data from our phones.

What is particularly worrying is that this data is going directly to the Chinese and is playing a part in their bid to obtain global domination. This may sound unrealistic, but President Xi Jinping has openly stated his desire to become a superpower. Calls have been made to try and limit how much of our data can be accessed and politicians are taking drastic steps to try and counteract the data rape of our phones.

To conclude, a complex matrix of data-gathering has spawned in our cyber world. Our phones are being data-raped, used to predict our behaviours and help China obtain global domination. We must act now to ensure our privacy is not further violated and this data-rape is stopped before it's too late.

What we see on TikTok is much different from what Chinese citizens experience, as they have an AI algorithm that encourages content creators to share videos of their day-to-day lives which are full of accomplishments and hard work. It is no surprise that this type of content is wholesome and heartwarming, but the Chinese version of TikTok also a section that promotes the political views of the Chinese Communist Party, a sharp contrast to what is seen in the West. The US and Canada are inundated with short clips of people, particularly girls, dancing to catchy beats making it all too easy for viewers to become mindlessly hypnotised by the screen before them. This fixation on the app has allowed an industry like the 'Only Fans' to flourish, where desperate young adults can pay money to talk to or watch creators deemed by the algorithm to be attractive. Investigations done by the The Wall Street Journal uncovered evidence that when users indicated their age to be 13, the app heavily recommended content that was of an adult nature and drugs. The power TikTok wields through its vast

AI capabilities can drastically shape how people lead their lives, as a majority of users are nonadults. Even more troubling is the spread of false political information and Chinese propaganda, something that require attention regardless of political stance since TikTok has infiltrated homes and offices to shape our understanding of world events. Ultimately, what is even more concerning is that behind this friendly and popular app is a ravenous data collection process aiming to create detailed files of every western user they can get their hands on.

As the use of apps like TikTok continues to engross the lives of people across the world, particularly the youth, and attention spans are starting to shrink, with relentless checking of mobile devices becoming a troubling norm, the culture in China is headed in a drastically different direction. China has now begun to craft their current and future generations with a view to prepare them for potential war, by imposing new codes of regulations and reshaping their social constructs. Princeling and business tycoon, Mr. Ren Zhiqiang, was famously censored after deeming the Chinese Government's coronavirus response 'incompetent', while the country has also limited the use of video games and have implemented compulsory reading of Sun Tzu's 'The Art of War' for students. Additionally, female and male students are being taught different skillsets in school to emphasise gender dichotomies, from computing to military technology. All of these measures are clearly sending a message that the Chinese Government is training their youth for war, and has been met with significant military and political propagandizing, a campaign that has caught the attention and criticism of the international community.

People everywhere are lured by the ostentatious display of unsavory visuals projected all over the internet. Videos that promote criminal activity and depravity have become increasingly prolific, and it appears that the Chinese have even taken the initiative to train their population to be more warrior-style in their approach. According to research conducted by the United States military, psychological warfare is a critical aspect of China's military strategy. It has been demonstrated on the international stage how China has utilized this strategy against a wide variety of opponents in order to acquire territory, resources, and ultimately, world power. In 2019, for example, China's influence in the South China Sea has grown substantially due to a series of "psychological campaigns" targeting its neighbors. China utilized such tactics to harass fisherman, land encroachments, and even military exercises in order to increase their own geopolitical standing in the region. Moreover, China has employed similar tactics in other areas, such as Africa and the Middle East, to strong-arm its adversaries and gain control over vital resources. Ultimately, manipulating the psychology of nations has become an increasingly effective tool in China's ever-expanding arsenal, and it will surely continue to be an integral part of their military strategy going forward.

As you absent-mindedly swipe away, please take a moment to ponder the weight of your actions. Your convenience may be weakening the very fabric of western society by contributing to rapid growth of technological domination by the People's Republic of China. Unbeknownst to you, precious information is being extracted from your every interaction thus allowing China an impressive insight into our collective habits and behaviours. Through the insidious presence of their Orwellian surveillance networks, the Chinese are capable of predicting elections, a feat that some have suggested is already occurring. The United States, for example, floundering in a sea of troubles including rising gas prices, border control issues, a skyrocketing inflation rate, the continuous threat of a burgeoning budget deficit and alarming crime rates at levels not seen for more than four decades, stands as a testament to the potential dangers which may be realised if one nation is allowed to wield such impressive technological power.

Moreover, Chinese influence has far-reaching implications beyond their predicted political control. By mining the data yielded by millions of interactions, they are now able to make well-informed predictions as to how and where you will spend your money and your time. Seemingly innocuous holiday plans or mundane shopping trips could potentially provide the Chinese with invaluable information which can be used against us in the future. In essence, it seems that China's malign agenda is often neglected thanks to the chaotic soup of troubles endured by the people of the United States. If you wish to protect your nation from the potential of Chinese interference, simply put down your screen and collectively ensure that our nations data is no longer surreptitiously stolen by the dragon.

The Stupid Continues

The past few years have brought about drastic changes to our lives. It seems that many of us are in a state of denial about the world we now live in, and yet, that same world presents a number of serious and pressing challenges that need to be addressed. Unfortunately, the people and institutions that are supposed to be most responsible for leading us out of this crisis are often found lying and deceiving us. This is both while they operate under the guise of 'looking out for our best interests' and yet fail to deliver on the promises they make. As an example, consider the recent reports concerning the US government and the way in which they have purposely misinformed and misled citizens about the origins, trajectory, and consequences of the Covid-19 pandemic. Whether it be through the systematic manipulation of media coverage or the dissemination of inaccurate or false data, it cannot be denied that we have entered into a period in which many of our very basic trust in our leaders has been compromised.

In order to combat this reality and move forward in a positive direction, it is essential that we demand truth from those in authority and strive to never accept anything less than complete honesty. In this way, we can start to restore the legitimacy of our government and other institutions, and thus break free from the range of negative consequences that can arise from a lack of trust and misinformation. By continuing my research and analysis as I have done throughout this book, I believe that we can begin to gain a clearer understanding of our current situation and the solutions that will lead us going forwards.

The Suppression of Science

How often have we heard the phrase "we are following the science" being used as a rallying cry for governments and private organizations? We've heard the same Communications Secretary of the United Nations claim that they in some way 'owned' the science when bragging about their close relationship with Google. Needless to say, such a statement has been nothing short of extremely annoying.

The last few years have seen science being compromised and suppressed in the name of "public interest" to an extent that is absolutely nothing short of criminal. It has all been for the purpose of gaining political power and financial benefits and is a grave injustice to both members of society and the scientific community. Now more than ever, when our very survival is under threat, we need science to be prioritized.

However, what we are instead seeing is a stark manipulation of science, especially with the advent of the pandemic, as politicians, media outlets, and celebrities have used it to serve their own interests while praising the efficacy of a vaccine. Unsurprisingly, this has resulted in a greater wedge between us and the truth, and in some cases, the suppression of science has been nothing short of deadly.

The issue of suppression of science has unfortunately been a common occurrence throughout recent history. In many instances, government advisers have been accused of unduly influencing the scientific community, withholding information which could otherwise be of immense value to the public. These insidious practices involve overstepping boundaries and can have grave consequences to public health and safety as well as provide little value to taxpayers.

One of the key individuals to call attention to these malignant practices is Dr. Kamran Abbasi, editor-in-chief of The BMJ. He claims that 'suppressing science, whether by delaying publication, cherry-picking favourable research, or gagging scientists, is a danger to public health and maladministration of taxpayer's money when entangled with commercial decisions...when good science is suppressed, people die.' This alarming warning must not be taken lightly and should be given due attention by policy makers and individuals alike.

We have examples of the suppression of science in multiple influential contexts. Perhaps the most infamous of these is 'Big Tobacco'. In this instance, cigarette manufacturers deliberately withheld research and information from the public which clearly showed the harms of smoking on physical health. As a result, thousands of lives were lost in the decades that followed and would have been spared had this information been made available to the general public.

It is now more important than ever to mitigate the effects of scientific suppression in order to secure the health, wellbeing and safety of citizens the world over. It is not just an ethical demand, but a moral obligation. It is for this reason that we must strive to prevent such occurrences at all times and should be extremely vigilant of any signs that such nefarious practices could be occurring in any realm.

Science is an invaluable foundation for modern society, acting as a source of information and knowledge on the natural world. From the writings of renowned scientists like Isaac Newton and Galileo to contemporary peer-reviewed articles, science provides answers to the unknown and encourages advancements in technology and industry. Yet, in the past few years, there have been increasing efforts to suppress science, overriding any sense of accuracy and trustworthiness it stands for. This manipulation of information is seen and felt everywhere, particularly surrounding politically and financially driven projects.

To illustrate, a notable example is the current public discourse regarding climate change. Although the scientific evidence demonstrating the severity of climate change is convincing, many politicians, corporations and industries have chosen to ignore it, instead prioritising their own agendas and profit margins. This 'cherry-picking' of science, which takes select pieces of evidence to support one's interests, is an unfortunate and misguided trend, and one that jeopardises the entire scientific community.

The coronavirus pandemic accelerated this process of pseudoscience, as individuals and groups sought answers to the pandemic, frequently in the form of conspiracy theories. By utilising social media outlets and the internet, these false allegations, although unsubstantiated and lacking any form of peer-review, have become widespread, representing a significant threat to the dissemination of true knowledge and facts. In order to restore faith in science and progress, it is important to eliminating any form of cherry-picked, untrustworthy information, and reiterate the need for proper and reliable research that meets the necessary standards of accuracy and verifiability.

Early on in the COVID pandemic, the CDC notoriously began a covert operation involving the manipulation of a scientific definition: one that had been accepted as true for nearly a century. This particular manipulation by the CDC, FDA, Pfizer, Moderna, the NIH, and various oligarchs like Bill Gates and Dr. Fauci would prove to be incredibly harmful and sinister. Initially, the definition of a "vaccination" was "the act of introducing a vaccine into the body to produce immunity to a specific disease." Stunningly, just like that, the word "immunity" was swapped out and replaced with the word "protection" - completely transforming the definition and allowing the aforementioned institutions to move forward with their plans. But it didn't end there - they decided to go above and beyond and altered the term "vaccine." Instead of being "a product that stimulates a person's immune system to produce immunity to a specific disease", the term was reworked to "a preparation that is used to stimulate the body's immune response against diseases." With these two alterations, they were set to carry out their crusade of global vaccination, utilizing a type of mRNA gene therapy that had never been used before on such a scale. By changing the terminology to sound like something more benign and relatable (namely, "vaccine") those who were ignorant to the covert operations chose to simply follow along, line up, and receive the jab. It is through such ruthless transformation of language, and sly manipulation of opinion, that those in positions of power are capable of perpetuating tyranny on a global scale.

When it comes to vaccines, the medical and legal definitions in place have been well-established and are accepted by many. Generally, a vaccine must make an individual immune to a virus and also prevent its further spread. The mRNA therapies currently being labeled as 'vaccines' however, don't meet these criteria, leaving many with concerns. Unfortunately, if anyone dares to share these doubts online, they are likely to face swift repercussions from tech giants such as Facebook and Twitter in the form of account suspensions.

Take for example the case of a doctor from Alaska, who had posted his thoughts about the new mRNA vaccines on Twitter. He was met with suspension of his social media account for daring to question the safety of the vaccines. Similarly, another doctor from California who regularly streamed his views of the gene therapy had his YouTube channel suspended for the same reason, showing how tech monoliths are clamping down on any voicing of scepticism relating to the new mRNA vaccines . It's clear that this is a concerning state of affairs and raises alarm bells about our freedom of speech.

The use of pseudoscience has long been an issue in the scientific community, with unsubstantiated claims and opinions often taken as fact. This was especially true when it came to Moderna and Pfizer's mRNA vaccines, which were viewed by many as a potential solution to the global coronavirus pandemic. In order to sway public opinion in favor of the vaccines, those behind the campaign rebranded them as something more palatable and less alarming: simply put, a vaccine. Such a label gave no cause for alarm or suspicion and allowed the campaign to expand quickly and easily, with the knowledge that using a term such as 'gene therapy' might have drawn resistance from the general public.

This relabeling of the vaccine has resulted in pseudoscience now overshadowing the factual and thoroughly researched science which has been learned over the past few generations from peer-reviewed studies. In this way, the promotion of pseudoscience has effectively clouded the conversation about the vaccines, with false claims being taken as true and the real science now becoming an afterthought.

Consider, for example, the case of a vaccine opposed to a particular virus. If a person was unaware of the science behind the vaccine, they would be much more likely to accept the claims of the proponents of pseudoscience than if they had seen a scientific explanation of the actual effects and processes of the vaccine. This is why it is so important to ensure that the labels and descriptions given to products and treatments are accurate and reflective of the truth; this way, the public can make informed decisions which are free of misinformation and false claims. If a biologically-based mRNa vaccine was labeled as 'gene therapy', then the public might be confused or misled into believing the inaccurate claims of pseudoscience, rather than understanding the real science behind the vaccine.

As the long awaited vaccine rolled out, it became a source of contention for many people globally. On one hand, it felt as though we were all up against a pandemic like never before and it was our responsibility to keep ourselves and those around us safe. We were even encouraged to view getting the shot as a way to contribute to the greater good - a somewhat patriotic act. But unfortunately, not everyone shared that sentiment as some were hesitant to get the jab. This led to some pushing a narrative that those who hadn't yet got the vaccine were in fact selfish, or even worse - that they were murdering those around them. This was a sentiment shared by US President Joe Biden when he said "Freedom. I have the freedom to kill you with my COVID.".

The idea that those who chose not to get the shot should be treated in such a harsh way continued to be perpetuated. Author Frank Schaeffer went as far as to call individuals who opposed vaccines and masks 'bioterrorists' and even suggested that drone strikes should be used on some of the worst offenders. To some this may seem a little over the top, however, it is easy to see why those opposed to the vaccine choose to make such a statement. This is because those who are hesitant are afraid; they fear that the risks associated with something new and untested may outweigh the potential benefits.

It is simply unfathomable to understand why humans are willing to harm one another, as we've seen time and time again throughout history, most notably during World War II in Germany. But this time, people have been conditioned to turn against each other if they refuse to comply with the controversial Pfizer and Moderna COVID-19 jab. Pressure to conform had been mounting as a result of the 'vaccine' label given to these jabs, coupled with the 'science' supporting them.

Though this push towards conformity is nothing new, it has become particularly concerning during COVID-19 as the situation has deteriorated into social ostracism for those who oppose certain measures such as the virus' associated 'vaccine'. In the UK, for example, many were left feeling isolated or excluded from events such as family gatherings or holiday plans due to their refusal to get the jab. Such punitive measures are particularly concerning when discussed in the context of a pandemic, as they can lead to an atmosphere of fear, distrust and paranoia that can further reduce the level of adherence to limit the spread of the virus.

In March 2021, Rachel Maddow of MSNBC provided viewers of her nightly newscast the momentous insight that science had proved the efficacy of COVID-19 vaccines. Speaking with a definitive blend of authority and resignation, Maddow declared that "Now we know that the vaccines work well enough that the virus stops with every vaccinated person. A vaccinated person gets exposed to the virus, the virus does not infect them, the virus cannot then use that person to go anywhere else. It cannot use vaccinated people as hosts to get more people." Her words were soon immortalized as the gold-standard of scientific understanding and shared millions of times throughout a variety of media platforms, including the ever-influential government channels, news networks and social media sites such as Facebook and Twitter.

This well-spoken statement wasn't just limited to digital reach, however. Maddow's words gave professionals and laypeople alike the peace of mind to go about their lives, with the knowledge that science was proving the efficacy of the vaccine. Since then, the research is resolute; studies from the World Health Organisation have demonstrated that by vaccinating populations against the virus, healthcare services are freed up and the risk of future surget in COVID-19 cases is diminished. This conclusion has been shared heavily by the Centers for Disease Control and Prevention, and similar bodies, who are embracing and promoting this resolute truth.

The impact of Maddow's statement cannot be understated and has resulted in the widespread implementation of vaccines across the globe. Today, many nations have already inoculated a sizeable proportion of their populations and the impacts of this development are clear. Ultimately, the words uttered by Maddow in March 2021 have allowed healthcare professionals and systems to focus on the urgent task of vaccinating as many people as possible, thus helping to finally bring some semblance of safety and security to the global community.

By mid-2020, there still hadn't been any conclusive proof that the mRNA vaccine produced immunity or prevented the transmission of COVID-19. Unfortunately, the FDA trials were carried out in secrecy, the criteria for vaccines was altered, and scientific data was carefully monitored and censored. As people took notice of the shady dealings, many began questioning what was happening, yet were quickly labeled as 'conspiracy theorists' for doing so. It's astonishing how a huge amount of money can influence science and push an agenda.

The underhanded tactics of those in power came under scrutiny when reports emerged of individuals becoming infected with the virus despite having been vaccinated. This raises serious issues when it comes to public trust in science, and highlights the fragility of our scientific institutions. Scientists fear that medical technology has been hijacked by those in power, and that the ethical standards of the medical profession could be seriously compromised.

These worrying developments present fundamental questions about the integrity of medical research and the independence of medical agencies. It demonstrates just how far individuals in positions of power are willing to go to advance their own agenda and make a huge profit in the process. We must stand together and fight to ensure science is done ethically and with integrity, and that those in power cannot exploit scientific knowledge for their own benefit.

The outbreak of Covid-19 championed a surge in scientific research, but despite the life-saving potential of discoveries, many doubted the gains of all scientific advancements. Fortunately, the events of a few weeks ago have made it harder to ignore the incredible success of science, as Janine Small, a senior executive from Pfizer, has openly admitted to the European Union that the vaccine has not been tested for its effectiveness as a barrier to disease transmission. This revelation has prompted a flurry of activity on social media, providing an outlet for criticism of mRNAs and public discussion about the vaccine and its efficacy. While the battle may have been won, the war of misinformation continues; it is now up to the public to use research and evidence to dismantle the waves of disinformation.

In the wake of what has seemed like a long struggle, it is a relief to see science reclaim its place at the forefront and bolster society with its successful advances. It has given us a newfound confidence in the power of scientific discovery and the lasting effects that profound research can have. Already, the implications of Janine Small's declaration have shown us the power of open dialogue and a commitment to staying informed in the ever-changing landscape of scientific advancement.

As we enter into the scorching heat of the election season, where tensions are already at a boiling point, the stakes could get even higher should the Republicans win majorities in the House and Senate in November's midterms. In anticipation of the potential result, Senator John Kennedy (R-LA) has gone on record with a warning issued to Dr. Anthony Fauci that he should use caution in his travels due to the likelihood of needing to face a barrage of questions from congressional committees.

Dr. Fauci is already facing immense scrutiny, with opinions across the political spectrum falling on both sides of the issue, however if the Republicans gain control going forward, the influx of questions and demands for answers would be extremely difficult for Fauci to navigate. Given that Kennedy has taken the preemptive measure of advising Fauci of the potential consequences should the Republicans win, the coming weeks could be filled with even more intensity.

As the days continue to pass, the effects of the midterm elections become more and more evident, with reverberations aftershocks potentially taking many different forms, just one of which is the spotlight that will be shone on Dr. Fauci if the Republican party come out on top. It therefore remains we shall see what the future brings, hopefully allowing the American people to make well-informed decisions come November.

Dr. Edeling's powerful statement highlights a key issue in our current global catastrophe - the tendency by many to make decisions based on authority, not facts. Indeed, we are witnessing a world increasingly dominated by extremes of opinion, without any logical underpinning to support them. This can be particularly dangerous as it encourages people to take important decisions based on personal views and emotions, rather than on an evidence-based body of knowledge.

A good example of this is the recent Covid-19 pandemic. As the virus spread across the world, so too did various opinions about how best to respond. From governments advocating for full lockdowns of entire communities to individuals espousing the benefits of herbal supplements and diets, it soon became clear that many people were relying heavily on their personal views and authority. Sadly, this resulted in a chaotic environment where factual evidence – such as those stemming from clinical laboratory studies – was often overshadowed by speculations built upon emotion.

In this way, Dr. Edeling's point rings true – it is imperative that we take a rational, evidence-based approach to decision-making in times of crisis. Rather than diving into a flurry of ungrounded opinions, we should strive to base our conclusions on established science that has been critically analysed and verified. In doing so, we can help ensure that we are responding to events in a way that is both effective and safeguarded from potential harm.

If we have learned anything, is that we can trust, but we must question and verify.

Deaths of Unknown Causes

Canadians across the country are alarmed by the sharp and drastic increase in the number of deaths attributed to "unknown causes". In 2018, this statistic was recorded as 150 in the province of Alberta alone, but only two years later in 2020, this number had ballooned to 1044, representing a staggering 700% increase in mortality due to causes that are still to be determined. In the time since 2018, this trend has only grown, with the number of deaths due to unknown causes shooting up by an unfathomable 4000% as of 2021.

It has been reported in some circles that in cases where an individual has been vaccinated against some ailment, an autopsy is not being performed to ascertain the cause of death. While the rationale behind this phenomenon is uncertain, it is fairly common knowledge that autopsies are the primary method for determining the root of a person's death. With the alarmingly high rate of deaths attributed to "unknown causes" in Canada, a much stronger emphasis needs to be placed on compulsory post-mortem examinations to ensure that the public is getting an accurate assessment of mortality cases in the country. Not only would this help provide answers as to why more and more Canadians are losing their lives to unknown causes; it would also assuage the fears of citizens and lend more trust to the health systems of the provinces affected by this alarming trend.

/

Throughout the decades, hospitals have been mandated to autopsy a certain portion of decedents to uphold their accreditation. A renowned examination of one hundred arbitrarily chosen autopsies and examinations across three distantly separated years (1960, 1970, and 1980) exemplified that substantial diagnoses were not immediately manifest in more or less twenty-two percent of all cases, even though state-of-the-art imaging techniques had been developed during that time. Fast forward to the present day, and financial gain has become almost non-existent for those performing the autopsies, and thus their rate of occurrence has dwindled to concerningly small levels of merely five percent. Nevertheless, if a more balanced approach was considered and if autopsies existed at the same frequency as before, we could assume that this rise in rate could yield many different findings of benefit to both parties involved. The invaluable nature of autopsies is not to be overlooked, as it has proven to be a reliable tool for verifying diagnoses in long-deceased patients where further tests may be impossible. In addition, autopsies hold a plethora of potential advantages, ranging from aiding the grieving process of a family to discovering new biological phenomena that were never known before.

It is incredibly surprising that neither the Centers for Disease Control and Prevention (CDC), the Food and Drug Administration (FDA), nor the Department of Health and Human Services (HHS) nor the government as a whole are demanding autopsies or testing of vaccine samples when so much public trust is placed in these medical interventions. Without autopsies, especially in cases where there is no established cause of death (COD) listed, there is no way to know if certain batches of vaccines are a contributing factor to deaths. Consequently it is vital that autopsies are conducted effectively, in a manner that rules out known possible adverse reactions such as blood clots and myocarditis.

It is undeniable that autopsies and testing of vaccine samples are necessary components of medical research. Autopsies should provide us with more insight into possible issues with certain vaccines and help us to better protect public health. When vaccine samples are tested for quality, it provides us with a better understanding of potential differences between batches, giving us greater peace of mind when vaccinating vulnerable individuals. Furthermore, it is only through proven scientific methods such as autopsies that products can be reliably declared safe and effective.

Manufacturers of pharmaceuticals and other related products are afforded a great deal of protection from product liability, due to the lopsided contracts that our governments have signed with major players such as Pfizer and Moderna. While this shielding of companies from legal consequences is troubling to some, it is also important to consider that our government agencies, employers, and educational systems are still accountable to us, when it comes to the decisions they make regarding experimental, and potentially dangerous products, be they drugs, vaccines, or otherwise. It would be alarming if, for instance, employers forced employees to take an experimental drug, or universities coerced students into taking a medical product that could be potentially lethal. It is the responsibility of our government bodies and organizations to ensure that these decisions are sound, and always prioritize the safety of the people.

In his comments on the current global crisis, Dr. Edeling astutely highlighted the prevalence of people relying on their positional authority as opposed to reasoned process when expressing opinions and giving instructions. He poignantly observed that, without backing their claims and demands with evidence, facts, scientific research, and logical argument, these statements are rendered baseless.

Throughout the world, many individuals are voicing strong opinions and issuing hard-lined commandments regarding issues such as the handling of the pandemic, the activities that should be condoned or prohibited and the manner in which public funds should be utilized. However, they are doing this without concrete and undeniable evidence to support their assertions. Countless times throughout history, dangerously erroneous decisions have been made with disastrous repercussions when opinions and commands issued from a position of authority have gone unchecked by an appropriate process of evaluation.

Dr. Edeling's comments serve as a reminder that to make sure that we are not repeating the same mistakes, each opinion and directive must be assessed on its merits. It is only through justified and logical reasoning, supported by reliable data and scientific analysis, that we can be confident that we are making the right decisions in these trying and uncertain times.

The cause of the current state of affairs, in which scientific evidence is being suppressed, can be very clearly highlighted by studying the principles of evidence-based medicine which have been replaced by opinions, suppositions and authoritative directives. Rather than reason, deduction and facts, policy is being determined without rigorous scientific scrutiny, and many of the decisions taken are extremely concerning. One such example is the continued utilization of mRNA gene therapy, despite the fact that there has been no proof of its efficacy or safety presented to the scientific and medical community. The warnings of the Vaccine Adverse Event Reporting System (VAERS) have been disregarded in favour of guidance from those with vested interests in the success of this technology, even though the outcome of this could be catastrophic. This profiteering at the cost of scientific principles and evidence shows just how deeply-rooted the suppression of science has become.

The unprecedented success of global vaccinations has largely been achieved without any transparency or adequate governance in place. As such, it is unsurprising that we have had to rely on the less reliable mechanism of Vaccine Adverse Event Reporting Systems (VAERs) as an early warning system. If a more comprehensive system had been established, perhaps one with the capacity of identifying patterns of concern regarding adverse reactions, it may have been avoidable for us to face articles like this introducing doubts about a huge development in the history of medicine. Furthermore, it is possible that more in-depth post-mortems may have been conducted to investigate any potential issues, which could have uncovered any potential danger ahead of time, potentially affording us a more reliable choice than having to weigh up potential health risks with potential benefits.

The issues surrounding vaccinations and the pandemic have been receiving more public scrutiny in recent months, as many of the current vaccines being employed for protection against Covid-19 are mRNA vaccines. This vaccine type differs from traditional vaccines, as mRNA is designed to prompt an artificial version of the desired immune response, instead of using weakened or inactivated pathogens akin to the majority of the immunizations administered in the past.

As a result of the rollout of these mRNA vaccines during the pandemic, some experts have reported that individuals battling Covid-19 might be impeded in their innate immune system responses due to the administration of mRNA, potentially allowing Covid to replicate further. In fact, 91 countries around the world have seen significant spikes in both Covid-19 cases and deaths coinciding with the rollout of vaccinations. Documents, containing graphs to illustrate the direct correlation, have been presented and analyzed to evaluate the potential side-effects of taking the mRNA vaccine.

Furthermore, studies also suggest that the individuals who have had the vaccinations experience a period of immune suppression approximately two weeks after being vaccinated. During this time, it has been observed that more people suffer from Covid-19 related deaths and a surge in those who have contracted the virus.

It is worth repeating here what has been discussed in many of my articles; these self-proclaimed 'vaccines' are not technically reckoned to be true vaccines. Vaccines are substances composed of either weakened or dead versions of organisms or pathogenic agents that are administered to a person, provoking an immune response to it. Vaccines have been used by humanity for centuries, and have been scientifically proven to be effective.

By these standards, the so called mRNA vaccines are drastically different. To put it simply, if vaccines are akin to cars, the mRNA is analogous to a flower – they are not even closely related in terms of their purpose. As opposed to being conventional vaccines, the mRNA vaccines are laboratory-engineered genetic messengers that are injected into the human body. When they are injected they propel the cells to produce coronavirus spike proteins.

The major difference between conventional vaccines and the mRNA vaccines is that the latter involves inducing cells to generate an antigenic spike protein instead of introducing the pathogenic agent to the body. In the world of vaccines, the process of making the cells generate the spike protein is entirely different from the typical process.

It is deeply concerning that the Centers for Disease Control seemed to covertly tweak the definition of a vaccine. If a regulatory agency such as the CDC can change the definition of a critical aspect of medicine, then a great deal of patience and vigilance must be utilized by the people when it comes to modern advances in science and technology.

The mRNA-based vaccines currently undergoing tests and trials have revolutionized the concept of vaccination by allowing scientists to trigger cellular processes from within the body itself. Through these vaccines, the body's cells are instructed to make a foreign antigen, the viral spike protein, which they were not designed to produce naturally. Both encouraging and concerning, this unprecedented production of molecular-level 'strangers' can set off a widespread attack on any cells that express it, triggering a destructive immune response in any organ of the body. This reaction can lead to dangerous and potentially fatal consequences, such as myocarditis, sudden cardiac arrest, and death. To make matters worse, recent microscopic analysis of tissue samples from various organs have provided substantial scientific evidence of an autoimmune-like disorder brought on by the vaccine. This discovery has further validated the fears of many individuals who have been raising similar concerns for quite some time.

In early 2020, three highly-qualified pathologists, two of whom were medical examiners, made a groundbreaking discovery when they published their findings in an early online release article entitled "Autopsy Histopathologic Cardiac Findings in Two Adolescents Following the Second COVID-19 Vaccine Dose" in the prestigious Archives of Pathology and Laboratory Medicine journal. The authors came to a pertinent and comprehensive conclusion, revealing that two teenage boys died at their respective homes mere days after receiving the second Pfizer-BioNTech COVID-19 vaccine dose. Upon further investigation, neither of the boys had tested positive for an active or previous COVID-19 infection, with negative results also present on their toxicology screenings for drugs or poisons in their body. In other words, the comparative autopsy evaluation suggested that the deceased boys passed away due to complications from the vaccine itself.

Most profoundly, their histopathological examination of the cardiac tissue unveiled an unconventional finding that can mainly be associated with a diverging form of heart injury known as toxic cardiomyopathy; a malady that is more closely connected to stress-induced catecholamine activity, as opposed to the more prevalent myocarditis pathology which typifies cardiotoxicity at large. As the authors of this article suggest, the myocardial injury displayed in these post-vaccine hearts is suggestive of a distinct kind of cardiotoxicity which is exclusive to a toxic cardiomyopathy, a medical topic that has been overlooked in the medical literature prior.

The recent surge in public uproar against the oligarchs is becoming increasingly difficult to ignore, as more and more courageous individuals are rallying together to demand that further autopsies be conducted to decisively confirm or deny the controversial findings attributed to the mRNA vaccine. Individuals from all walks of life - from conscientious healthcare workers to socially engaged parents and grandparents - are joining forces to express their deep concern about the potential long-term effects of the vaccine, how it may affect the health of their loved ones, and what it could mean for the future of medicine.

This is certainly not an isolated issue, as medical experts and policy makers continue to laud the mRNA vaccine as an efficient and necessary preventative against the novel coronavirus, raising the potential dangers of such a dramatic shift in the standard of care. Reports of adverse effects, ranging from minor aches and pains to more serious fatigue, headaches, seizures, and even paralysis, have put many on high alert, prompting a wave of both grassroots and governmental initiatives that call for further investigation into the matter.

Ultimately, those brave enough to stand up against the oligarchs understand the paramount importance of pursuing a heightened level of due diligence in their efforts, ultimately hoping to ensure that the public is left with a full understanding of the potential risks and benefits of the mRNA vaccine. As such, it is clear that these demands for further autopsies may set in motion a much-needed level of transparency in the medical community, providing the public unprecedented access to information regarding treatment and prevention.

The efficacy of mRNA gene therapy has been widely studied by reputable scientists, which has yielded plentiful evidence highlighting numerous potential hazards of the treatment. Numerous autopsies performed on those who passed away after receiving the mRNA gene therapy have revealed alarming findings. For instance, a recent autopsy conducted at the University of Arizona revealed numerous irregularities in the individual's cardiovascular system, including irregularities in the heart valves, inflammation in the arteries, and other anatomical anomalies that would not have otherwise been visible during life. These findings suggest that mRNA gene therapy may be implicated in some cases of cardiac disorder complications. Furthermore, other autopsies have revealed that some individuals who received mRNA gene therapy experienced adverse reactions in their immune system, doing away with the possibility of the successful implementation of the treatment in the foreseeable future.

Ultimately, these autopsy results emphasize the need for additional research into the potential risks and benefits of mRNA gene therapy. Although there have been some promising developments, we must continue to evaluate the safety of mRNA gene therapy to ensure the safety of those who resort to it for medical treatment.

The exciting scientific approach of using mRNA vaccines to combat the devastating SARS-CoV-2 virus has generated an immense amount of public and scholarly interest. Unlike traditional vaccine formulations, which generally remain at the injection site, mRNA vaccines are designed to travel throughout the body and accumulate in various organs. As is the case with mRNA-based COVID vaccines, these vaccines trigger long-lasting expression of the SARS-CoV-2 spike protein. But, because this spike protein looks generally similar to proteins found in many human organs, its presence within them can initiate an autoimmune-like inflammatory response that could ultimately lead to serious organ damage, such as that of blood vessels, and in the worst cases, death.

This phenomenon has been documented in instances where military personnel and healthcare workers have received the mRNA vaccine and then reported vascular problems, stroke, and death. It's possible that for some individuals, the autoimmune response that results from the spike protein's presence could have a more deleterious than beneficial overall effect. Therefore, it is of utmost importance to continue to monitor the safety and efficacy of mRNA vaccines as they are deployed worldwide.

Pathology remains the ultimate arbiter in establishing causation of disease, and research involving autopsy materials from patients who had passed away between several days and several months after receiving vaccinations has been conducted with some noteworthy findings. To cite a few specific examples, there was a study in which researchers examined autopsy materials from 790 individuals who had died after receiving a smallpox vaccination. The researchers discovered that when the patients had pre-existing predisposing conditions which made them more vulnerable to severe adverse events, there was a causal link between the smallpox vaccination and the ultimately fatal outcomes. Then, there was a study of autopsy materials from 822 people who had received the inactivated poliovirus vaccine, which ascertained that when the administered dose was higher than that recommended by the manufacturer, the mortality rate was significantly higher and there was a clear association between the vaccine dose and the adverse outcome. Finally, there was a research project into analysis of autopsy materials from 4,711 patients who had been vaccinated with a BCG intramuscular vaccine which determined that there was a higher-than-expected rate of mortality in those individuals who had pre-existing medical conditions. Although further study is needed, these studies and others demonstrate that there is a definite need to more closely monitor and assess patients who receive vaccinations in order to ensure that if any untoward effects occur, they can be managed effectively and promptly.

Professor Burkhardt, a renowned pathologist from Reutlingen, Germany, is exceptionally experienced in the field, and his colleague, Professor Walter Lang, has together undertaken detailed analysis of numerous cases of individuals who passed away either shortly after - within days - or months, receiving vaccines. In each instance, their cause of death had officially been certified as either 'natural' or 'unknown', however the loved ones of these citizens had doubts related to the diagnosis and asked for an additional evaluation. It is noteworthy that after Burkhardt's examination, the majority of these deaths were determined to be related to the vaccinations that the aforementioned individuals had been given.

For instance, after his assessment of the case of a four-month-old infant who had tragically died shortly after being immunized with a pack of three vaccinations, Burkhardt and Lang were able to find evidence of immediate-type hypersensitivity to the pertussis component in the vaccine, a condition that is known to be fatal in some cases, though it was initially categorised as 'natural' by the attending medical personnel. This is just one example of how Professor Burkhardt has made a genuine difference in providing bereaved families with closure, as well as furthering his years of research and analysis on the subject.

Researchers Burkhardt and Lang conducted an exhaustive study in which all four major vaccine manufacturers - Pfizer, Moderna, Johnson & Johnson and AstraZeneca - were taken into account. Their survey revealed that the majority of patients surveyed had received an mRNA vaccine from either Pfizer or Moderna. However, many of the deceased patients in the sample had ventured outside of mRNA-based vaccinations, opting instead to receive a viral vector-based vaccine as well, on a separate occasion. This raises the possibility that a combination of both types of vaccines, either in tandem or in swift succession, had a more multifaceted and deleterious effect on certain individuals.

It goes without saying that more research is needed in this arena to ascertain the implications behind mixing vaccine types. Such research could play a crucial role in the development of comprehensive guidelines for vaccinology, especially in this era of unprecedented global efforts to massively roll out precautionary measures. In other words, examining the pros and cons of both mRNA and vector-based vaccines, as well as their interactions with each other, is essential to ensuring the safety and efficacy of new vaccines – and accordingly, for mitigating the pandemic.

Pfizer conducted animal experiments that helped ascertain how quickly their mRNA vaccine can be distributed throughout the body, which is crucial for it to cause potentially lethal damage. By observing the movement of a radiolabeled model mRNA vaccine in rats for up to 48 hours, Pfizer was able to determine that the vaccine was able to quickly move from the injection site into the bloodstream, and subsequently into various organs. This decisive experimentation, performed by Pfizer and reported to Japanese authorities, played a significant role in the approval process of their mRNA vaccine in the country [5].

The importance to uncover the rate of distribution of the vaccine throughout the body, particularly to other organs, is paramount. Accordingly, these animal experiments conducted by Pfizer yielded a beneficial outcome: by determining the efficacy of their mRNA vaccine, they were effectively able to demonstrate the reliability of their product, thus gaining approval in Japan. As high-stakes scenarios such as those posed by the coronavirus pandemic persist, it is important to recognize the powerful role that the understanding of a vaccine's rate of distribution can have on the journey to approval.

Autopsies of individuals who had received postmortem mRNA treatment have revealed an unexpected panoply of results. Although a low number of autopsies have been completed, the outcomes are already demonstrating that this kind of treatment can have varying degrees of efficacy and outcomes. Notably, some of the postmortem autopsies concluded that the mRNA treatment established a platform from which a more successful course of restoration of the patient's tissue and function could be developed; however, in other cases the post-mortem tests revealed results that differed significantly from the outcomes or expectations that individuals or the provider had previously held. In one case, the researchers noted that post-mortem mRNA treatment had "simply not worked, with the tissue damage continuing beyond the duration of the treatment". This case demonstrated that while there may be a potential therapeutic benefit of postmortem mRNA treatment, this kind of treatment is not universally efficacious and that the outcomes should always be interpreted with caution. Consequently, even with the positive results seen in the postmortem autopsies, more research is needed to understand the full risk profile of the treatment and its potential to produce successful outcomes.

The results of our investigation confirmed a novel phenomenon of myocardial injury postvaccine, presenting distinct characteristics from other forms of myocarditis. As opposed to typical myocarditis, the intramyocardial inflammation seen in sample postvaccine hearts shares more similarities to catecholamine-induced stress cardiomyopathy. Stress cardiomyopathy is a type of pathological condition induced by excessive levels of catecholamine hormones (such as adrenaline, epinephrine and norepinephrine) which is produced in response to trauma or intense psychological stress. Through our research, we gain insight into the feedback mechanism between an overwhelming cytokine storm and catecholamine surge.

Having this knowledge is essential to providing better screening and therapy to alleviate the effects of postvaccine myocardial injury. For example, identifying at-risk groups that are more prone to a cytokine storm or exacerbation of inflammatory responses can help inform priority of vaccination or direct contra-indication. Furthermore, timely interventions aimed at controlling the catecholamine level can prove beneficial to patients experiencing postvaccine cardiac complications.

In conclusion, our insight into this unique form of postvaccine myocardial injury opens up opportunities for medical practitioners to better manage and treat their patients. With our understanding of the feedback loop between cytokines and catecholamine hormones and deeper insight into patient risk profiles, health care providers can implement safe and effective interventions and treatments for postvaccine myocardial injury.

Upon first glance, the scientific jargon surrounding catecholamine-mediated stress (toxic) cardiomyopathy appears to be accurate and reliable; yet when the individual components of the phrase are examined, a more complex picture becomes evident. Ultimately, this unfamiliar clinical phrase describes a novel heart failure syndrome that is thought to be caused by intense emotional or physical trauma. Unlike a traditional myocardial infarction, no plaque rupture or coronary thrombosis is present, and this condition is thus considered to be reversible.

This condition is known as stress cardiomyopathy (SCM). In many cases of SCM, the individual's plasma catecholamine levels are significantly higher than usual. SCM has been observed in other contexts of catecholamine excess, such as central nervous system injury and pheochromocytoma. Within the scope of SCM, an additional factor often comes into play: Abnormal myocardial blood flow resulting from sympathetic microvascular dysfunction, which causes coronary flow reserve to diminish and can trigger a so-called 'cytokine storm'. However, the phrase 'cytokine storm' does not have a uniform definition, leading to a substantial amount of media speculation surrounding the phrase and the phenomenon itself.

Indeed, the medical data does not appear to unequivocally validate the term 'cytokine storm', leaving many to question its legitimacy and validity. While 'cytokine storm' poses a dramatic image for the public and the media, it does not seem to be consistent with the findings of the autopsies. Therefore, it appears that the use of this term is merely a tool used by some to present unfounded findings in a manner that appeases the public and obscures the facts present in nature.

After having conducted my research several times, I could not help but notice a peculiar pattern that started to emerge as I continued my analysis. At first, I was unconcerned and unconvinced, confident that nothing irregular was transpiring. Nevertheless, as I began to read further into the text, I noticed certain words being used continuously and even some that seemed to be exaggerated, worrying me to the point where I decided to delve deeper into the observation. Therefore, I was able to uncover the detailed intricacies of the issue which I had originally passed over and determined that it, in fact, was concerning.

I am gravely concerned as to why there is a paucity of postmortem examinations being conducted for the deceased whose cause of death remains unknown, given the proliferation of unfounded suspicions that abound in the discourse surrounding this issue. It has truly come full circle for me - the need for postmortems is clearer than ever. Without such assessments, how can we confirm a cause of death or dispel many of the dangers conjectures that are routinely raised? From merely a scientific standpoint, the importance of finding out the real cause of death should be thoroughly recognized. But beyond that, how can we remain complacent and content when the truth of a perishing is unknown? Would it not be more beneficial for society to bear the hard truth, however uncomfortable it may be, than to remain in ignorance?

Things Just Keep Getting Worse

Just when we thought that the damaging effects of messenger RNA (mRNA) "vaccines" could not get any more extreme, reports have now emerged of cattle dying within 24 hours of being vaccinated with it. While most of the discourse around mRNA is focussed on its impact on humans, the implications for other animals that are being injected with it are deeply concerning. In Australia, a mass vaccination of an entire herd of cattle led to 35 of the 200 animals dying once the mRNA vaccination had been administered. Shockingly, this is not an isolated event and similar problems have been occurring across the globe. Governments have been mandating that these mRNA "vaccines" be given to cattle, and the results have been catastrophic. Although those animals that survive are able to produce milk and cheese, the safety of consuming these products is deeply uncertain. After all, nobody knows what the impact of drinking milk and eating cheese from a "fully-vaccinated" dairy cow would be, as the mRNA vaccines currently being used in the medical field remain experimental in nature. This is incredibly concerning, as mRNA technology is also being tested and experimented with in the food industry. In a previous article, I explored how mRNA is being added to our produce, and now, the same disregard for safety is being extended to livestock farmers, too. Going forward, consumers may have to start seeking out hormone-free and mRNA-free meats and produce in order to ensure that they are safe and natural.

Despite the fact that there is no current evidence that cattle can catch COVID, the question remains as to whether they can be a potential risk to humans. The reality of the situation is that many cattle are already treated with hormones and antibiotics, so it can be a bit confusing when we start discussing genetically modified organisms (GMOs) created through mRNA manipulation. This form of genetic engineering was enforced upon us, and it has since been revealed that this act was perpetrated as an egregious crime. It's painfully clear that powerful organizations, such as Bill Gates and the World Economic Forum, have been pushing for a global agenda that would require people to give up their cars, forego eating meat, and find ways to endure the harsh winters with minimal rations. Oftentimes this means resorting to "government handouts" such as social credits in an effort to remain afloat. Suffice to say, in light of this situation, much more research and information is necessary before we can fully comprehend the potential implications of using GMOs in the food industry.

Recently, a major deal was made between the United States-based biotechnology company, Tiba Biotech and two of the leading vaccine manufacturers, Moderna and Pfizer. This highly-lucrative agreement has generated a significant amount of wealth for the company, yet the irony remains that in early 2021, during a senate committee hearing on vaccines in Texas, a doctor testified that animal trials of the mRNA injections had been discontinued due to them resulting in the deaths of many test subjects. This can serve as a stark reminder of the consequences that can occur when attempting to develop a vaccine and should provide pause for reflection for all those who look upon this news. The danger of these vaccines and their associated risks are undeniable, and thus should be taken seriously. Taking these factors into consideration, it is apparent that this is a situation that should not be taken lightly by anyone.

If we perform some basic mathematical calculations and aggregate the deaths from the European Union Drug Regulating Agency (EUDRA) and the United States Vaccine Adverse Event Reporting System (VAERS), we can start to obtain an initial understanding of the scale of fatalities attributed to the COVID vaccine. Upon inspection of the published Underreported Factor (URF), which stands at 41, we can adjust our understanding by multiplying the previously mentioned total by this factor, yielding an estimated global figure of 19.9 million directly related to the COVID vaccine. Seeing as this URF is an average of 8:1, it can be extrapolated that the aforementioned total corresponds to an average of around 8 ripple effect deaths stemming from every single single coronavirus vaccine related fatality. This indicates that a tragically vast number of people across the world may have had their lives, and those of their families, changed indelibly by the COVID vaccine.

Research and statistics have revealed that an estimated 2.2 billion serious adverse reactions have been observed annually from the same sources of data. This figure is alarming in and of itself and is even more concerning considering the reports of an increasing spectrum of deaths classified as 'unknown causes' across the globe. These fatalities, now considered the leading cause of mortality, are causing numerous individuals, doctors and experts to fear that the already ubiquitous numbers may rise even more exponentially.

It is now more important than ever to take preventative measures when it comes to health and safety as the unknown cause of death continues to consume a greater portion of the data. An example of such precautionary actions can be seen in Japan, where they have implemented a comprehensive network of sensor technology to detect and predict the emergence of environmental factors which put the population at risk of infection or other ailments. In other parts of the world, specifically the United States, numerous regulations and laws have been implemented to ensure a safe and protected environment in specialized patient care.

The plague of the unknown cause of death is unavoidable and in order to combat its overwhelming effect, the implementation of effective prevention and enforcement measures are paramount to safeguarding the population from its dangers. It is clear that further research is needed if we hope to understand this phenomenon more clearly and uncover trends that help us learn more about its increasing prevalence and ultimately prevent further deaths.

CDC Massive Coverup

It took two lawsuits, an abundance of convincing of an inflexible judge, and an astonishing amount of time to be devoted to legal debates for us to finally receive the information concerning adverse reactions to the COVID-19 vaccinations reported to the CDC through their V-Safe application. Presenting this data nudged the CDC out of the shadows; it's an indiscretion that instantly invalidates the concept that the injections are 'safe'. Most of us are already aware of this, however, yet more people will hopefully recognise this soon thanks to the evidence at hand.

The legal proceedings took a concentrated amount of effort to win, as the CDC seemed unwilling to bow to the public demand for transparency. It was a prolonged period where we fought hard against a defiant government force, with the intention to expose the truth of the matter. The mitigation of the potentially severe health relaters of the vaccine should be common knowledge. It's not only due to the CDC's complacence that this information has not been publicised, but now, with these revelations, we can confidently state that the vaccinations are not without risk.

Unfortunately, while the magnitude of the adverse effects caused by the mass vaccine rollout has been brushed aside by larger outlets of media and news, a few more conservative and alternative sources have provided a more accurate account of the situation. Of course, the headline of the article in this case does not quite accurately lend to the full impact of the vaccine rollout—it does not even come close to giving proper context to the real issue at hand.

Indeed, what has been perhaps most worrying about the whole affair is that—at least initially—the narrative that was being put forward by those in the know severely understated what was really going on: a reported 71 MILLION symptoms caused by the vaccines themselves, according to the CDC. This startling figure, however, was only made known much later after the fact, after over 230 million Americans had already received their first dose of the various available vaccines. It's quite concerning that this crucial detail was conveniently left out of the narrative at the very beginning, and only ended up coming to light much later on.

For this reason, it is imperative that news outlets utilize greater caution when disseminating facts and information regarding the current state of affairs. More emphasis needs to be placed on providing readers with the full story—one that is presented in a truthful, unbiased and accurate manner. This will help to ensure that those looking for factual, reliable news coverage do not get misled or taken advantage of, and can make informed decisions based on the weight of the evidence.

As the number of vaccinated Americans steadily increases, so does the number of cases of those seeking medical attention after receiving a COVID-19 inoculation. A staggering amount of individuals have found themselves in need of medical attention post-vaccination, with The Centers for Disease Control and Prevention (CDC) revealing on October 3rd, 2022 that hundreds of thousands of Americans have experienced adverse reactions following their immunization.

From mild to severe, post-vaccination symptoms ranging from fatigue and sleepiness to vomiting, wheezing and low oxygen levels, have caused many of those given the jab to subsequently seek medical care. Unsurprisingly, the greatest increase in the number of patients has been those over the age of 65, who are some of the most at risk individuals when it comes to the coronavirus pandemic.

Given the breadth of potential afflictions resulting from a COVID-19 vaccine, it has become even more critical that individuals are closely monitored, as well as properly screened before receiving the shot. With more and more Americans receiving their vaccinations, the steady increase in medical assistance sought post-vaccination is an alarming trend that warrants close attention.

According to recent data, a staggering number of individuals have sought medical attention, or have been hospitalized, in the wake of receiving a COVID-19 vaccine. Approximately 782,900 individuals reported needing medical attention, including seeking assist from their local emergency room or hospitalization, after receiving the inoculation. And, worse still, as many as 2.5 million more people worldwide have had to miss attending school, work, and other typical activities due to an adverse event they experienced following their inoculation.

These numbers are alarming, and remind us of the importance of making sure individuals are educated on the potential side effects they could experience post-vaccination. Furthermore, knowing the signs and symptoms of a severe reaction are instrumental in helping to keep individuals safe, as they seek to protect themselves and those around them from the virus. Fortunately, the majority of people experience mild side effects following their vaccine, which generally subside within a few days of receiving their dose. However, for those few that might face a more complex reaction, it is important to understand that support is available and a healthcare professional can help create a plan for managing any side effects one could experience.

The Centers for Disease Control and Prevention (CDC) recently implemented their V-safe program - an innovative vaccine safety monitoring system to which people can report issues associated with vaccines they have received. Through the V-safe program, individuals who have been vaccinated can report side effects or other issues using a smartphone app. This ambitious undertaking is an important component in tracking and monitoring the safety of vaccines and provides valuable feedback for health officials and policy makers to consider when making decisions about vaccine distribution.

The reports generated through the V-safe program will provide real-time data to support public health decision-making, a process that has typically been time-consuming and complex. By having access to real-time feedback from individuals who have received vaccines, authorities have the opportunity to quickly and effectively evaluate potential risks associated with vaccines and ultimately prioritize the health and wellbeing of the public.

This groundbreaking project illustrates the importance of citizens taking an active role in promoting the health of their communities. By sharing their experiences, individuals can help create a safer, healthier environment for all of us. The reports submitted to the V-safe program will also be invaluable in understanding and improving vaccine safety going forward.

PFIZER ADMITS TO WHAT WAS ONCE DEEMED AS MISSINFORMATION

Pfizer's admission that they had not tested their vaccine for effectiveness for blocking the transmission of the disease exhibited a flagrant disregard for the truth. Senior executive Janine Small went even further, claiming that the vaccine had saved over four million lives. However, the evidence directly contradicts this bold, and largely unsubstantiated assertion. It appears highly likely that, rather than being based on scientific data, this statement is instead derived "from very old mortality predictions, or perhaps is just plain unscientific mathematical predictions, which have proven time and time again to be inaccurate".

At the time of the vaccine's release in December 2020, a majority of the world had already been exposed to COVID-19, with seroprevalence studies indicating that the virus had been in circulation since mid-2019. Furthermore, multiple studies have conclusively determined that natural immunity is more effective than any short-term vaccine-induced immunity, rendering the chances of damage from a subsequent infection much lower.

This estimated death toll has since been found to be drastically exaggerated. This is due to the fact that those at the greatest risk of serious complications or death had already lost their lives, and additionally because of numerous misreporting errors such as incorrect COD labelling and the tendency to list almost any death as COVID-related. It has become clear that the actual number of lives saved by the vaccine is far lower than the four million suggested by Pfizer.

The coronavirus outbreak has highlighted the danger of mutating viruses and how quickly they can spread. While the virus first seemed to be a very severe strain, it actually became less virulent as it mutated during the course of the pandemic, resulting in fewer fatalities and serious illnesses. However, there has been an unfortunate increase in death rates from 'unknown causes.' Reports suggest that those deaths can be attributed to delayed or inadequate medical care caused by the pandemic, with limited or closed healthcare facilities leading to more serious health problems, ultimately leading to more deaths than previously seen. Examples of this include a spike in infection-related deaths in the United States, particularly among the elderly. Other areas of the world, such as Africa, have been severely impacted by the pandemic, with health officials noting that the continent experienced the highest number of pandemic-related deaths per capita.

This highlights the need for increased awareness of not only the illness itself, but also the ripple effects it can have in terms of healthcare access and the number of deaths.

Overall, the mutating nature of viruses is a concerning reality that must be addressed, not just in terms of stemming the spread of the illness, but in terms of its continued effects. The fact that there have been more deaths from 'unknown causes' correlated with the pandemic is a reminder that the impact of an outbreak can be far greater than the initial effects of the virus itself.

The truth of the matter is that, in many cases, people who had contracted Covid-19 may have recovered fully from the virus, only to face an unrelated medical issue that caused their death; however, due to the positive PCR result in their medical history, their death was, in fact, mistakenly attributed to Covid-19. An illustrative example of this phenomenon was that of a patient in New York City; this individual had come down with Covid-19 in early April, showing symptoms such as a fever and difficulty breathing. After receiving appropriate medical treatment, the patient recovered, being completely asymptomatic and finally testing negative for the virus, to the relief of their family. Unfortunately, only a few weeks after being discharged, the patient passed away due to an undisclosed medical condition unrelated to Covid-19. Nonetheless, it was erroneously recorded as a death caused by Covid-19.

Sadly, this is a highly pervasive problem with no clear answers at present. As of now, it is essential for medical workers and epidemiologists to take special steps to differentiate between deaths that were actually caused by Covid-19 and other deaths that were misattributed. However, this can be a complex process and one that is difficult to ensure is done thoroughly in every case.

In the Canadian province of Ontario, initial forecasts had predicted over 25,000 deaths from COVID-19 in one year. However, as of today that figure has been firmly revised down to approximately 1,300 fatalities. In fact, it is possible that this number is still overly inflated. Recent studies that employed advanced forensic analytics, have estimated the final death toll to be even lower: less than 10 percent or as little as 5 percent of the original estimates.

In the face of this development, numerous public health initiatives have been given credit for keeping the death toll much lower than previously thought. A combination of increased empathy, robust support and collaboration among the local governments and compliance by citizens to the guidelines have been shown to have helped bring down the number of casualties in Ontario. Furthermore, the medical professionals and scientists should also be acknowledged for the huge effort made in controlling the pandemic. Their research and dedication, in trying to better understand the virus, have played a vital part in slowing the spread of the virus and thus lessen the total amount of reported deaths.

Though the overall numbers of confirmed deaths in Ontario have not reached the catastrophic levels predicted, it is important to remember that the province has nevertheless suffered. Each individual loss is the result of a tragedy that has affected entire families and communities. In addition, the mental health of the citizens has been greatly affected by the drastic changes enforced by the pandemic, such as lockdowns and social distancing measures. It is thus of the utmost importance to continue tracking the impact of the virus, in order to effectively deal with the challenges presented by its spread.

The world is witnessing an unprecedented manufactured pandemic emerge on a global scale, fueled by propaganda-style news reporting that is set to soon be met with substantial investigation and judicial inquiries. These will then uncover the lies, coercion, and profiteering by those who have proclaimed themselves to be the saviors of mankind in this age. Such a situation is nothing short of a crime against humanity.

The extent to which information has been manipulated to the advantage of the more powerful and influential has been used to construct a narrative that serves their interests and has been forced upon the unsuspecting public. This has been especially pertinent in the current situation, with governments, institutions, and media companies across the world pushing a predetermined agenda.

This has caused substantial amounts of suffering to populations based on this fallacious reporting, with citizens being denied access to basic health care services, resources and knowledge that could have tremendous implications for our future. The cost of this deception can be seen in the economic forecasts of our nations and the health of those who have lost their lives in what many are referring to as a human tragedy.

The very people who have been responsible for this series of events must be held accountable. Proactive steps to prevent additional deceit in the future must be taken, and justice must be sought in order to ensure that the truth is no longer viewed as disposable. When the injustice of what is unfolding today has been levied against the perpetrators, only then can we ensure that the victims of this manufactured pandemic will receive the recognition and compensation that is justly deserved.

Pfizer Contracts

It has long been known that Pfizer contracts between Canada and the United States contain a caveat that is often tucked deep into the text and not easily noticed: a statement reading "Pfizer's efforts to develop and manufacture the Product' are 'subject to significant risks and uncertainties". What does this mean?

To put it simply, this type of statement is in place to alert consumers and business partners alike to the potential risks associated with Pfizer's production and development of the Product. While Pfizer takes extensive measures to ensure the safety and reliability of their products, unforeseen issues like a sudden shortage of raw materials, unexpected changes in the marketplace, or even a natural disaster could create external pressures that jeopardize their production process and lead to significant losses. As a result, this statement serves as a reminder to take these risks into consideration when engaging with Pfizer products and services.

In addition to the protections afforded by Pfizer's risk statement, Canadian and American law also provides some additional safeguards that are worth noting. For instance, both countries' Consumer Protection Acts give consumers the right to return certain goods within a set amount of time if they are found to be defective or unsatisfactory in any way. Furthermore, Canadian and U.S. product liability laws hold Pfizer liable if they don't take due care and caution in manufacturing their products and impose strict warranty regulations to ensure they're held to a higher standard by consumers around the world.

By protecting the interests of their customers and other stakeholders with both words and legislation, Pfizer's risk statement sends a strong message that their commitment to product safety and quality is top-priority, making them a trusted name in the development and manufacturing of the world's most innovative products.

The evidence is incontrovertible that safer and more reliable medication could have drastically reduced mortality rates in the country in question by 81%. But, sadly, the government has exhibited an egregious lapse in judgement and an apathetic disregard for public safety by signing a contract that, while good for the government, was devoid of both transparency and public input. At the expense of their own people, the government has chosen to put their own financial interests above the well-being of their citizens.

The lack of visibility and knowledge of who signed this agreement has been a cause of great concern among the population. This has resulted in immense public distrust, and has led to massive outcry against it. Even more alarmingly, this contract was negotiated with taxpayers' money, with absolutely no accountability or transparency as to how it was spent. This egregious decision has been made despite the fact that we have evidence pointing to a safer, far more effective medication for the population.

In light of this situation, it has become apparent that governments need to act in the best interests of their citizens. They must not ignore scientific evidence and proven solutions in favour of a contract that fails to meaningfully contribute to the public's welfare. Rather, governments should prioritize the safety and protection of the people they are elected to serve and make decisions that reflect their commitment to this.

Then there are these little gems of contractual statements that never should never been agreed to.

The Purchaser recognizes that the Vaccine and its related items, components and substances are being quickly developed in response to the urgent state of the COVID-19 pandemic and that their study is predicted to continue after their provision to the Purchaser as stipulated in this agreement. Furthermore, the Purchaser acknowledges that the long-term health effects and performance of the Vaccine remain unknown at present, as do any potential short or long-term side effects. With the accelerated development of the Vaccine and its related items, there is uncertainty as to any long-term success or risks associated with their use. It is therefore essential to continually monitor their effectiveness and any associated risks so as to prepare for an emergency situation and adequately inform our population of the potential implications before the vaccine's wide-scale application.

The expansive language of the contract between Pfizer and their client makes it evident that Pfizer understands any potential risks that may arise from the use of the self-proclaimed "experimental drug". Thus, the contract goes out of its way to protect Pfizer from any legal repercussions due to any harm or damage incurred by the drug. This means that, should the drug inadvertently cause harm or injury to anyone, regardless of their status or knowledge, Pfizer will be freed from any legal or monetary repercussions.

For example, if an experimental drug originated from Pfizer was mistakenly given to a patient and resulted in a complication, any judicial action taken would occur solely between the patient and the medical facility. Pfizer would remain fully indemnified due to the language of the contract, protecting them from any liabilities that may arise.

In other words, the contract between Pfizer and their client firmly shields Pfizer from any legal claims or associated costs due to harm or damage sustained by the "experimental drug". In this way, Pfizer is able to undertake complex scientific research without being weighed down or held back by the prospect of potentially ruinous legal action or legal fees.

Vaccine Overreach

When hundreds of millions of people took the mRNA vaccine for COVID, they were assured it was safe and would immunize them and their loved ones from the virus. Unfortunately, it was soon discovered that the vaccine did not guarantee immunity or protect them fully from the virus and only mitigated the clinical symptoms instead. In hindsight, we can see that traditional FDA approved medications would have been much safer to use. Worryingly though, scientists around the world have had time to uncover some truly horrific details about the vaccine, which has not yet been marked as safe by the FDA, and has caused tens of thousands of unexplained deaths and millions of adverse reactions. The actual payload of the vaccine is now coming to light, and it's clear that the government, media and all other relevant authorities have failed to provide us with adequate protection once again.

DNA is the foundational code of life—making up the genetic composition of each and every one of us. Every human carries within their genetic sequence a unique set of instructions, supplying crucial guidance regarding our body's development and overall health. Put simply, the information contained in one's DNA dictates everything from the colour of their eyes to their susceptibility to certain health conditions such as cancer.

Components of DNA, known as genes, are the ultimate determiners of an individual's physical attributes. Genes determine everything from hair and eye colour to height and muscle tone. Additionally, inherited genetic malfunctions can result in an increased likelihood of diseases such as diabetes, heart disease, and even mental health disorders.

The complexity and coding of DNA is astonishing. Each gene contains thousands of base pairs, which together form the genetic building blocks that make each individual human unique. Every time a cell divides, a totally new DNA structure is created. In fact, the combined number of base pairs in one human body is estimated to be more than 5 octillion—an unfathomable figure!

Ultimately, it's safe to say that we are, quite literally, products of our genes.

Recent reports are raising questions about the implications of the presence of mRNA in the widely administered vaccine, BNT162b2, which is thought to turn back into DNA through a reaction called reverse transcription. Following vaccination with the BNT162b2 vaccine, it has now been demonstrated by doctors that mRNA uptake into the liver can be detected as soon as six hours after inoculation, as testified by a newly released scientific study. The statement elucidating the researchers' findings stated: "We also show that BNT162b2 mRNA is reverse transcribed intracellularly into DNA in as fast as six h upon BNT162b2 exposure."

The ability of mRNA to reverse transcribe into DNA in a relatively short amount of time plants the seeds of a myriad of questions in the minds of many – is this role of the mRNA impacting other genetic influencers within the body and, if so, could enzymatic pathways in the body be adversely affected? Moreover, can other mRNA components that follow the transformation of mRNA to DNA influence the production of proteins and, if so, in what ways?

The conversion of mRNA to DNA within the body could also result generate an array of other unintended influences and these need to be explored in order to gain a more robust understanding of the potential long-term implications of reverse transcription. Studies conducted on the long-term effects of mRNA to DNA conversion are now crucial for the safety and efficacy of the vaccine over the long-term, and the research to date suggests the need for further exploration of the topic.

The widespread distribution of the COVID-19 vaccine has sparked debate over its effectiveness and long-term side effects. One fear is the possibility of mRNA that has been inserted into the vaccine to be transcribed back into DNA and potentially integrated into the human genome. While the initial risk of developing an autoimmune disease or early death from taking the vaccines seemed to be an afterthought or no thought at all, this is a realistic possibility that must be considered as the vaccine continues to be administered.

The lack of clinical trials and research conducted to assess the long-term side effects of taking the vaccine has made it difficult to accurately assess any form of risk. Although the vaccine has passed through all of the necessary safety checks and has been authorized for use by various safety commissions, it is still a pressing concern.

The inability to predict the long-term effects of the vaccine is deeply concerning, but further research and clinical trials must be done in order to draw any definitive conclusions. In the meantime, individuals should do their own research and make an informed decision regarding taking the vaccine. What is important to remember is that the vaccine is safe and effective, and by continually following health advice, as well as continuing to wear a mask and practice social distancing, people can continue to stay safe from both the virus, and from any potential long-term adverse effects of taking the vaccine.

The significance of studying whether mRNA from the vaccine can be transcribed back to DNA and potentially integrated into the human genome was essential for vaccine developers to consider. With the current type of vaccines developed, which are based on mRNA, this question is of even greater significance. Despite its importance, this topic has not been thoroughly researched and thus we are currently unable to answer this question. Therefore, it must be commended that a group of researchers chose to explore this issue to better understand the potential risks and rewards of mRNA vaccines.

Scientists must take such steps to ensure the safety of the entire population. To illustrate the importance of this research, it can be compared to a hypothetical situation in which a software developer considers the possibility that their code may be embedded into a computer's hardware, as opposed to just running on top of the hardware. In this case, the researcher would need to be mindful to consider the implications of any changes made to the code that may be unknowingly embedded into the hardware, thereby introducing unexpected modifications in the hardware. Such a situation serves to demonstrate why it is essential that vaccine developers should consider studying the possibility that mRNA can be transcribed back to DNA and potentially integrated into the human genome.

In light of this, it is important to recognize the work of these researchers for taking the initiative to explore this topic, and for ultimately providing the scientific community with a better understanding of vaccines and their potential risks and rewards. It is only through such attempts to gain knowledge that we can make informed decisions pertaining to public health and to ensure the safety of us all.

The Centers for Disease Control and Prevention (CDC) has been accused of deceiving the public through false statements regarding the nature of mRNA and viral vector COVID-19 vaccines. For example, some have alleged that the CDC has misrepresented the facts by telling the public that "the genetic material delivered by mRNA vaccines never enters the nucleus of your cells, which is where your DNA is kept." However, this is not entirely accurate.

It is true that mRNA vaccines do not deliver genetic material to the cell nucleus; instead, the mRNA vaccines provide instructions for our cells to produce viral proteins which build protection against the virus. However, viral vector COVID-19 vaccines are different; these actually do deliver genetic material to the cell nucleus in order to produce an immune response. Although these vaccines may allow our cells to take on characteristics of the virus, the vector virus does not hold the capability to integrate its genetic material into our own DNA, therefore it cannot alter our DNA in any meaningful way.

To conclude, the CDC may have misled the public regarding mRNA and viral vector vaccines, but the truth remains that neither type of vaccine pose any threat to our DNA.

The recent investigation into the Centers for Disease Control and Prevention's persistent misinformation surrounding the COVID-19 pandemic has revealed just one of many deceptions the organization has been complicit in. It's no wonder that so many people express feeling of surprise and betrayal upon learning this stark truth, which ultimately begs us to ponder: What other shady lies and untruths are the CDC keeping from us?

The lack of faith instilled in the organization's credibility has unleashed a cascade of scrutiny from scientists, medical professionals, and members of the public alike, as it appears that the CDC has stooped to irresponsible and deceitful behaviors in order to control the narrative. To make matters worse, instances such as this have been happening for some time now, with a disturbing pattern of dissembling and misguiding the public about their health and safety.

This leaves us to wonder what other details are being purposely withheld from us that may have a profound effect on our lives. This is a legitimate question that deserves answers. Until trust and transparency can be restored and the CDC can be held accountable for their past transgressions, it is critical that we remain vigilant when it comes to our health and safety, as knowledge really is power.

Reinfection or reverse transcription?

With each passing day, it seems that more and more reports of people testing positive for COVID-19 after receiving all their recommended booster shots are surfacing. While there is no denying that some of these claims can be corroborated by investigation, as evidenced by the presence of various strains of the SARS-CoV-2 virus from two separate episodes, it seems unlikely that the majority are actually reinfections. At this point in time, no replication-competent virus has been isolated from any of the reported cases, further leading to the conclusion that these individuals simply tested positive again due to a continued presence of the virus in their bodies.

While it is still unclear as to why some people that have already defended themselves against the virus have tested positive again, what is certain is that proper health protocols and measures must be taken. This means that while vaccinations provide a necessary layer of protection, it is equally imperative to observe the basics; namely regular handwashing, wearing masks, maintaining physical distance and avoiding crowded places, which have time and time again been clinically proven to minimize the risk of reinfection.

The present investigation has examined the transposition of SARS-CoV-2 RNA, which is the cause of the COVID-19, into the human genome through reverse transcription. This can lead to positive polymerase chain reaction (PCR) tests owing to the continual transcription of viral RNA into DNA and integration into the genetic makeup of the human organism. This phenomenon is of considerable medical and scientific import due to the considerable morbidity and mortality associated with SARS-CoV-2 infection and its implications for the future of viral control.

Reverse transcription is the process of converting RNA into DNA, and is carried out primarily by enzymes known as reverse transcriptases. These enzymes have various features and functions, such as helping to replicate certain viruses and being used as a diagnostic tool. In this way, they are considered to be intimately associated with virus-host interaction and the development of diseases. Moreover, as was demonstrated in this study, these reverse transcriptases can lead to the integration of viral RNA into the human genome, thus creating positive PCR tests from only a single infection and potentially creating health risks for many more individuals than the single person in whom it was initially detected.

This study emphasizes the potential for a piece of the virus to become part of our permanent genetic code, and the importance of understanding the implications of this occurrence. Indeed, if this reverse transcription is allowed to occur, it may produce a future of viral disease control challenges that must be addressed in order to ensure the health and wellbeing of people worldwide. As such, this investigation not only serves to inform public health policy with regard to SARS-CoV-2, but also holds important implications for researchers and clinicians in managing the challenges of treating and assessing other viral infections.

Despite the vast amount of research that has been conducted, many have failed to become concerned about the potential implications of the Pfizer/BioNTech Covid injection (BNT162b2). A study conducted in Sweden has confirmed that mRNA from the Pfizer Covid vaccine can enter cells and be transcribed onto human DNA within just 6 hours, potentially leading to modifications of our own DNA. This is an incredibly profound and potentially concerning finding which has sparked some alarm.

Not only this, but preclinical studies involving animal testing have displayed reversible effects on the livers of rats, including liver enlargement, creation of vacuoles, and increased levels of certain enzyme markers (γGT, AST, ALP). Furthermore, the documentation provided to the European Medicines Agency by Pfizer itself demonstrates that some of the "vaccine" is not kept at the injection site, and can end up in the liver in considerable proportions, up to 18% of the total dose.

Overall, there is a plethora of evidence out there that we should all be cognizant of and be questioning directly. Unfortunately, many individuals may be ignorant to the situation until they are directly confronted with it; as by this point, it may be too late. It is our responsibility to remain informed and ask the necessary questions to ensure that we make educated decisions regarding our own individual health and safety.

My Thoughts on the Emergency Measures Act

The recent controversy surrounding the Prime Minister's decision to declare an emergency has raised questions surrounding Canada's adherence to the guidelines of the Emergency Measures Act and the Canadian Security Intelligence Service Act. As it stands, it appears that the Prime Minister has extended his own authority beyond the limitations imposed on him by these laws, thus undermining the spirit of the legislation and putting Charter, Bill of Rights and International Covenant of Civil and Political Rights protections at risk.

The emergency measures in question could lead to drastic and extreme limitations placed on civil liberties, including significant limitations on freedom of movement, speech, assembly and more. Even before the emergency has been declared, many Canadians have expressed their deep concern about the possibility that their fundamental rights could be violated.

The Canadian Charter of Rights and Freedoms, the Canadian Bill of Rights, and the International Covenant on Civil and Political Rights provide clear cut guidance for citizens on what types of restrictions should be allowed in a time of national emergency and have been included as an integral part of the Emergency Measures Act and the Canadian Security Intelligence Service Act for that very purpose. It thus stands to reason that the declaration of an emergency shouldn't undermine basic human rights, and any attempts to do so should be seen as a violation of both the spirit and the letter of the law.

In order to fulfill its obligation to Canadians and ensure the appropriate respect for fundamental liberties and human rights, the government must rise to the occasion and guarantee that the Emergency Measures Act and the Canadian Security Intelligence Service Act are upheld. These laws should be seen as an important tool for protecting Canadians during times of national emergency and not be flouted if the government wishes to act in good faith. This is not only a matter of upholding the Due Process of law, but is also a matter of preserving Canadians' basic human rights.

Although not set out in the Preamble, these rights are:

• the right to life.

• the protection against cruel, inhuman or degrading treatment or punishment

• the protection against slavery.

• the protection against imprisonment for debt.

• the protection against acts made retroactively into crimes

• the right of every individual to be recognized as a person under law.

• the freedoms of thought, conscience and religion.

A public order emergency is an exceptionally serious threat to the security of Canada that has been officially classified as a national emergency. The severity of these threats is determined in accordance with Section 2 of the Canadian Security Intelligence Service Act, which states that a threat must be so serious as to be considered a national emergency according to Section 14 of the same Act.

The term 'public order emergency' is not limited to security threats, however. It also includes other emergencies that can impact public safety, such as natural disasters, civil unrest, or environmental degradation. For example, in the province of Ontario, a 'state of emergency' was declared in 2019 due to record flooding, which caused many people to be evacuated from their homes and wildlife to be affected. In other cases, a public order emergency may be declared as a result of an epidemic, allowing governments to implement quarantine measures to contain its spread.

In order for a public order emergency to be declared, it must be determined that the situation poses a significant threat to the safety and security of Canada. This process requires collaboration between the federal and provincial governments, as well as other key stakeholders such as the Canadian Security Intelligence Service, the RCMP, and emergency services personnel. The emergency declaration can only be lifted when the threat has been mitigated.

threats to the security of Canada means

The act of espionage or sabotage directed against Canada or which is detrimental to its interests shall constitute a major violation and will not be tolerated. This includes any activities that are in support of such espionage or sabotage. For instance, the contents of confidential government information being shared with foreign entities, or the malicious destruction of government property for the purpose of sabotage, would both be viewed as serious threats to Canadian interests. In addition, the deliberate withholding of information, or the spread of misinformation intended to damage or discredit the nation or its policies to foreign nations, would also be considered equally serious offenses. These activities have the potential to put Canada's sovereignty and security at risk and therefore must be taken very seriously.

Canada has recently experienced issues arising from foreign influenced activities carried out within or relating to the country, activities that undermine the interests of Canada and its people. In particular, these activities are typically clandestine, deceptive, or involve a threat to the security of individuals. For instance, foreign agents have been known to use covert techniques to collect confidential information from government officials, intelligence agencies, and corporations for the purposes of espionage. Additionally, hostile foreign entities have been linked to cyberattacks on critical infrastructure and other attempts to disrupt the peaceful operation of Canadian society and its economy. These activities may also be related to attempts to unduly influence public opinion or interfere in the democratic process, or spread disinformation to sow discord and discord among Canadian citizens. As such, it is essential that Canadian intelligence and law enforcement agencies remain vigilant and take all appropriate measures to identify and neutralize any such nefarious activities.

Activities within or related to Canada that aim to further a political, religious, or ideological objective at home or abroad, as well as the threat or use of seriously violent acts against people or property, are considered to be terrorism within Canadian law. This includes numerous examples, such as an individual writing a blog post urging readers to utilize violence to further a specific cause, or a group of people planning an attack on a government building in order to further their political tactics. Terrorism can come in any form, but the intent must always remain to further a political, religious, or ideological objective through serious violence.

Covert activities and unlawful acts that are intended to undermine, or which ultimately lead to the destruction or overthrow by violent means of the constitutionally-mandated system of government in Canada, are strictly prohibited and condemned. Such actions include, but are not limited to, espionage, sabotage, subversion, coups, instigations of revolution, and acts of terrorism. By way of example, efforts to illicitly spy on the government, or to disseminate deliberately misleading information in order to influence public opinion are both illegal, and detrimental to the maintenance of the democratic system of government in Canada. Similarly, the use of violence, either through demonstrations or by other means, to attempt to overthrow the Canadian government constitutes a serious threat to stability, and the security of all citizens of this nation. It is of the utmost importance that all Canadians oppose any and all efforts to undermine the rule of law and democratic systems of governance in our nation.

Canadian security involves protecting the nation and its citizens from threats, both foreign and domestic. Unfortunately, menaces that threaten the nation's security are all too real. Consequently, it is of paramount importance that the country's safety is safeguarded through a combination of belligerent action, such as protecting the nation's physical borders, as well as through less overt means, like the use of law, education and advocacy.

When it comes to protecting the security of Canada, a range of activities come into play. These include monitoring of individuals, groups and organizations suspected of criminal activities that threaten the country's security; suppressing the use, possession and procurement of weaponry, military-grade technology and other dangerous materials; and cracking down on terrorism, espionage and other unlawful activities.

Additionally, reducing the ease of access to prohibited materials, such as automatic weapons and explosives, helps to deter acts of violence, terrorism and sabotage in Canada.

It is essential to remember, however, that lawful advocacy, protest or dissent are not to be excluded from the discussion. In fact, these measures can often supplement the aforementioned security measures, and play a key role in creating a secure nation. Responsible protest and working peacefully to engage with law enforcement, the government and international partners to promote security and human rights are just some of the ways in which citizens can get involved in the fight to keep Canada safe. It is only when adverse activities are carried out in conjunction with any of the aforementioned activities that worries regarding national security arise.

Only if the above criteria have been met, the Governor in Council could by order or regulation:

It is important for local governments to enforce laws regulating public assembly to protect public tranquility and safety. Such legislation may prohibit certain types of gatherings when there is reasonable cause to believe that they could lead to disturbances, harm to people or property, or other forms of disruption. For example, existing laws often restrict public assemblies at late hours, particularly in residential areas, to protect against noise disturbances. Similarly, laws can be enacted to limit the size or duration of gatherings in public parks or other shared spaces in order to minimize the potential for violence or vandalism. Municipal authorities may also use regulations to limit public movement or the use of certain properties, such as prohibiting public marches or protests on private properties. In this way, local governments can effectively manage public assembly in order to maintain public order and safety.

- Designate and secure protected places;
- Assume the control, restoration and maintenance of public utilities and services;
- Authorize or direct the provision of essential services and the provision of reasonable compensation therefore; and

Under Section 19(1), any individual found in violation of an order or regulation is liable to punishment according to the severity of the infraction. Upon summary conviction, an individual may be imposed with a fine of no more than 500 dollars or imprisonment of no more than six months, or both. In the event of an indictment, one found guilty may be fined up to $5000 or could face prison sentence of up to five years, or potentially both. For example, a person found to be recklessly discharging a firearm in a prohibited area might be subject to a summary conviction, which could result in a fine of no more than 500 dollars or a prison sentence of up to six months. Alternatively, a person found guilty of a more serious offence such as murder or manslaughter could be sentenced to a maximum five-year jail term, along with a hefty five-thousand dollar fine if found guilty in an indictment.

Search and seizure powers would apply only to inquiries in relation to black market practices (s. 30(1)(c) and (d)).

In Friday, the Government of Ontario, led by Premier Doug Ford, declared a state of emergency in the province and invoked a set of new emergency measures to further punish protestors who do not comply. Under the new measures, offenders may face a maximum fine of $100,000 or up to a year in prison for non-compliance with the Emergency Management and Civil Protection Act (EMA).

These penalties represent a significant escalation of the government's response to civil disobedience and undoubtedly represent a major deterrent to those who are considering taking part in protests in the future. Though Ford's office has defended their decision by emphasizing the need to protect public safety, critics have roundly criticized the move as heavy-handed and disproportionate when compared to similar measures in other provinces.

For Canadians interested in staying up-to-date with the latest news from Parliament Hill and the Ontario government, Capital Dispatch is an essential source of information. Sent each weekday morning, Capital Dispatch provides subscribers with a comprehensive overview of the goings-on in the political arena, plus updates on the status of the EMA and other important developments. With thorough coverage, accurate reporting and timely updates, Capital Dispatch is a must-read for anyone concerned about the trajectory of the province during this state of emergency.

TERRIFYING NEWS AND BILL GATES IS BEHIND IT

The FDA Advisory Committee Recommends Annual Boosters!

The FDA and CDC recommended on January 26th to skip the first dose of the vaccine and instead receive annual or bi-annual shots. It is important to note that the current vaccine is not FDA-approved and still under emergency use authorization. This raises concerns about the potential influence of the pharmaceutical industry on the FDA and CDC. Additionally, the current pandemic situation does not meet the criteria for emergency use authorization, as it has been stated that the pandemic is over. The recommendation for bi-annual shots also raises questions as the variants targeted by the vaccine have become extinct. Most of the population has already achieved immunity through natural means or the vaccine. These recommendations by the FDA and CDC, funded by the pharmaceutical industry, for using an unapproved treatment raise valid concerns.

The recommendation by a committee funded by the pharmaceutical industry to use a treatment that is not FDA-approved raises questions about its safety and efficacy. The limited testing of the vaccine, only on eight mice, raises concerns about its application to the wider population, especially children and pregnant women, who typically require extensive data on a drug before it is considered for use. The lack of clinical trials for the vaccine further compounds these concerns.

This situation is unacceptable and becoming increasingly outrageous. There is a blatant disregard for the valid concerns raised, allowing big pharma to profit from the administration of non-FDA-approved injections on a virus strain that no longer exists right before our eyes. This situation has reached a point of absurdity.

Thailand nullifies the Pfizer contracts – Princess in Coma.

Thailand is cancelling Pfizer and now demands billions of dollars back from Pfizer. They will also seek money to pay people for the damages and suffering caused by what they believe is the result of the Pfizer vaccine. This all happened quickly after Thailand's cherished princess collapsed due to a heart problem 23 days after receiving a booster and remains in a coma weeks later. The royal family was originally told that the princess had an infection, which was the cause of her condition, but soon, it has been found to be a lie and a coverup.

What Thailand is now reporting is that the jab was the cause of the princess's fate. Reports now show she started having severe arrhythmia of the heart, which is very irregular heartbeats, that can cause fainting and, in some cases, not allow sufficient oxygen to get to the brain. What I am thinking from similar instances, was a blood clot cutting off blood reaching the brain is the cause.

This is only the beginning of what will be more countries to follow and will result in an uprising never seen before.

BREAKING!

A PFIZER SENIOR EXECS ADMITS PLANS TO MANIPULATE THE COVID VIRUS!

JORDAN TRISHTON WALKER, DIRECTOR OF RESEARCH AND DEVELOPMENT

It has been recorded on video by Project Veritas that Pfizer had intentions or may have performed, through gain of function, manipulated the COVID virus to remain highly infectious for years to come. This would ensure an ongoing massive revenue stream. With this latest bombshell, there must be an inquest into the COVID matter and the profiteering, coverup, and gene manipulation (gain of function) that we were led to believe never happened by Dr. Anthony Fauci. It's time we find out!

BREAKING!

THE CDC IS INVESTIGATING POSSIBLE LINKS BETWEEN THE VACCINE AND HEART CONDITIONS SUCH AS MYOCARDITIS.

We Told You So

Something many of us has been talking about for over a year, the mainstream media are now talking about. Their news headlines read like this is something discovery and sudden, when in fact, full well knowing, this is not new at all.

Now that the CDC is looking into this, perhaps under pressure from the newly forming US government, I sense there will be some upset people in the not-so-distant future as this entire COVID situation starts to unravel to expose some pretty harsh truths.

How about a big a$$ bowl of "I told you so soup."

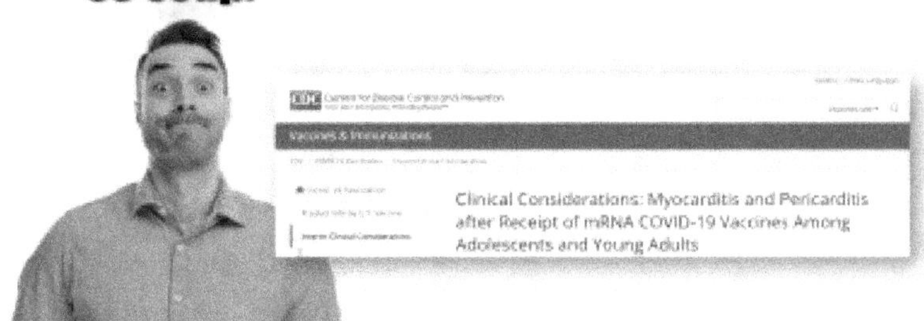

www.ingramcontent.com/pod-product-compliance
Lightning Source LLC
LaVergne TN
LVHW052213200926
848481LV00001B/92
9798223247159